Contents

REVISED
for the 17th Edition IEE
Wiring Regulations

Electrical Installations

Learner

Dave Allen

John Blaus

Nigel Harman

Brian Tucker

LEVEL 2

2330 Technical Certificate

www.heinemann.co.uk

✓ Free online support
✓ Useful weblinks
✓ 24 hour online ordering

01865 888118

Heinemann

Heinemann is an imprint of Pearson Education Limited, a company incorporated in England and Wales, having its registered office at Edinburgh Gate, Harlow, Essex, CM20 2JE. Registered company number: 872828

www.heinemann.co.uk

Heinemann is a registered trademark of Pearson Education Limited

Text © EAS and JTL 2008

First published 2008

12 11 10 09 08
10 9 8 7 6 5 4 3 2

British Library Cataloguing in Publication Data
A catalogue record for this book is available from the British Library
.
ISBN 978 0 435401 09 2

Typeset by HL Studios
Original illustrations © Pearson Education Limited 2008
Illustrated by Illustration Ltd. (Stuart Holmes), Ken Vail Graphic Design and HL Studios
Cover photo/illustration © Getty Images
Printed in Spain by Graficas Estella

The websites used in this book were correct and up-to-date at the time of publication. It is essential for tutors to preview each website before using it in class so as to ensure that the URL is still accurate, relevant and appropriate. We suggest that tutors bookmark useful websites and consider enabling students to access them through the school/college intranet.

Endorsement and acknowledgements

Original material by JTL. Additional content supplied by Nigel Harman and Brian Tucker on behalf of Electrical Assessment Services UK Ltd.

JTL endorsement

The electrotechnical industry has determined that the minimum requirement for skilled craft status is the NVQ level 3. However it is also acknowledged that some entrants to the industry may wish to commence with level 2 before progressing to the level 3 technical certificate and NVQ level 3 award. JTL therefore endorses the contents of this first of two reference books in so far as it supports such candidates and will encourage them to proceed to achievement of full electrical craft status.

This material has been developed from JTL resources with kind permission.

Acknowledgements

The author and publishers would like to thank the following individuals and organisations for permission to reproduce photographs:

AAF-Creatas – page 344

Alamy Images – page 37, 184 [scissor], 191, 224

Arco – pages 10 [goggles/helmet], 11 [leggings], 12, 13, 15

Art Directors and Trip – pages 10 [mask], 11 [trainers], 347 [smoke]

Construction photography – pages 182, 184 [boom]

Construction photography/Jean-Francois Cardella – page 244

Construction photography/Paul McMullin – pages 1, 319

Corbis/Comstock – page 314

Digital Vision – page 39

Dreamstime – page 321

Getty Images – pages 209, 311

Getty Images/Photodisc – page 57

Getty Images/Stone – page 40

Ginny Stroud-Lewis – pages 14, 347 [break glass], 348

Introduction

What is this book?

This book is designed to specifically support the syllabus of the level 2 of the City and Guilds 2330 certificate. The Certificate is designed to set a quality standard across the industry for learning and training in the electro-technical sector. It is available both to new starters and experienced workers wishing to update their qualifications.

The Technical Certificate is the primary means of providing the underpinning job knowledge and understanding that supports your ability to actually do the job. Apprentices require both the Level 2 and Level 3 qualification.

However, this book also serves as an introduction to Electro-technical Technology and provides a preliminary foundation to candidates preparing to start work on the C&G 2356 NVQ in Electro-technical Services – Electrical Installation (Buildings and Structures).

The NVQ contains National Occupational Standards that describe what a competent person in an industry can be expected to be able to do. The NVQ is therefore awarded when a candidate has shown they can perform the job within the workplace. The minimum standard that must be obtained is the level 3 qualification, which this book provides the initial basis for study towards.

Successful completion of the above will enable you to become a level 3 qualified electrician!

This book aims to support that process, by introducing you to the theory and giving guidance on practical activates you can do to put your skills into practice.

How can this book help?

This book includes several features designed to help you make progress and reinforce learning:

- **Illustrations and photographs** – drawings and photographs showing essential information in a clear and concise way.

- **Tables, bullet points and flowcharts** – crucial information delivered in a clear and concise format.

- **Margin notes** – short hints to aid good practice and reinforce information. These also provide tips on health and safety and further research that can be carried out. These margin notes include:

    - **Did you know** – useful additional information you may not have known

- **Remember** – reminders about information you have already covered which is relevant

- **Find out** – opportunities and suggestions for further research

- **Safety tips** – those all important guides to safe practice both for you and your co-workers.

- **On the job scenarios** – typical situations from work, testing your ability to make decisions in scenarios you might encounter.

- **End of chapter activities** – these allow you to put some of the skills you have learnt into practice.

- **FAQs** – these provide answers to regularly asked questions.

- **Knowledge checks** – use these to test yourself on the information you have learnt.

This book is designed to give you all the knowledge you need to progress and gain success in the 2330 Level 2 certificate.

From here you will have the opportunity of progressing in a varied and interesting career – well done for choosing such a great start!

17th Edition IEE Wiring Regulations

This book has now been updated in accordance with the BS7671:2008 (17th Edition, IEE Regulations) standards.

A fundamental change from the 16th to the 17th edition is that phase conductors are now known as line conductors. It is important to note that this terminology refers to the name of the conductor itself. Particularly when dealing with a.c. theory, the terms 'phase' and 'line' are still used to avoid confusion when dealing with specific concepts and values of voltage and current.

Note: references to the On-site guide are to the BS7671: 2001 edition. At the time of writing this book the 2008 On-site guide had not been published.

Remember whenever carrying out any electrical work please consult the current edition of the Regulations.

chapter 1

Legal responsibilities of employers and employees

Unit 1 Outcome 1

All employers, including the self-employed, have duties under the Health and Safety at Work Act 1974 to ensure the health and safety of themselves and others affected by what they do. This includes people working for employers and the self-employed (e.g. part-time workers, trainees and subcontractors), those who use the workplace and equipment they provide, those who visit the premises and people affected by their work (e.g. neighbours or the general public).

The aim of this chapter is to enable you to identify the health and safety legislation and to be able to work safely within the electro-technical industry.

On completion of this chapter the candidate will be able to:

- **state the legal responsibilities of employers and employees under the Health and Safety at Work Act (1974), and list current relevant Health and Safety legislation**

- **identify regulatory requirements which are statutory and non-statutory**

- **explain the use of personal protective clothing**

- **state the need for isolation before work begins**

- **identify types and meaning of safety signs**

- **follow accident and emergency procedures, and deal with electric shock victims**

- **identify types and application of fire extinguisher.**

Legal responsibilities

On completion of this topic area the candidate will be able to identify statutory and non-statutory legislation that affects the workplace, and state the main responsibilities of employers towards maintaining health and safety.

Legislation

Unfortunately, work is a dangerous activity and each year many hundreds of thousands of hours are lost to industry through time off for injuries. Through the years, countless governments have tried to reduce both the loss in time and, of course, the loss of lives, as well as the pain and suffering resulting from accidents at work.

Governments have mainly tried to use self-regulation and persuasion. However this has not been successful, so they have had to resort to legislation (laws) to try to reduce this waste. Early examples of these laws are:

- The Health and Morals of Apprentice Act of 1802

- The Factories Act of 1833

- The Electricity (Factories Act) Special Regulations, 1908 and 1944.

As you can see from the dates of these laws, many of them have now been outdated by events. The main Act that will concern you at work is the Health and Safety at Work Act 1974. This law opened the floodgates to a whole raft of safety legislation over the following years.

Statutory and non-statutory legislation

When a law is passed by a government it is said to be placed on the statute book, and is therefore enforceable in law. Non-compliance with the laws can lead to prosecution by the courts, which may result in financial penalty or imprisonment. Three such statutory acts or regulations that affect us in the electro-technical industry are:

- Health and Safety at Work Act 1974

- Electricity at Work Regulations 1989

- Electricity Safety, Quality and Continuity Regulations 2002 (formerly the Electricity Supply Regulations 1989).

Statutory laws and regulations define what the law is – in other words they make the law. What they do not do is tell us how to comply with the law. This duty is fulfilled by non-statutory regulations and codes of practice. These are written for every aspect of industry and enable us to comply with the statutory acts. Statutory Regulations that affect the electro-technical industry are:

- Management of Health and Safety at Work Regulations 1999

- Provision and Use of Work Equipment Regulations 1998

Did you know?

The Factories Act 1833 was introduced to reduce the number of hours that children were allowed to work in the textiles industry. It stated that children could work a maximum of nine hours a day with two hours of education

Did you know?

The Health and Safety at Work Act 1974 is often called HASAWA for short

Remember

Non-statutory regulations are not enforceable by law themselves, but if you do not follow their conditions you will not meet the standards needed to fufill the statutory laws

- Control of Substances Hazardous to Health Regulations 2002
- Personal Protective Equipment at Work Regulations 1992
- BS 7671 Requirements for Electrical Installations 2008
- Workplace (Health, Saftey and Welfare) Regulations 1992
- Construction (Health, Saftey and Welfare) Regulations 1996.

The following are non-statutory documents:

- HSE ES 38
- IEE Code of Practice for In-Service Inspection and Testing of Electrical Equipment
- Building Regulations.

Health and Safety at Work Act 1974

The Health and Safety at Work Act 1974 (HASAWA) is what is called an 'Enabling Act'. It sets out the basic principles by which health and safety at work is regulated. It outlines a number of duties that employers have to their employees and vice versa. The Act also covers a series of Regulations that explain the practical details of maintaining a safe working environment.

Duties of employers to their employees (section 2)

The Act requires employers to ensure, so far as is reasonably practicable, the health, safety and welfare of their employees at work. The things it covers include:

- making the workplace safe and without risk to health
- keeping dust and fumes under control
- the provision and maintenance of plant and systems that are safe
- safety in the use, handling, storage and transport of articles and substances
- the provision of information, instruction, training and supervision as is necessary to ensure the health and safety at work of employees
- the provision of access to and egress from the place of work that are safe and without risk
- the provision of adequate facilities and arrangements for welfare at work
- the provision of Personal Protective Equipment (PPE).

The employer must also:

- draw up a health and safety policy statement if there are five or more employees
- carry out an assessment of risks associated with all the company's work activities
- identify and implement control measures
- inform employees of the risks and control measures

- periodically review the risk assessment
- record the assessment if more than five persons are employed
- consult a safety representative, if one is appointed by a recognised trade union, about matters affecting your health and safety
- if requested in writing to do so by any two safety representatives, establish a safety committee within three months of the request being made
- display a current certificate, as required by the Employers' Liability (Compulsory Insurance) Act 1969
- undertake precautions against fire, providing adequate means of escape and fighting fire
- provide adequate first aid facilities
- report certain injuries, diseases and dangerous occurrences to the enforcement authority.

Duties of employees (section 7)

As an employee you have legal duties to:

- take reasonable care at work for the health and safety of yourself and others who may be affected by what you do or do not do
- not intentionally or recklessly interfere with or misuse anything provided for your health and safety
- co-operate with your employer on health and safety matters, and assist the employer in meeting their statutory obligations
- bring to your employer's attention any situation you consider serious and an imminent danger
- bring to your employer's attention any weakness in their health and safety arrangements.

Electricity at Work Regulations 1989

Made under the HASAWA 1974, the Electricity at Work Regulations (EAWR) came into force on 1 April 1990. They impose general health and safety requirements to do with electricity at work, on employers, the self-employed and employees.

Every employer and self-employed person has a duty to comply with the provisions of the Regulations, in so far as they relate to matters that are within their control.

Every employee has the duty to co-operate with their employer, so far as is necessary to enable the Regulations to be complied with. Because these are **statutory** Regulations, penalties can be imposed on people found guilty of malpractice or misconduct.

The Regulations refer to a person as a 'duty holder' in respect of systems, equipment, and conductors. Every duty holder must comply with the provisions of these

Regulations in so far as they relate to matters that are within their control. The regulations define the following:

Employer: any person or body who:

a) employs one or more individuals under a contract of employment or apprenticeship, or

b) provides training under the schemes to which the HASAWA applies.

Self-employed: an individual who works for gain or reward other than under contract of employment, whether or not he or she employs others.

Employee: Regulation 3(2)(b) repeats the duty placed on employees by the HASAWA, which are equivalent to those placed on employers and self-employed persons where these matters are within their control. This includes trainees like you who are considered employees under the Regulations.

The Regulations are designed to take account of the responsibilities that many employees in the electrical trades and professions have to take on as part of their job.

We will look more closely at EAWR in chapter 8.

Electricity Safety, Quality and Continuity Regulations 2002

The Electricity Safety, Quality and Continuity Regulations 2002 are **statutory** regulations. They replace the Electricity Supply Regulations 1989, and specify safety standards that are aimed at protecting the general public and consumers from danger. In addition, the Regulations specify power quality and supply continuity requirements to ensure an efficient and economic electricity supply service for consumers.

Management of Health and Safety at Work Regulations 1999

The Management of Health and Safety at Work Regulations 1999 generally makes more explicit what employers are required to do to manage health and safety under the Health and Safety at Work Act. Like the Act, they apply to every work activity. These regulations are **statutory**.

Provision and Use of Work Equipment Regulations 1998 (PUWER)

PUWER 1998 replaced the Provision and Use of Work Equipment Regulations 1992 and carried forward those existing requirements, with a few changes and additions. Examples of this include the inspection of work equipment and specific new requirements for mobile work equipment. These regulations are **statutory**.

The Regulations require risks, to people's health and safety from equipment that they use at work, to be prevented or controlled.

Control of Substances Hazardous to Health Regulations 2002 (COSHH)

Using hazardous substances can put people's health at risk. COSHH therefore requires employers to control exposure to hazardous substances to protect both employees and others who may be exposed from work activities.

Hazardous substances are anything that can harm your health when you work with them, if they are not properly controlled, e.g. by using adequate ventilation.

Hazardous substances are found in nearly all work places, e.g. factories, shops, mines, construction sites and offices and can include:

- substances used directly in work activities, e.g. glues, paints, cleaning agents
- substances generated during work activities, e.g. fumes from soldering and welding
- naturally occurring substances, e.g. grain dust, blood, bacteria.

Employers must ensure employees' and others' safety wherever practicable. COSHH is again **statutory**.

Personal Protective Equipment at Work Regulations 1992

The Personal Protective Equipment at Work Regulations set out the principles for selecting, providing, maintaining and using personal protective equipment (PPE). This is equipment designed to be worn or held to protect against a risk to health or safety. It includes most types of protective clothing and equipment such as eye, hand, foot and head protection, safety harnesses, lifejackets and high visibility clothing all of which, under the Health and Safety at Work Act, employers must provide free of charge. These regulations are **statutory**.

BS 7671: Requirements for Electrical Installations 2008 (IEE Wiring Regulations 17th Edition)

The BS 7671 Wiring Regulations are **non-statutory**. They relate to the design, selection, erection, inspection and testing of electrical installations, both permanent and temporary, in and about buildings generally. This includes agricultural and horticultural premises, construction sites, caravans and caravan sites.

Although **non-statutory**, non-compliance with BS 7671 will almost certainly mean that **statutory** regulations, such as the Electricity at Work Regulations, will not have been complied with. BS 7671 only applies to installations operating voltages up to 1000 volts a.c. and does not apply to equipment on board ships, aircraft, mines and quarries, railways and motor vehicles, and any equipment that is mobile or fixed on offshore installations.

Remember

In the UK electrical industry, and throughout this book, reference to 'Regulation' or 'Regulations' means BS 7671 the *IEE Wiring Regulations*, unless stated otherwise

The Wiring Regulations were incorporated into the **British Standards** as part of the recognition of the harmonisation of these regulations with those of **European Standards**. BS 7671 is divided into seven parts. These are:

- scope, object and fundamental principles
- definitions
- assessment of general characteristics
- protection for safety
- selection and erection of equipment
- inspection and testing
- special installations or locations.

We will look at BS 7671 in greater depth in chapter 2.

Workplace (Health, Safety and Welfare) Regulations 1992

These Regulations apply to all workplaces, where a workplace is defined as being any premises or part of premises which are not domestic premises, and are made available to any person as a place of work, including:

- any place within the premises to which such person has access while at work
- any room, lobby, corridor, staircase, road or other place used as a means of access to or egress from that place of work or where facilities are provided for use in connection with the place of work other than a public road.

They do not include a modification, an extension or a conversion of any of the above until such modification, extension or conversion is completed.

The Regulations do not apply to places of work that relate to ships, aircraft, locomotive or rolling stock, construction sites and sites where extraction of mineral resources is carried out, temporary work sites and workplaces in agricultural or forestry land.

The requirements of these Regulations are imposed on every employer or any person who has to any extent control of a workplace, in respect of the following matters:

- working environment
- safety
- facilities.

These regulations are **statutory**.

Construction (Health, Safety and Welfare) Regulations 1996

These Regulations are of interest to electricians, as a contractor will spend much of his or her time on construction sites. The temporary nature of construction sites makes them one of the most dangerous places to work. These **statutory** regulations apply to the construction work in any building, or civil engineering work, including construction, assembly, alterations, conversions, repairs, upkeep, maintenance or dismantling of a structure.

The general provisions of the regulations set out minimum standards to promote safety on site. These regulations specify the requirements for guardrails, working platforms, ladders, emergency procedures, lighting and welfare facilities such as washing facilities, sanitary conveniences and protective clothing. There is now a duty for all those working on construction sites to wear head protection. Obviously this includes electricians working on site as subcontractors.

Here are some guidelines for safe working. Take a minute to look at these and think about how you work. Are you taking any unnecessary risks?

Work tidily and cleanly. Do not leave objects lying on the floor where they may cause accidents. Clean all materials and debris away from the site at the end of the working day and ensure that when working overhead on scaffolds, trestles, ladders and steps, you do not lay anything down in such a position that it may fall on anyone or anything below.

Observe all rules and work instructions provided. When you start work your employer will make you aware of the company's rules and expectations with regard to safety. Most employers will have a health and safety policy or statement, and it is part of your job to acquaint yourself with the contents and to ensure that all of your activities comply with the stated requirements.

Running or hurrying can cause accidents. Never run or take short cuts, even if you are in a hurry. You may collide with someone, trip over an obstruction or run into a protruding object causing an injury. It is always better to walk and arrive safely. Construction sites are particularly dangerous places to work if you do not take common sense precautions.

Keep all machinery and equipment well maintained and in good condition. Never use damaged machinery, tools or equipment, and make sure that any damage you may cause is reported or repaired promptly so that it does not endanger the next user.

Secure all loose clothing and repair any torn articles immediately. Overalls should always be fastened with no flaps or torn pieces hanging off that may become tangled with rotating machinery etc. If your hair is long, it should be covered by a 'snood' cap (hair net) or tied up so that it is not a hazard, and all jewellery should be removed where a safety hazard exists.

Advise supervisors immediately if you observe any unsafe practices or notice any defects in any of the equipment that is provided for use by yourself and others.

Follow all manufacturers' instructions and recommendations when using items of equipment.

Examine all electrical tools and equipment very carefully before use to ensure that they are in good working order and show signs of having been recently inspected.

Let others know when you are working overhead or nearby when your activities may pose a particular danger to them.

You are responsible for the safety of yourself and others with whom you work or who may be affected by your work. Do not leave things lying around. Everything that you do must be of the very highest commercial and safety standards so that it DOES NOT present any significant danger to you or other people who may be affected by your actions.

Remember

Accidents do not just happen; they are caused and are invariably the result of human failing. However, with knowledge and the application of sensible thoughts and procedures, there is no reason why you should ever have an accident

Personal protective equipment (PPE)

On completion of this topic area the candidate will be able to explain the use of personal protective clothing appropriate to the task being undertaken.

Personal Protective Equipment at Work Regulations 1992

The Personal Protective Equipment at Work Regulations 1992 set out principles for selecting, providing, maintaining and using personal protective equipment (**PPE**). This is equipment designed to be worn or held to protect against a risk to health or safety. It includes most types of protective clothing and equipment such as eye, hand, foot and head protection.

Employers have a duty to reduce the workplace risk to as low as is reasonably practicable; where some risk remains, then PPE is issued as a last resort. PPE should then be issued free of charge by the employer and must be suitable to protect against the risk for which it is issued.

Employers	Employees
Must train employees and give information on maintaining, cleaning and replacing damaged PPE	Must use PPE provided by their employer, in accordance with any training in the use of the PPE concerned
Must provide storage for PPE	Must inform employer of any defects in PPE
Must ensure that PPE is maintained in an efficient state and in good repair	Must comply with safety rules
Must ensure that PPE is properly used	Must use safety equipment as directed

Table 1.01 PPE responsibilities of employers and employees

Areas requiring protection, together with related examples of PPE, are shown on the following pages. However, as formal risk assessment defines the type of PPE required in any given situation, no attempt has been made to quantify the circumstances in which they should be used.

Eye protection

Every year thousands of workers suffer eye injuries, which result in pain, discomfort, lost income and even blindness. Following safety procedures correctly and wearing eye protection can prevent these injuries. There are many types of eye protection equipment available, e.g. safety spectacles, box goggles, cup goggles, face shields, and welding goggles. To be safe you have to have the right type of equipment for the specific hazard you face.

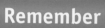

Remember

Your eyes are two of your most precious possessions. They are among the most vulnerable parts of your body to injury at work

Safety goggles

Helmet with visor face screen

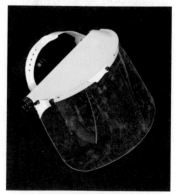

Half-face mask

Foot and leg protection

Guarding your toes, ankles, feet and legs from injury also helps to protect your whole body from injury caused by improper footwear, for example an injury caused by electric shock.

Protective footwear can help prevent injury and reduce the severity of injuries that do occur.

PPE		Usage
Safety shoe, boot or trainer		Basic universal form of foot protection Safety clogs can also be worn. These are used particularly as protection against hot asphalt
Spats	Often made of leather and worn over the shoe	Protect the feet from stray sparks during welding
Gaiters	Worn over the lower leg and top of the shoe	Used to give protection against foul weather/ splashing water
Leggings		Protect general clothing

Table 1.02 Types of leg and foot PPE

Did you know?

Toe and foot accidents account for a large proportion of reported accidents in the workplace each year, and many more accidents go unrecorded

Hand protection

This involves the protection of two irreplaceable tools – your hands, which you use for almost everything: working, playing, driving, eating etc. Unfortunately hands are often injured. One of the most common problems other than cutting, crushing or puncture wounds is **dermatitis**.

Skin irritation may be indicated by sores, blisters, redness or dry, cracked skin that is easily infected.

To protect your hands from irritating substances you need to:

- keep them clean by regular washing using approved cleaners
- wear appropriate personal protection when required
- make good use of barrier creams where provided.

Rigger gloves

Gauntlet gloves

Head protection

Head protection is important because it guards your most vital organ – your brain! Head injuries pose a serious threat to your brain and your life. Head protection can help to prevent such injuries.

Here is a list of good safety practices:

- know the potential hazards of your job and what protective gear to use
- follow safe working procedures
- take care of your protective headgear
- notify your supervisor of unsafe conditions and equipment
- get medical help promptly in the case of head injury.

There are several types of protective headwear for use in different situations; use them correctly and wear them whenever they are required.

Definition

Dermatitis
– inflammation of the skin normally caused by contact with irritating substances

Remember

A single injury can handicap a person for life or even be fatal

Safety helmets

Here are a few important rules:

- adjust the fit of your safety helmet so it is comfortable
- all straps should be snug but not too tight
- do not wear your helmet tilted or back to front
- never carry anything inside the clearance space of a hard hat, e.g. cigarettes, cards and letters
- never wear an ordinary hat under a safety helmet
- do not paint your safety helmet because this could interfere with electrical protection or soften the shell
- only use approved types of identification stickers on your safety helmet, e.g. First Aider
- do not use sticky or Dymo tape because the adhesive could damage the helmet
- handle the helmet with care; do not throw it or drop it etc.
- regularly inspect and check the helmet for cracks, dents or signs of wear, and if you find any, get your helmet replaced
- check the strap for looseness or worn stitching; also check your safety helmet is within its 'use-by' date.

Bump caps

For less dangerous situations, where there is a risk of bumping your head rather than things falling, or where space is restricted, bump caps, which are lighter than safety helmets, may be acceptable.

If you have to work outside in poor conditions, and a safety helmet is not a requirement, consider using a Souwester and cape.

Did you know?

An estimated 80% of industrial head injuries are sustained by people who are not wearing any protective equipment

Safety helmet

On the Job: Head protection

Fazal is a first-year apprentice who is starting his first day on a construction site. The electrician in charge gives Fazal a hard hat to wear. Fazal notices that the hat has a crack down one side and reports this to his supervisor. The supervisor tells Fazal that he must wear this hat because there are no more in the store and he needs to get on with his work.

1. What should Fazal do in this situation?

2. If an object hit Fazal on the head and he was injured because of the crack in the hat, who would be responsible for this injury?

Safety tip

If you have long hair and do not want it ripped off by something, tie it up or use a hair net

Hearing protection

You may sometimes have to work in noisy environments and, if you do, your employer should provide you with suitable hearing protection. Like any other sort of PPE it is important that you wear hearing protection properly and check it regularly to make sure it is not damaged. Typical types are ear plugs that fit inside the ear, or ear defenders that sit externally, such as headphones.

Ear plugs

Ear defenders

Lung protection

We all need clean air to live, and we need correctly functioning lungs to allow us to inhale that air. Fumes, dusts, airborne particles (such as asbestos) or just foul smells, (such as in sewage treatment plants), can all be features of construction environments.

A range of respiratory protection is available, from simple dust protection masks to half-face respirators, full-face respirators and powered breathing apparatus. To be effective these must be carefully matched to the hazard involved and correctly fitted. You may also require training in how to use them properly.

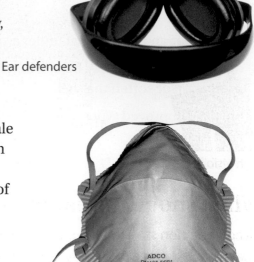

Disposable dust respirator

Whole body protection

To complete our 'suit of armour' against all things harmful, we need to protect the rest of our body. Usually this will involve overalls, donkey jackets or similar, to protect against dirt and minor abrasions. However, specialist waterproof or thermal clothing will be needed in adverse weather conditions; high visibility clothing is required on sites or near traffic; and chemical resistant clothing, such as neoprene aprons, is necessary if working near or with chemical substances. When working outdoors, sunscreen is now also a consideration.

High-visibility jacket   Overalls   High-visibility waistcoat

When should I wear PPE?

All construction sites that you work on will require you to wear a similar basic level of PPE.

PPE	When worn
Hard hats	• where there is a risk of you either striking your head or being hit by falling objects
Eye protection	• when drilling or chiselling masonry surfaces • when grinding or using grinding equipment • when driving nails into masonry • when using cartridge-operated fixing tools • when drilling or chiselling metal • when drilling any material that is above your head
Ear protection	• when working close to noisy machinery or work operations
Gloves	• whenever there is a risk to the hands from sharp objects or surfaces • when handling bulky objects to prevent splinters, cuts or abrasion • when working with corrosive or other chemical substances
Breathing protection	• when working in dusty environments • when working with asbestos • when working where noxious odours are present • when working where certain gases are present

Table 1.03 When to wear PPE

Remember

All items of PPE that are provided for your protection must be worn and kept in good condition

Other items

When working involves long periods of kneeling or having to take your weight on your elbows, you may be issued with specialist protectors for these areas. Other items you may use could include safety harnesses.

Safe isolation of electrical supplies

On completion of this topic area the candidate will be able to state the need for isolation before any work is carried out on an electrical installation or system.

Many fatal electrical accidents occur during the proving of isolation. You must be on guard, as you may have no idea whatsoever of the type of supply you are dealing with. Before beginning work on any electrical circuit you should make sure that it is completely isolated from the supply by following recognised procedures drawn up by the Joint Industry Board for the Electrical Contracting Industry:

1 identify sources of supply

2 isolate

3 secure the isolation

4 test voltage indicator

5 test system dead

6 retest voltage indicator

7 begin work.

Identify sources of supply

Not only is it important to identify the source of supply, but you also need to know what type it is. You need to be aware that a Band I supply covers extra-low voltage, that is a voltage normally not exceeding 50 volts a.c. Fifty volts should not be enough to kill you by passing/forcing a current through your body, but a shock could cause you to lose balance and fall from height if you were working on ladders or scaffold, and that could result in a personal injury.

Band I supplies also cover telecommunications, signalling, bell, and control and alarm installations where the voltage is limited for operational reasons.

Band II supplies contain low voltage, which is a voltage normally exceeding extra-low voltage but not exceeding 1000 V a.c. between conductors or 600 V a.c. between conductors and earth. Band II also includes the voltages for supplies to households and most commercial and industrial installations. Band II will therefore contain both single-phase and three-phase supplies. The actual voltage of the supply may differ from the nominal value by a quantity within normal tolerances, which are +10 per cent to −6 per cent.

After identifying the type of supply, you now need to identify the source of supply including any stand-by supplier that may switch in automatically. It is not unusual for rooms or buildings to be fed from more than one source.

The lighting circuit may be fed from a different distribution board than the ring circuit. You may also find that a radial circuit has been added to the installation and is fed from a completely different source.

All these sources need to be located and isolated before safe work can proceed. Schematic drawings will help you to identify the source of supply for isolation.

Isolate

Now that the source and type of supply have been identified they need to be isolated. Regulations require that a means of isolation must be provided to enable electrically skilled persons to carry out work on or near parts that would otherwise normally be energised.

Isolating devices must comply with British Standards, and the isolating distance between the contacts must comply with the requirements of BS EN 60947-3 for an isolator. The position of the contacts must either be externally visible or be clearly, positively and reliably indicated.

If it is installed remotely from the equipment to be isolated, the device must be capable of being secured in the open position using a lock and key that are unique to that isolator.

You must appreciate the basic difference between switching off and isolating.

- Switching off may involve the breaking of normal load current (or even higher current due to overload or short circuit).

- Isolation is concerned with cutting the already dead circuit so that re-closing the switch will not make it live again.

A semi-conductor device, e.g. a dimmer, cannot be used for isolation. While a switch must be able to break currents, an isolator need not be designed to do so because it should only be operated after the current has been broken. Typically a single-phase installation would be isolated using a double-pole device and an installation supplied with a three-phase four-wire system would be isolated using a three-pole switch.

Secure isolation

To prevent any unauthorised or unexpected re-closing of the contacts the **isolator** must be provided with a means of locking in the 'off' position, with a padlock and a key that is unique to that padlock and can be kept by the person at work on the circuit or equipment (Reg. 537.2.2.4).

Test that the equipment and system is dead

All circuits to be worked on must be tested to ensure they are dead. First, test the voltage indicator on a known supply or proving unit before use. Then test between line and neutral, line and earth, and neutral and earth for single-phase supplies. For three-phase supplies, test between brown and black, brown and grey, grey and black, brown and blue, black and blue, grey and blue, and brown and green/yellow, grey and green/yellow, black and green/yellow, blue and green/yellow.

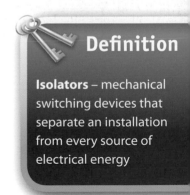

Definition

Isolators – mechanical switching devices that separate an installation from every source of electrical energy

On 1 April 2004 the insulation colours changed. However, for many years to come you will find existing installations with the old colours. These were:

- single-phase – red, black
- three-phase – red, yellow and blue.

Retest the voltage indicator on a known supply or proving unit after use. You must also erect warning notices stating 'Danger: Electrician at Work'.

Begin work

Some companies make use of a paper-based 'Permit to Work' system for all isolation procedures. Make use of the warning notices 'Danger: Electrician at Work' when you are working, and you should ensure that your name and the time and date you started work are included on the notice. This will help anyone to establish why power has been turned off. The charts on pages 19 and 20 show you standard isolation procedures.

Test equipment

All test equipment must be regularly checked to make sure it is in good and safe working order. You must ensure that your test equipment has a current calibration certificate indicating that the instrument is working properly and providing accurate readings. If you do not do this, test results could be void.

When checking the equipment yourself the following points should be noted.

- Check equipment for any damage, and see if the case is cracked or broken. This could indicate a recent impact, which could result in false readings.
- Check that the batteries are in good condition and have not leaked (local action), and check that they are all of the same type.
- Check that the insulation on the leads and probes is not damaged but complete and secure.
- Check the operation of the meter with the leads both open and short-circuited.
- Then zero your instrument on the ohm scale. If you have any doubt about an instrument or its accuracy ask for assistance.

Remember

All live conductors must be isolated before work can be carried out. As the neutral conductor is classified as a live conductor this should also be disconnected. This may mean removing the conductor from the neutral block in the distribution board. Not all distribution boards are fitted with double pole isolators so the connecting sequence for neutral conductors needs to be verified and maintained

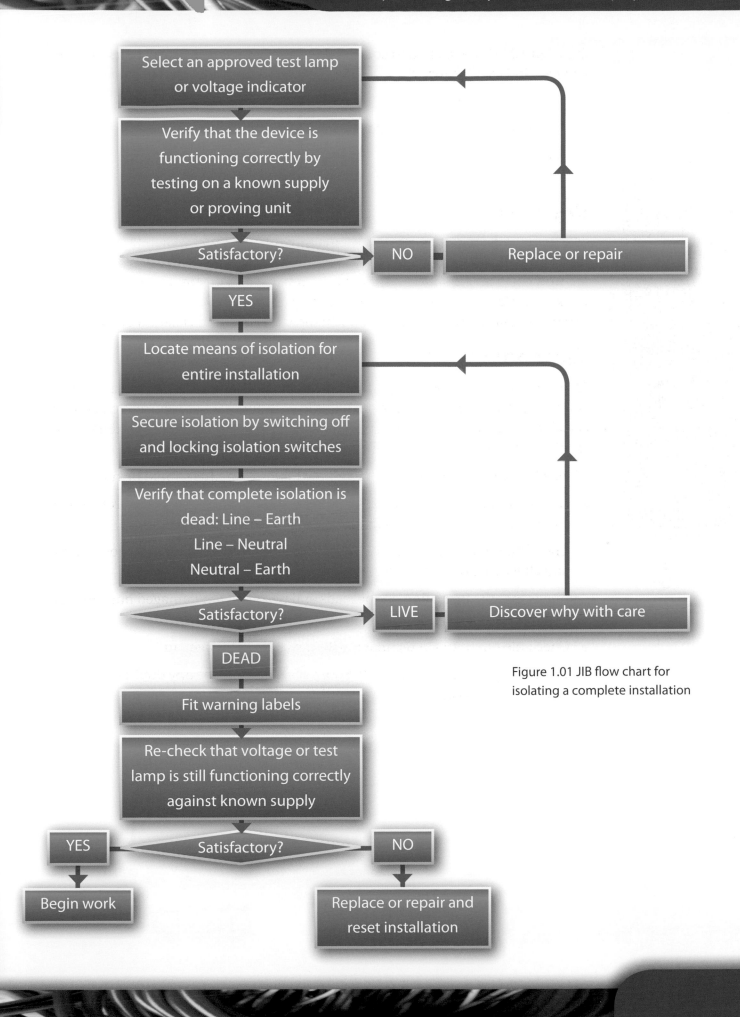

Figure 1.01 JIB flow chart for isolating a complete installation

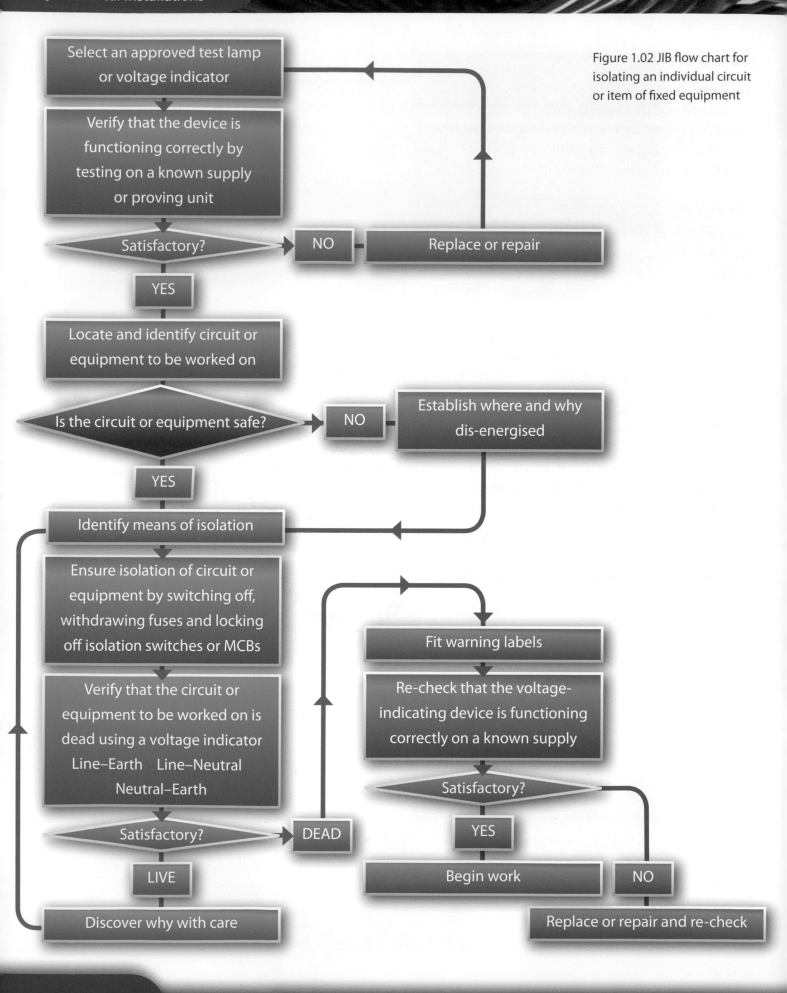

Figure 1.02 JIB flow chart for isolating an individual circuit or item of fixed equipment

Select an approved test lamp or voltage indicator

Verify that the device is functioning correctly by testing on a known supply or proving unit

Satisfactory? → NO → Replace or repair

YES

Locate and identify circuit or equipment to be worked on

Is the circuit or equipment safe? → NO → Establish where and why dis-energised

YES

Identify means of isolation

Ensure isolation of circuit or equipment by switching off, withdrawing fuses and locking off isolation switches or MCBs

Fit warning labels

Verify that the circuit or equipment to be worked on is dead using a voltage indicator
Line–Earth   Line–Neutral
Neutral–Earth

Re-check that the voltage-indicating device is functioning correctly on a known supply

Satisfactory? → DEAD

Satisfactory?

LIVE

YES

DEAD

Begin work

NO

Discover why with care

Replace or repair and re-check

Guidance Note GS 38

Published by the HSE, this document covers electrical test equipment used by electricians and gives guidance to electrically competent people involved in electrical testing, diagnosis and repair. Electrically competent people may include electricians, electrical contractors, test supervisors, technicians, managers or appliance repairers. Advice is given in the following areas.

Test probes

Test probes and leads used in conjunction with a voltmeter, multimeter, electrician's test lamp or voltage indicator should be selected to prevent danger. Good test probes and leads will be as described below.

Probes

- Should have finger barriers or are shaped to guard against inadvertent hand contact with the live conductors under test.

- Are insulated to leave an exposed metal tip not exceeding 2mm measured across any surface of the tip. Where practicable it is strongly recommended that this is reduced to 1mm or less, or that spring-loaded retractable-screened probes are used.

Recommended test probe

- Should have suitable high-breaking capacity fuse or fuses with a low current rating usually not exceeding 500 mA or a current limiting resistor and a fuse.

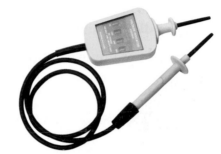

Non-recommended test probe

Leads

- Are adequately insulated.

- Are coloured so that one lead can easily be distinguished from the other.

- Are flexible and of sufficient capacity for the duty expected of them.

- Are sheathed to protect against mechanical damage.

- Are long enough for the purpose while not being so long that they are clumsy or unwieldy.

- Do not have accessible exposed conductors other than the probes or tips, nor live conductors accessible to the person's finger should a lead become detached from a probe. The test lead or leads are held captive and sealed into the body of the voltage detector.

- Are not cracked or damaged in any way.

Voltage-indicating devices

Instruments that are used solely for detecting a voltage fall into two categories as described below.

1. Detectors that rely on an illuminated lamp (test lamp) or a meter scale (test meter)

Test lamps are fitted with a 15 W lamp and should not give rise to danger if the lamp is broken. They should also be protected by a guard.

These detectors require protection against excess current. This may be provided by a suitable high breaking capacity fuse with a low current rating usually not exceeding 500 mA or by means of a current-limiting resistor and a fuse. These protective devices are housed in the probes themselves. The test lead or leads are held captive and sealed into the body of the voltage detector.

2. Detectors that use two or more independent indicating systems (one of which may be audible) and limit energy input to the detector by the circuitry used

An example is a 2-pole voltage detector, i.e. a detector unit with an integral test probe, an interconnecting lead and a second test probe.

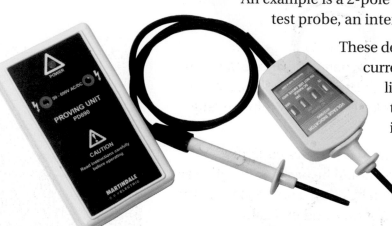

These detectors are designed and constructed to limit the current and energy that can flow into the detector. The limitation is usually provided by a combination of the circuit design, using the concept of protective impedance, and current-limiting resistors built into the test probes. These detectors are provided with in-built test features to check the functioning of the detector before and after use. The interconnecting lead and second test probes are not detachable components.

2-pole voltage detector

These types of detector do not require additional current-limiting resistors or fuses to be fitted provided that they are made to an acceptable standard and the contact electrodes are shrouded as previously mentioned.

It is recommended that test lamps and voltage indicators are clearly marked with the maximum voltage that may be tested by the device and any short-time rating for the device if applicable. This rating is the recommended maximum current that should pass through the device for a few seconds; these devices are generally not designed to be connected for more than a few seconds.

Systems of work

The use of test equipment by electricians falls into three main categories.

1. Testing for voltage.

2. Measuring voltage.

3. Measuring current, resistance and, occasionally, inductance and capacitance.

Item 1 forms an essential part of the procedure for proving a system dead before starting work, but it can be associated with simple tests to prove presence of voltage. Items 2 and 3 are more concerned with commissioning procedures and fault-finding.

Before testing begins it is essential to establish that the device – including all leads, probes and connectors – is suitably rated for the voltage and currents that may be present on the system under test. Where a test is being made simply to establish the presence or absence of voltage, the proprietary test lamps or 2-pole voltage detectors suitable for the working voltage of the system, are preferred to a multimeter.

For voltage detection or measurement, fuse-protected test leads are recommended when voltmeters and, in particular, multimeters are used. Some multimeters have electromechanical overload devices, but these are often inadequately rated to deal with short-circuit energy present on electrical power systems.

It is usually necessary to use leads that incorporate high breaking capacity fuses even if the multimeter has an overload trip. If thermal clips are provided for connection to test points they should be adequately insulated and arranged to be suitable for use with the test leads as a safe alternative to the use of test probes.

It is important that a multi-function or multi-range meter is set to the correct function and/or range before the connections are made. Where there is doubt about the value of voltage to be detected or measured, the highest range should be selected at first provided that the maximum voltage possible is known to fall within the range of the instrument.

Progressive voltage detection or measurement is often used to prove circuit continuity. The dangers from exposed live conductors should be borne in mind when using this method. In many cases continuity testing can be carried out safely with the apparatus dead using a self-contained low voltage d.c. source and indicator.

Keep records of inspection and testing of the equipment, particularly when faults are found. These will help decide how often visual inspection or testing will need to be carried out. Test instruments should be re-calibrated at regular intervals and the calibration certificate retained for reference purposes.

Safety tip

All items of test equipment, including those items issued on a personal basis, should receive a regular inspection and where necessary a test by a competent person

Did you know?

Accident history shows that the use of incorrectly set multimeters or makeshift devices for voltage detection has often caused accidents

FAQ

Q Is it possible to use any multimeter to measure voltage on site?

A No, as there is a risk that you could set the instrument range incorrectly, causing damage to the instrument and yourself. You should use an approved GS 38 voltage-indicating device.

Safety signs

On completion of this topic area the candidate will be able to identify types and meanings of safety signs.

Working on construction sites, in factories or elsewhere, you will see a variety of signs and notices. It is up to you to learn and understand what they mean and to take notice of them. Most warn of a possible danger and must not be ignored.

Figure 1.03 Types of safety sign

Each type can be recognised by its shape and colour. Some signs are just symbols, while others have words or other information such as heights and distances.

Accident and emergency procedures

On completion of this topic area the candidate will be able to follow the accident and emergency procedures that can be found in the workplace.

Accident prevention

How do we prevent accidents? So far everything we have looked at has been based around the legal requirements and fairly obvious material problems. These tend to be referred to as 'environmental causes', as they relate to the environment that you work in. Such causes can be unguarded machinery, defective tools and equipment, poor ventilation, excessive noise, poorly lit workplaces, overcrowded workplaces and untidy or dirty workplaces.

However there are other causes of accidents and they invariably involve people. It would be nice to think that we all possess common sense, but the 'human causes' of accidents include:

- carelessness
- bad and foolish behavior
- improper dress
- lack of training

- lack of experience
- poor supervision
- fatigue
- use of alcohol or drugs.

Remember

Prevention is better than cure!

We each need to take personal responsibility for our actions. By thinking about what we are doing, or are about to do, we can avoid most potentially dangerous situations. This can be by reporting potentially dangerous situations, hazards or activity to the correct people. Sometimes the hazard is something you can easily fix yourself, for instance you might move a cable that was causing a trip hazard. In these cases, though, you must still report it to your supervisor. Remember the fact that the cable was there at all might be indicative of someone not doing their job properly and it could happen again.

Accident procedures

In the event of you discovering someone who has had an accident it is important to:

- check for further danger to yourself and the casualty
- make an assessment of the casualty – are they bleeding, are they breathing, are they unconscious
- summon help
- only move the casualty if it is really necessary; for serious injuries the casualty should remain still until professional medical help has arrived.

Remember

The first person to alert about a health and safety issue should be your Site Supervisor or Safety Officer

In the event of an accident it is important to record details of the accident, so that lessons can be learned to prevent other accidents, and to make sure that, in the event of a claim for compensation, the facts of the accident can be presented for evaluation.

All accidents must be reported to the employer and details recorded into the accident book. To do so you will need to record:

- details of the person injured
- details of what happened
- details on the injury sustained

- details of any witnesses
- date and time of the accident
- name and signature of the person reporting the accident.

The Reporting of Injuries, Diseases and Dangerous Occurrences Regulations 1995 (RIDDOR) places a duty upon the employer to report some work-related accidents, diseases and dangerous occurrences to the Health and Safety Executive. Such a report would be made in the event of a death or where the casualty loses a limb, loses consciousness, or loses more than three days from work. Reports to the Health and Safety Executive should be made by completing form F.2508 or by calling the Incident Contact Centre on 0845 300 9923.

Before starting work on any site it is important to establish the emergency procedures such as the emergency exits, the emergency assembly points, how to raise the fire alarm and the position of the first aid boxes, as well as how to summon a first aider.

Usually first aid boxes can be found in the site foreman's office or in the offices of the various contractors on site. It is also common practice for most company vans to have a first aid box in them. On most sites the main contractor will have a qualified first aider. On a large site this person may be employed in this role full time, while on smaller sites the first aider is usually one of the workforce. The first aider is contacted through the site foreman. This is done either by radio or mobile phone. This can vary from site to site, so you would need to check arrangements when you first arrive.

First aid for electric shock

On completion of this topic area the learner will be able to state the basic action to be taken in the event of discovering a casualty receiving an electric shock.

Electric shock

Electric shock occurs when a person becomes part of the electrical circuit. The severity of the shock will depend on the level of current and the length of time it is in contact with the body.

The lethal level is approximately 50 mA, above which muscles contract, the heart fibrillates and breathing stops. A shock current above 50 mA could be fatal unless the person is quickly separated from the supply. Below 50 mA an unpleasant tingling or painful sensation may be experienced. However, this may cause you to fall from a roof or ladder, which could lead to a serious injury.

In the event of an electric shock

☑ First of all, check for your safety to ensure that you would not put yourself at risk by helping the casualty.

☑ Next, break the electrical contact to the casualty by switching off the supply, removing the plug (if it is undamaged) or wrenching the cable free (only attempt this if the cable and plug etc. are undamaged). If this is not possible, break the contact by pushing or pulling the casualty free using a piece of non-conductive material, e.g. a piece of wood.

☑ If the casualty is conscious, guide them to the ground, making sure that further injuries are not sustained, e.g. banging their head on the way down. If the casualty is unconscious, get help straight away, then check the casualty's response. Talk to and gently shake the casualty to gauge their level of response. If the casualty appears unharmed, they should be advised to rest and see a GP.

☑ If there is no movement or any sign of breathing summon help immediately. If there is someone with you, tell them to get help, i.e. ring 999; if you are on your own with the casualty you will have to leave them for a moment while you get help yourself.

☑ As soon as you return to the casualty you need to begin CPR (cardio-pulmonary resuscitation).

Airways, Breathing, Circulation (ABC)

When a casualty is unconscious, you need to know to what degree they are unconscious. Ask a question or give a command – for example, 'What happened?' or 'Open your eyes'. Speak loudly and clearly, close to the casualty's ear. Carefully shake their shoulders. A casualty in a serious state of 'altered consciousness' may mumble, groan or make slight movements. A fully unconscious casualty will not respond.

Remember

Remember your ABC of resuscitation: Airways, Breathing, Circulation

Open the airway. An unconscious casualty's airway may be narrowed or blocked, making breathing difficult and noisy, or impossible. The main reason for this is that muscular control in the throat is lost, which allows the tongue to sag back and block the throat. Lifting the chin and tilting the head back lifts the tongue away from the entrance to the air passage.

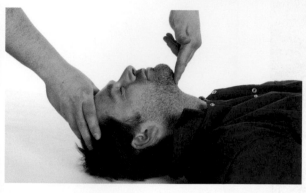

To open the airway, remove any obvious obstruction from the mouth. Place two fingers under the point of the casualty's chin as shown here and lift the jaw. At the same time, place your other hand on the casualty's forehead and tilt the head well back.

To check for breathing put your face close to the casualty's mouth. Look and feel for breathing as well as looking for chest movements. Listen for the sound of breathing. Feel for breath on your cheek. Look, listen and feel for 10 seconds before deciding that breathing is absent.

Opening the airway

Check for signs of life and a pulse. Because of the time being lost trying to find the carotid pulse in the neck, the National Institute of Clinical Excellence currently does not advise this method of checking for a pulse. Current thinking is to listen to the patient's chest instead. Check for about 10 seconds to see whether a pulse is present. If the casualty is breathing, place them in the recovery position (see page 30). If not, deliver CPR (cardio-pulmonary resuscitation) in the following manner.

Mouth-to-mouth ventilation is given with the casualty lying flat on their back:

- first, remove any obvious obstruction including broken or displaced dentures from the mouth

- open the airway by tilting the head and lifting the chin

- close the casualty's nose by pinching it with your index finger and thumb

- take a full breath, then place your lips around their mouth, making a good seal

- blow into the casualty's mouth until you see the chest rise; take about 2 seconds for full inflation

- remove your lips and allow the chest to fall fully

- repeat

- if there is a response from the casualty, place them in the recovery position. If not, begin chest compressions.

Chest compressions

- With the casualty still lying flat on their back on a firm surface, kneel beside them, and find one of the lowest ribs using your index and middle fingers.

- Slide your fingers upwards to the point in the middle where the rib margins meet the breastbone.

- Place your middle finger over this point and your index finger on the breastbone above.

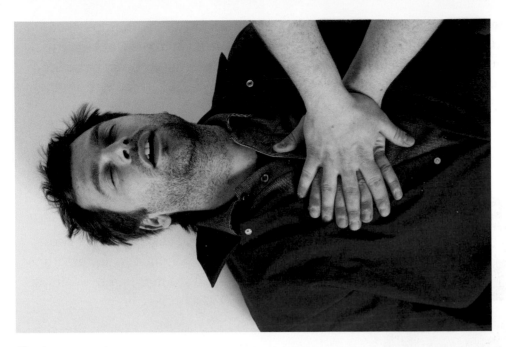

Chest compressions

- Place the heel of your hand on the breastbone, and slide it down until it reaches your index finger. This is the point at which you will apply pressure.

- Place the heel of your first hand on top of the other hand, and interlock fingers.

- Leaning well over the casualty, with your arms straight, press down vertically on the breastbone to depress it approximately 4–5 cm, then release the pressure without removing your hands.

- Repeat the compressions 30 times, aiming for a rate of approximately 100 compressions per minute.

- Return to the head and give two more ventilations, then 30 further compressions.

- Continue to give two ventilations to every 30 compressions until help arrives or you are unable to continue. Do not interrupt CPR to make pulse checks unless there is any sign of a returning circulation. With a pulse confirmed, check breathing and if it is still absent continue with artificial ventilation. Check the signs of life/pulse after every 10 breaths, and be prepared to re-start chest compressions if it disappears. If the casualty starts to breathe unaided, place them in the recovery position. Re-check breathing and pulse every three minutes.

The recovery position

Any unconscious casualty whose breathing is not impaired (see Opening the airway page 28) should be placed in the recovery position. This position prevents the tongue from blocking the throat, and, because the head is slightly lower than the rest of the body, it allows liquids to drain from the mouth, reducing the risk of the casualty inhaling stomach contents.

The head, neck and back are kept in a straight line while the bent limbs keep the body propped in a secure and comfortable position. If you must leave an unconscious casualty unattended, they can be left safely in the recovery position while you get help.

The recovery position

The technique for turning is described below and assumes that the casualty is found lying on their back from the start.

- Kneeling beside the casualty, open the airway by tilting the head and lifting the chin; straighten the legs. Place the arm nearest you out at right angles to the body, elbow bent, and with the palm uppermost.

- Bring the arm furthest from you across the chest, and hold the hand, palm outwards, against the casualty's nearer cheek. (Remember to turn any rings with stones in them to face the palm so that there is no contact between the ring and the casualty's face. Also remember to check pockets for keys etc. that could cause more injury when the casualty is turned on to his or her side.)

- With your other hand, grasp the lower thigh just above the knee furthest from you and pull the knee up, keeping the foot flat on the ground.

- Keeping the casualty's hand pressed against their cheek, pull at the knee to roll the casualty towards you and on to their side.

- Tilt the head back to make sure the airway remains open. Adjust the hand under the cheek, if necessary, so that the head stays in this tilted position and circulation to the hand is not restricted.

- Adjust the upper leg, if necessary, so that both the hip and the knee are bent at right angles.

Treatment for burns, shock and breaks

An electric shock may result in other injuries as well as unconsciousness. There may be burns at both the entry point and exit point of the current. The treatment for these burns is to flood the site of the injury with cold water for at least 10 minutes. This will halt the burning process, relieve pain and minimise the risk of infection.

You would then need to treat for shock, which is the medical condition where the circulatory system fails and insufficient oxygen reaches the tissues. If this shock is not treated quickly the vital organs can fail, leading ultimately to death. To treat shock you need to:

- stop external bleeding if there is any
- lay the casualty down, keeping the head low
- raise and support the legs but be careful if you suspect a fracture
- loosen tight clothing, braces, straps or belts to reduce constriction at the neck, chest and waist
- keep the casualty warm by wrapping them in a blanket or coat
- continue to check and record breathing, pulse and level of response
- be prepared to resuscitate if necessary.

Once you are satisfied that the casualty is in a stable condition, you should cover any burned skin with a loose, lint-free dressing, or even with sheets of cling film (do not wind any dressing around the injured area). If the casualty has sustained a broken bone, then your first aim is to prevent movement at the site of the injury.

Do not move the casualty until the injured part is secured and supported, unless they are in danger, i.e. from further electric shock. You must arrange for the casualty's immediate removal to hospital, maintaining comfortable support during transport.

Treatment for smoke and fume inhalation

If you find someone suffering from the effects of fume inhalation or asphyxiation, provided that it is safe to do so, get them outside into the fresh air as soon as possible. Loosen any clothes around their neck or chest that may impair their breathing. Call the emergency services and refer to the first aid section of this book regarding the ABC procedure, pages 27–29.

Remember

This is only a temporary measure. The casualty must reach hospital as soon as possible, preferably by ambulance

Remember

Be prepared to treat for shock at any time

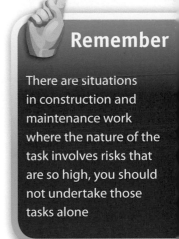

Remember

There are situations in construction and maintenance work where the nature of the task involves risks that are so high, you should not undertake those tasks alone

Fire safety

On completion of this topic area the candidate will be able to identify types and applications for various fire extinguishers.

As an electrician, you may well end up working in what are known as **explosive atmospheres**, and fire is always a risk when dealing with electrical installation. So what exactly is a fire?

What is a fire?

Essentially, fire is very rapid **oxidation**.

Rusting iron and rotting wood are common examples of slow oxidation. Fire, or **combustion**, is rapid oxidation as the burning substance combines with oxygen at a very high rate. Energy is given off in the form of heat and light. Because this energy production is so rapid, we can feel the heat and see the light as flames.

Fire-fighting equipment

Normally available fire-fighting equipment includes portable appliances such as extinguishers, buckets of sand or water and fire-resistant blankets. In larger premises you will find automatic sprinklers, hose reels and hydrant systems.

Fire extinguishers

There are many types of fire extinguisher, each with a specific set of situations in which they may or may not be used. Four common types of fire extinguisher are covered below.

In the past, the whole extinguisher was coloured according to its type. Black ones, for instance, were carbon dioxide (CO_2). Today this colour-coding is not immediately obvious as all extinguishers are red. In the heat of the moment, be careful to pick up the right extinguisher for the type of fire you are trying to put out. Powder and foam extinguishers both come in two types, and none is totally effective on every kind of fire. Before buying one, it is vital to look carefully at what kinds of fires it can be used on.

Fire extinguisher types	
Standard/Multi-purpose dry powder	
Colour	Blue
Application	The powder 'knocks down' the flames. Safe to use on most kinds of fire. Multi-purpose powders are more effective, especially on burning solids; standard powders work well only on burning liquids.
Dangers	The powder does not cool the fire well. Fires that seem to be out can re-ignite. Does not penetrate small spaces, like those inside burning equipment. The jet could spread burning fat or oil around.
How to use	Aim the jet at the base of the flames and briskly sweep it from side to side.

Water fire extinguisher

Water	
Colour	Red
Application	The water cools the burning material. You can only use water on solids, like wood or paper. Never use water on electrical fires or burning fat or oil.
Dangers	The water can conduct electricity back to you. Water actually makes fat or oil fires worse – they can explode as the water hits them.
How to use	Aim the jet at the base of the flames and move it over the area of the fire.
CO$_2$	
Colour	Black
Application	Displaces oxygen with CO$_2$ (a non-flammable gas). Good for electrical fires as they do not leave a residue.
Dangers	Pressurised CO$_2$ is extremely cold. DO NOT TOUCH. Do not use in confined spaces.
How to use	Aim the jet at the base of the flames and sweep it from side to side.
Foam/AFFF (Aqueous Film Forming Foam)	
Colour	White or cream
Application	The foam forms a blanket or film on the surface of a burning liquid. Conventional foam works well only on some liquids, so it is not good for use at home, but AFFF is very effective on most fires except electrical and chip-pan fires.
Dangers	'Jet' foam can conduct electricity back to you, though 'spray' foam is much less likely to do so. The foam could spread burning fat or oil around.
How to use	For solids, aim the jet at the base of the flames and move it over the area of the fire. For liquids, do not aim the foam straight at the fire – aim it at a vertical surface or, if the fire is in a container, at the inside edge of the container.

Table 1.04 Some fire extinguisher types

Dry powder fire extinguisher

Carbon Dioxide fire extinguisher

Foam fire extinguisher

When considering using a fire extinguisher remember the following points:

- never use a fire extinguisher unless you have been trained to do so
- do not use water extinguishers on electrical fires due to the risk of electric shock and explosion
- do not use water extinguishers on oils and fats as this too can cause an explosion
- do not touch the horn on CO$_2$ extinguishers as this can freeze burn the hands
- do not use a CO$_2$ extinguisher in a small room as this could cause suffocation
- read the operating instructions on the extinguisher before use.

Activity

1. Have a look around your place of work and see if you can spot any 'improvised' equipment such as home-made test probes and connectors. Do you think they comply with PUWER?

2. Under supervision, carry out a safe isolation procedure on the following:

   • an individual item of equipment

   • a single final circuit

   • an entire installation.

FAQ

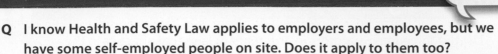

Q I know Health and Safety Law applies to employers and employees, but we have some self-employed people on site. Does it apply to them too?

A Yes. Self-employed people have to comply with Health and Safety Law just like employed persons.

Q Does the Electricity at Work Regulations 1989' mean it doesn't apply to domestic properties?

A No. While you are working at a domestic property it becomes your place of work, so the regulations will apply to you.

Q I keep being told that BS 7671 is non-statutory. Does this mean that I don't have to comply with it?

A Strictly speaking, deviations from the Regulations are allowed, but they must not result in a lesser degree of safety than that provided by the Regulations and you cannot claim that the installation conforms to the Regulations if you do depart from them. In practice, deviating from the Regulations is not a good idea and may well mean that you have contravened a statutory requirement.

Q I hate wearing PPE; it gets in the way and is uncomfortable. Do I have to wear it?

A You're quite right, PPE does get in the way, but it is only doing its job! The purpose of PPE is to put a barrier between you and the hazard to stop you being harmed by it. The best form of protection is to remove the hazard, then you won't need to wear PPE but, if the hazard remains, then so must the PPE.
If you find PPE uncomfortable, help yourself by ensuring that it fits properly, is correctly adjusted and is clean. Modern PPE is lighter and more comfortable than it used to be – you can even buy steel toe-capped boots that don't hurt your feet!

Q Can I use my multimeter to test for voltage during a safe isolation procedure?

A Multimeters are not the best instrument to use when isolating supplies because they have lots of switches and buttons which may be moved during the procedure. You could be fooled into believing a supply is dead when, in fact, you have simply selected the wrong range. A simple two-lead voltage indicator is best – there is less that could go wrong.

Knowledge check

1. List the three main 'statutory instruments' or laws that affect us in the electrical industry.

2. For each of the examples below, state which would be the most specific and relevant law breached in each case.

   • Failure to provide a means of reporting an accident.

   • Using a screwdriver as a wood chisel.

   • Storing a hazardous substance in a fizzy drinks bottle.

   • Customising your hard hat with stickers.

   • Working live on a domestic shower circuit.

3. Make a list of the different type of PPE that you could be expected to wear during your normal work and describe their function.

4. Describe the circumstances in which you would be required to wear PPE.

5. How many keys do you think a padlock used for safe isolation purposes should have? Explain the reasons for your answer.

6. Why is it important to test from L–N, L–E and N–E when isolating an electrical supply? Why could you not just test between L–N?

7. State three features of a 'good' set of test probes, according to GS 38.

8. Why is it no longer recommended that you check for a casualty's pulse at the carotid artery (the neck)?

9. Which is the best position for an unconscious casualty to be placed in and why?

10. List the main types of fire extinguisher and their applications.

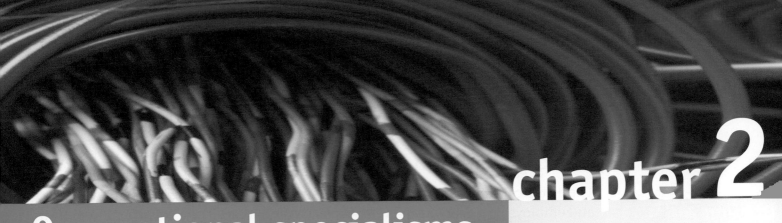

chapter 2

Occupational specialisms in the electro-technical industry

Unit 1 Outcome 2

The electro-technical industry is a large and diverse sector, taking in a number of specialisms. These range from public lighting installations to electrical maintenance and repair. All these areas have their own specific skills that you will need to develop in order to work in them.

Specialist skills will be needed in many different organisations – for example you will need to have a different expertise for factory work than you would for working with the local council. The tasks undertaken will also change with the job title you may have. In all cases, though, it is important to remember the principles of safe working.

On completion of this chapter the candidate will be able to:

- describe briefly the service provided within specialisms

- state organisations that carry out electro-technical activities and the roles of personnel within these organisations

- identify professional bodies, trade and employer associations, trade union and regulatory bodies associated with specific specialist companies.

Specialisms in the electro-technical industry

On completion of this topic area the candidate will be able to briefly describe the service provided within key specialisms.

The construction industry

The electrical contracting industry is just one sector within the construction industry, and employs well over one million people. Organisations in the construction industry range from 'sole traders' (for example, a jobbing builder), to large multinational companies employing thousands of workers.

The work done by these companies is very varied, but we can broadly think of it under three headings:

- building and structural engineering
- civil engineering
- maintenance.

Building and structural engineering covers the construction and installation of services for buildings such as factories, offices, shops, hospitals, schools, leisure centres and, of course, houses.

Civil engineering involves the construction and installation of services for large structures such as bridges, roads, motorways, docks, harbours and mines.

Maintenance covers the repair, refurbishment and restoration of existing buildings and structures.

Larger companies will be able to undertake work in all these areas, but smaller companies may specialise in one area.

Nearly every project, large or small, will involve a variety of different trades, such as bricklaying, plastering, plumbing and joining, as well as electrical work. The ability to work well with others and establish good professional relationships with colleagues and people in other trades is very important for the successful completion of a project.

Scope of the electro-technical industry

Most people are unaware of the vast range of activities and occupations that make up the electrical industry. Their understanding is usually limited to a person who installs lights and sockets in their house. In reality, during his or her career an electrician could be involved with the installation, maintenance and repair of electrical services (both inside and outside) associated with buildings and structures such as houses, hospitals, schools, factories, car parks, leisure centres and shops. There are also specialist areas that call for additional knowledge and training.

Electricians may be employed within the electrical contracting industry, but they may also be employed directly by organisations needing their skills.

These include:

- factories and manufacturing plant
- processing plant (e.g. chemical works, refineries)
- local councils
- commercial buildings
- leisure centres
- shopping complexes
- universities
- panel builders
- railways
- hospitals
- motor rewind and repair companies
- the armed forces.

The rail industry is one major sector that employs electricians

The largest sector is associated with building services. We will look at these in more detail followed by some of the specialist areas.

Building services

Some of the electrical services that can be found in and around our buildings and structures are:

- lighting and power installations
- alarm, security and emergency installations
- building management and control installations
- communication, data and computer installations.

Lighting and power installations

This covers all the varied types of lighting we see in houses and offices, from normal overhead lights through to the spotlights in car parks and floodlights in leisure centres and stadiums. Power installations cover everything from the installation of sockets in a house through to the installation of supplies for machines in factories.

Alarm, security and emergency installations

Many buildings have access control and security systems to prevent loss of, or damage to, valuable assets and information. If a fire or other emergency occurs, there must be a way of quickly alerting people to the danger. Most buildings contain some sort of fire alarm system, from a basic smoke detector in a house to complex systems in offices and shops (these are often linked directly to the emergency services).

Did you know?

Emergency lighting systems, capable of operating when the mains power fails, can help to guide people safely out of a building

Building management and control installations

There is an increasing use of complex electronic systems to control building services and installations, such as:

- 'intelligent' lighting (which switches on only when someone is present)
- air conditioning, heating and climate control systems
- energy management
- alarms and access control
- machine and motor control systems
- computer systems.

These developments place further demands on the skills and responsibilities of an electrician. More complex installations may require specialist contractors.

Communication, data and computer installations

Communications today, even within individual buildings, go well beyond simple speech and the telephone. The widespread use of computers enables people to communicate via email, to access data directly from other computers and the Internet, and to manage many aspects of an organisation from their desks. This interconnection of computers is called **networking**, and uses a structured cable installation. There are several different types of cable in use, such as Cat 5, Cat 5e, Cat 6 and fibre optic. In many cases, voice, data, video and control signals for building services are carried on a single cable.

Definition

Networking
– connecting computers together so that they all have access to the same common information

Networking computers can involve several different cable layers

Specialist areas

The installation electrician could deal with all of the previous systems. However, the following areas within the electro-technical sector are more likely to be undertaken by specialist contractors.

Cable jointing

This work involves connecting main power cables (high or low voltage, live or disconnected) to buildings, or within power stations and substations. It includes:

- making the joint
- placing insulating material around it
- testing the performance of the cables and insulation.

It must often be done outdoors. Sometimes non-specialist electricians can make small low voltage joints using cast resin kits.

Highway electrical systems

Electricians undertake some public lighting work, but this category of work also includes more complex areas like street lighting, and traffic control systems such as motorway signs and traffic lights. These tend to be catered for either by specialist sections within a local council or by electricians within specialist contractors who have received additional training.

Electrical machine drive installations

A growing number of countries are keen to encourage their industries to use electricity more efficiently. One way of doing this is to offer financial incentives for installing more energy-efficient motors and related equipment, such as electronic **variable speed drives** (also known as adjustable speed drives). This is an important measure because, with a high proportion of electricity being converted by motors into mechanical work, a small improvement in the efficiency of electric motors and motor-driven systems can lead to quite dramatic electricity savings. Motor systems can be either simple (e.g. a motor driving a fan) or highly complex systems used in process plants such as refineries. These will include motors, control systems, power supplies and motor-driven machines such as pumps and compressors. Such systems can consume around 40 per cent of industrial drive energy.

Variable speed drive

Improving the overall design and operation of the process, rather than simply improving the drive motor efficiency, usually gives the biggest energy savings. Performance improvement therefore focuses first on reducing process inefficiencies, before selecting new drive components and control strategies to match the new reduced load. In practice, however, production machinery is not always replaced so systematically, and existing motors often have to drive loads that are much higher, lower or more varied than those for which they were first specified. Therefore, minimising energy losses in the existing system becomes the priority.

Did you know?

Motor system efficiency drops quickly when motors operate below 40% of full-rated load

Pumps, compressors and fans etc. operate at optimum efficiency only within a narrow range of operation (flow and pressure), and so it is important that they are kept within that range. In some systems, load control can be helpful, but the biggest improvement is often obtained simply by matching the motor size to the actual load. Energy losses in drive trains (gearboxes, drive chains etc.) can also be reduced. In some cases, effective maintenance will improve the situation.

Organisations specialising in electrical machine drive installations consider all the above factors to produce the most energy efficient system.

Panel building

Control panels are used to control electricity distribution or control systems and equipment. Most panels are built in a factory, although some companies will build their products on site. There are three types of panels: switchgear, control panels and motor control circuit systems.

Instrumentation

This involves installing process control and measurement systems for a wide variety of industries, including water companies, chemical, pharmaceutical and process plant, and manufacturing sites. System components include:

- flow meters
- pumps
- pressure and temperature gauges
- valves and level switches.

Electrical maintenance

In the past, many installation electricians used their experience of installation to transfer to maintenance activities. Today, this job is usually seen as a specialist role, as it often has to cover a wide range of expertise related to the location, including electronics, safety issues and preventive maintenance, as well as electrical and mechanical repair and installation. Maintenance electricians are usually employed in:

- factories
- hospitals
- universities
- manufacturing plants.

They work on a wide range of systems and equipment, and do minor installation work when necessary.

We will look at other areas connected with this in greater detail later in this book.

Organisations in the electro-technical sector

On completion of this topic area the candidate will understand the main organisations in the electro-technical sector.

Employer structure

There are about 21,000 electrical contracting companies registered in the UK. They range from single-person organisations to large multinational contractors, but the majority employ fewer than ten people. The structure of electrical companies varies considerably between firms, depending on the number of employees and the type and size of the business. For ease we will think of them in two groups: small firms and large firms.

There are many tasks to be dealt with in electrical contracting, including:

- handling initial enquiries
- estimating cost
- issuing quotations
- dealing with suppliers and subcontractors
- supervising the contract
- carrying out the work
- financial control
- final settlement of the account.

A large firm will have specialists who concentrate on one or two of these tasks. In a small firm, one person might do several (or even all) of them.

Small firms

In a small company, the tasks listed above (with the possible exception of actually carrying out the work) are often the responsibility of one person. This type of company structure is known as a vertical structure. The principal advantage is that lines of communication are short: everyone knows who is responsible for different aspects of a project. When someone from outside (another tradesperson, contractor or customer) needs information, they know who to talk to, and are not passed from one department to another.

Figure 2.01 Small firm organisational flowchart

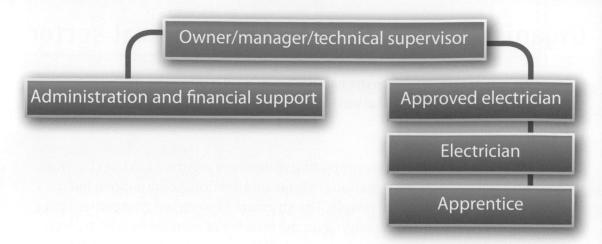

However, if the person dealing with a project is ill, on holiday, or leaves the company, vital information may be missing or lost and it may be very difficult for someone else to continue handling the project smoothly.

Large firms

In the larger company, the tasks are allocated to different people, for example estimators (who handle the initial enquiry and produce an estimate) and contracts engineers (who see that the work is done). This type of structure is known as a horizontal structure. The advantage is that individuals become specialists in a particular task and can work more efficiently. Companies of this size often belong to a trade organisation such as the Electrical Contractors' Association (ECA) and, because of their structure, can offer good career development prospects.

Figure 2.02 Large firm organisational flowchart

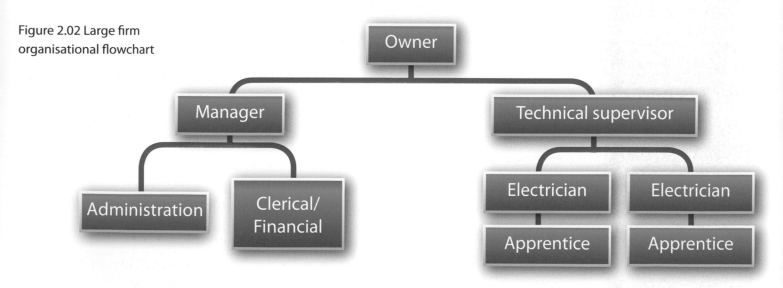

Project roles and responsibilities

On completion of this topic area the candidate will be able state the role of different personnel within an organisation that has electro-technical activities basing this on a construction project.

Several different people are involved in a construction project, from its initial design through to construction and completion. As an electrician, you will be dealing with many of these people, and it is important that you understand their job function and how they fit into the overall project. We will look at the people involved in the three main stages of a typical construction project:

- design
- tendering
- construction.

Definition

Bill of quantities – a list of all the materials required, their specification and the quantities needed

Design stage

Client	<ul><li>Person or organisation that wants the work done and is paying for it.</li><li>Specifies the purpose of the building.</li><li>Usually gives an idea of number of rooms, size, design etc., and any specific wants.</li><li>May also give an idea of the price they are willing to pay for the work.</li></ul>
Architect	<ul><li>Designs the appearance and construction of the building so that it fulfils its proper function.</li><li>Advises the client on the practicality of their wishes.</li><li>Ideally provides a design solution that satisfies the client and also complies with the appropriate rules and regulations for the type of building.</li><li>For small projects, an architect may draw up a complete plan. For larger, more complex buildings they will consult specialist design engineers about technical details.</li></ul>
Consulting engineers (Design engineers)	<ul><li>Act on behalf of the architect, advising on and designing specific services such as electrical installation, heating and ventilation etc.</li><li>Create a design that satisfies the client and architect, the supply company and regulations.</li><li>Ensure that cable sizes have been calculated properly, that the capacities of any cable trunking and conduit are adequate, and that protective devices are rated correctly.</li><li>Produce drawings, schedules and specifications for the project that will be sent out to the companies tendering for the contract.</li><li>Answer any questions that may arise from this.</li><li>Once the contract has been placed, they will produce additional drawings to show any amendments.</li><li>Act as a link between the client, the main contractor and the electrical contractor.</li></ul>
Quantity surveyor (QS)	<ul><li>Responsible for taking the plans and preparing an initial **bill of quantities** for a project; contractors who are tendering for the project use this information to prepare their estimates.</li><li>During construction, the quantity surveyor monitors the actual quantities used, and also checks on claims for additional work and materials.</li></ul>

(continued over)

Clerk of works	• Checks that the quality of the materials, equipment and workmanship used on the project meet the standards laid down in the specification and drawings. • On big contracts there may be several clerks of work, each responsible for one aspect, such as electrical, heating or ventilation. • Effectively employed by the client. • Inspects the job at different stages. • Is also present to check any tests carried out. • May also be given the authority by the architect to sign day worksheets and to issue architect's instructions for alterations or additional work.

Table 2.01 Design stage

The architect, consulting engineer, quantity surveyor and clerks of works are traditionally part of the architect's design team. However, installations may not always be tackled like this.

For example, for a small job the client may approach an electrical contractor directly and ask it to carry out the work. The client provides a few basic details and requirements, and expects the electrician to ensure that the installation is properly designed and carried out.

Larger electrical contractors, and in particular those that are multi-disciplined, often offer a design service for customers which eliminates the need to use an external architect and makes things simpler (and possibly cheaper) for the client. However, if a dispute arises over design aspects of the job, the client no longer has anyone to arbitrate.

Tendering stage

In competition with others, tendering is the process by which a contractor works to the drawings and specifications issued by the consulting engineer and submits in writing a total cost for carrying out the work (i.e. materials, tools, equipment and labour).

Most invitations to tender have strict guidelines about the information to be supplied and a fixed deadline by which the tender must be received.

Estimator

The estimator's task is to calculate the total cost that will be given in the tender. At the start of the tendering process, a consulting engineer or building contractor usually issues an 'Enquiry to submit a tender' (or Invitation to Tender). This contains various documents, such as:

• a covering letter, giving a broad description of the work, including start and finish dates

• the form of contract that will be applicable to the project – for example, whether it is to be a fixed or fluctuating price

Did you know?

Some larger electrical contracting firms have their own design/consulting engineers

- drawings and specifications for the project
- a tender submission document that must be used
- a day work schedule.

This type of tender always has a time limit, so that the contractor is not held to a fixed price if the project is delayed. A fluctuating price contract allows the contractor to claim back the difference between costs included at the time of estimate and actual costs incurred at the time of installation.

When the Invitation to Tender is received, the estimator will check with his or her management (sometimes the contracts manager) to see whether the company wishes to tender on the basis of the submitted documents. If it does, the estimator reads the specifications carefully to understand the requirements.

Using this information and the scaled drawings, the estimator calculates the amount of materials and labour required to complete the job within the time specified by the client. This information is recorded on a 'take-off' sheet. A typical sheet is shown below.

Did you know?

A fixed price contract is exactly what it says. The contractor agrees to complete the job for the price quoted in the tender, even if the material or labour costs go up before the project is completed

Item	Qty	Description	Unit cost	Discount	Material cost £	Hours to install	Hourly rate £	Total labour £
1	200m	20mm galvanised conduit	1.2/m	0	240.00	70	9.50	665.00
2	30	Earthing couplings	.24	0	7.20	0	0	0
3	30	20mm std. brass bushes	.10	0	3.00	0	0	0
4	150	20mm distance saddles	.48	0	72.00	0	0	0
5	150	1.5 x 8 brass screw and plugs	.03	0	4.50	0	0	0

Figure 2.03 Take-off sheet

Nowadays this work is normally done using dedicated computer software. The final tender is based on the results of these calculations.

We will cover the specification in more detail throughout chapter 9.

Construction
The contract

Contract law is very complex, and it cannot be covered fully on this course. To minimise the risks, you can use a standard form of contract. The Joint Contracts Tribunal (JCT) 'Standard Form Of Contract' (normally for projects of a complex nature or in excess of 12 months' duration) or the 'Intermediate Form Of Contract' are typical of contracts used in the industry.

If you are involved in any kind of contract it is always advisable to seek professional legal assistance before making or accepting an offer. Once contracts are agreed, construction and installation work can begin and many more people become involved.

Roles during the construction stage

Main contractor	• Usually the builders, because they have the bulk of the work to carry out. • Has the contract for the whole project. • Employs subcontractors to carry out different parts of the work. • In refurbishment projects, where the amount of building work is small, the electrical contractor could be the main contractor. • Paying and co-ordinating subcontractors.
Nominated subcontractors	• Named (nominated) specifically in the contract by the client or architect to carry out certain work. • Must be used by the main contractor. • Normally have to prepare a competitive tender. • Subcontractors will include electrical installation companies.
Non-nominated subcontractors	• Companies chosen by the main contractor (i.e. not specified by the client). • Their contract is with the main contractor.
Nominated supplier	• Supplier chosen by the architect or consulting engineer to supply specific equipment required for the project. • Main contractor must use these suppliers.
Non-nominated supplier	• Selected by main contractor or subcontractors. • For electrical supplies, this will be a wholesaler selected by the subcontractor who can provide the materials needed for the project.
Contracts manager	• Oversees the work of the contracts engineers. • May also be responsible during tendering for deciding whether a tender is to be submitted and the costs and rates to be used.
Contracts engineer	• Employed by the electrical contractor to manage all aspects of the contract and installation through to completion. • Responsible for planning labour levels, ordering and organising materials required. • Ensures the contract is completed within the contract timescales and on budget. • Liaises with suppliers to ensure planned delivery dates and builders' work programme are acceptable. • May negotiate preferential discounts with suppliers. • Attends site meetings.

Project engineer	<ul><li>*Role definitions vary within the industry but generally the role is similar to that of a contracts engineer.*</li><li>Responsible for day-to-day management of on site operations relative to a specific project.</li><li>Often based on the site.</li></ul>
Site supervisor	<ul><li>Contractor's representative on site.</li><li>Oversees normal day-to-day operations on site.</li><li>Experienced in electrical installation work, normally an Approved Electrician.</li><li>Responsible for the supervision of the approved electricians, apprentices and labourers.</li><li>Uses the drawings and specification to direct the day-to-day aspects of the installation.</li><li>Liaises with contracts engineer to ensure that the installation is as the estimator originally planned it. Ensures materials are available on site when required.</li><li>Liaises with the contracts engineer where plans are changed or amended to ensure additional costs and labour/materials are acceptable and quoted for.</li></ul>
Electricians, apprentices and labourers	<ul><li>The people who actually carry out the installation work.</li><li>Work to the supervisor's instructions.</li></ul>
Electrical fitter	<ul><li>Usually someone with mechanical experience.</li><li>Involved in varied work including panel building and panel wiring and the maintenance and servicing of equipment.</li></ul>
Electrical technician	<ul><li>*Job definition varies from company to company.*</li><li>Can involve carrying out surveys of electrical systems, updating electrical drawings and maintaining records, obtaining costs, and assisting in the inspection, commissioning, testing and maintenance of electrical systems and services.</li><li>May also be involved in recommending corrective action to solve electrical problems.</li></ul>
Service manager	<ul><li>*Similar role to contracts manager (and in some cases the roles are combined) but focuses on customer satisfaction rather than contractual obligations.*</li><li>Monitors the quality of the service delivered under contract.</li><li>Checks that contract targets (e.g. performance, cost and quality) are met.</li><li>Ensures customer remains fully satisfied with the service received.</li></ul>
Maintenance manager	<ul><li>*Once the building has been completed.*</li><li>Keeps installed electro-technical plant working efficiently.</li><li>May issue specifications and organise contracts for a programme of routine and preventive maintenance.</li><li>Responsible for fixing faults and breakdowns.</li><li>Ensures legal requirements are met.</li><li>Carries out maintenance audits.</li></ul>

Table 2.02 People involved in the construction stage

The electrical contracting industry structure

On completion of this topic area the candidate will be able to identify those organisations associated with a specific specialist company in the electro-technical industry.

Electrical contracting industry bodies

There is a variety of organisations that you should be aware of within the industry.

The Electrical Contractors' Association (ECA)

The ECA represents the interests of electrical installation companies in England, Wales and Northern Ireland and is the major association working within the electrical installation industry. It was founded in 1901 and has more than 2000 member companies, ranging in size from small traders with just a few employees to large multinational organisations. The aim of the ECA is to ensure that all electrical installation work is carried out to the highest standards by properly qualified staff. Consequently, firms that wish to become members of the ECA must demonstrate that they have procedures, staff and systems of the highest calibre.

The National Inspection Council for Electrical Installation Contracting (NICEIC)

Find out

Who is your local NICEIC inspector, and where are their offices?

The NICEIC is an accredited certification body set up in 1956 to protect users of electricity against the hazards of unsafe and unsound electrical installations. It is the industry's independent electrical safety regulatory body and is not a trade association.

The NICEIC maintains a roll of approved contractors who meet the council's rules relating to enrolment and national technical safety standards, including BS 7671 (*IEE Wiring Regulations*). The roll is published annually and is regularly updated on the NICEIC website so that consumers and specifiers can select contractors who are technically competent.

The council also employs 46 inspecting engineers who make annual visits to approved contractors to assess their technical capability and to inspect samples of their work.

Amicus (AEEU)

Find out

Who is your local Amicus representative?

For many years workers within the electrical installation industry have enjoyed good labour relationships with their employers. This was largely due to the excellent co-operation and relationships that existed between the industry's principal trade union at the time (EETPU) and the ECA. In 1992, the EETPU merged with the AEU to become the Amalgamated Engineering and Electrical Union (AEEU).
However, Amicus is the name of the new amalgamation that came into being on 1 January 2002, with Amicus–AEEU being the UK's largest manufacturing union, having approximately 730,000 members in the private and public sectors.

The Joint Industry Board (JIB)

Formed in 1968, the Joint Industry Board for the Electrical Contracting Industry (JIB for short) came into existence as the result of an agreement between the ECA and the EETPU. Effectively the industrial relations arm of the industry, the JIB has as its main responsibility the agreement of national working conditions and wage rates.

The Health and Safety Executive (HSE)

The UK's Health and Safety Commission (HSC) and the Health and Safety Executive (HSE) are responsible for the regulation of almost all the risks to health and safety arising from work activity in the UK. Their mission is to protect people's health and safety by ensuring that risks in the workplace are properly controlled. They look after health and safety in nuclear installations and mines, factories, farms, hospitals and schools, and offshore gas and oil installations; they are further responsible for the safety of the gas grid and the movement of dangerous goods and substances, railway safety, and many other aspects of the protection both of workers and the public.

Did you know?

Local authorities are responsible to the HSC for enforcement in offices, shops and other parts of the services sector

SummitSkills

SummitSkills is the Sector Skills Council for the building services engineering sector, representing the electro-technical, heating, ventilating, air conditioning, refrigeration and plumbing industries.

SummitSkills has been created for employers to identify skills shortages and deliver action plans to address them. The organisation will also provide career information for all industries within the sector as well as dealing with all training standards and policy matters previously handled by the former National Training Organisations (NTOs).

The Institution of Lighting Engineers (ILE)

The Institution of Lighting Engineers (ILE) is the UK's most influential professional lighting association, dedicated solely to excellence in lighting. Founded in 1924 as the Association of Public Lighting Engineers, the ILE has evolved to include lighting designers, architects, consultants and engineers among its 2,500-strong membership. The key purpose of the ILE is to promote excellence in all forms of lighting. This includes interior, exterior, sports, road, flood, emergency, tunnel, security and festive lighting as well as design and consultancy services.

The Institution is a registered charity, a limited company and a licensed body of the Engineering Council.

The Institute of Electrical Engineers (IEE)

The Institute of Electrical Engineers (IEE) was founded in 1871. It has now merged with other technical institutions to become the Institution of Engineering Technology (IET) but publications will continue to be referred to under their IEE designation until there is a need to reissue them.

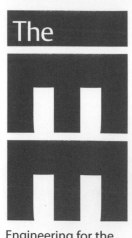

The IEE

Engineering for the future

It is the largest professional engineering society in Europe and has a worldwide membership of just under 130,000. As well as setting standards of qualifications for professional electrical, electronics, software, systems and manufacturing engineers, the IET prepares regulations for the safety of electrical installations for buildings, the *IEE Wiring Regulations* (BS 7671) now having become the standard for the UK and many other countries.

Activity

Find out who has the following roles within your company:

- designer
- installer
- inspector and tester
- supervisor(s)
- manager(s)
- quantity surveyor
- apprentice(s).

FAQ

Q I really like working with computers. Is there a place for me in the industry?

A Definitely. Computers are becoming increasingly used to control our environment and reduce energy consumption and emissions. These systems all need designing, installing, operating and maintaining.

Q I been offered on-the-job training with a small firm. Is it worth taking or should I try with a bigger company?

A This is a difficult one. There are good and bad firms, both small and large. You should ask about the range of work the firm does and how much time it is prepared to give you to train properly. There are some very good small firms that provide excellent one-to-one training, so bigger isn't always better.

Q Why do I need to train for so long when I just want to connect up cables?

A Being an electrician isn't just about connecting up cables. To be a good electrician you need to study and practise hard to build up your knowledge, skills and experience, and all this takes time.

Knowledge check

1. Describe in your own words 'electrical services'.

2. List five types of building services that use electronics or computers in their control systems.

3. List the main functions of the clerk of works.

4. Explain the difference between the terms 'electrician', 'electrical fitter' and 'electrical technician'.

5. What is the difference between the main roles of the Electrical Contractors Association (ECA) and the National Inspection Council for Electrical Installation Contracting (NICEIC)?

chapter 3

Storage and retrieval of technical information

Unit 1 Outcome 3

Whatever the environment you will be working in, you will always need to know how to have access to essential technical information. As such, all electricians need to know not only how to store information, but also the best method of retrieving the information.

As there is a huge variety in the types of technical information that will be stored by an individual or an organisation, there is naturally a large range of options open to electricians. These can range from paper and filing to more modern electronic methods.

Once information is retrieved you will need to know how to apply it. Many of the technical issues will be presented in graphs and diagrams. You must make sure that you understand how to interpret the information that these contain.

On completion of this chapter the candidate will be able to:

- state the sources of technical information
- explain how this information may be retrieved
- identify types of drawing and diagram
- identify and describe how different scales can be used to produce drawings, diagrams and plans
- recognise and explain BS EN 60617 symbols
- state the necessity to use drawings and diagrams in relation to specifications.

Sources of technical information

On completion of this topic area the candidate will be able to state various sources of technical information available in the electro-technical sector and explain how that information may be retrieved.

Storage media

We have seen that lots of different people may be involved in an electrical installation project. For them to work together they must communicate with one another, using or transmitting technical information.

There are many sources of this information, including drawings, diagrams, charts and data, British Standards, Codes of Practice, specifications, manufacturers' instructions and manuals, catalogues and reports.

The information can come from hundreds of different sources, including the British Standards Institute (BSI), BS EN standards, equipment and component manufacturers, consulting engineers, trade associations, HMSO book stores, libraries and the Internet. The material can be on paper, microfilm, CD-ROM or electronically downloaded from a remote source.

Nowadays, technical information is stored and communicated using a wide variety of methods. Many of these require special equipment (e.g. computers) to get at the information. The same information is often available in several different formats.

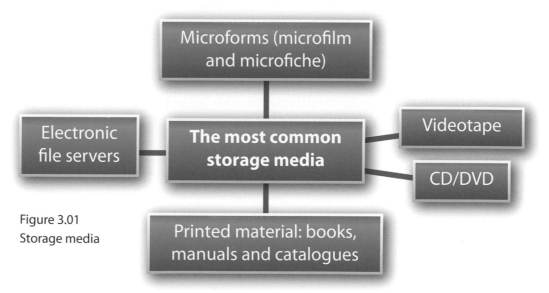

Figure 3.01
Storage media

Printed materials

This is still the most common form of storage for general information – for example, books, newspapers, leaflets, datasheets and catalogues. Their major advantage is that they need no special equipment for reading. Increasingly, many of these items are also available in the other formats described below. This is particularly true of technical information, where electronic methods have the advantage of lower costs and increased delivery speed.

Microforms

Microforms (also known as microfiches) store large amounts of information such as text, documents and photographs using photographic techniques. Their main advantage is long life, and therefore they are popular for storing information that needs to be preserved for many years or even centuries, for example in libraries or newspaper offices. Microforms are based on polyester film, and have a life expectancy of about 500 years when stored correctly. Microfilm is a length of 16mm or 25mm film; microfiche is a larger piece of photographic film (around 150mm × 100mm). Both formats store information as tiny pages, and each type can hold hundreds of pages on one microform. Special optical reading devices are used to read the images. They have a temperature-controlled light source and a magnifying device to show the images on a screen similar to a computer display.

Videotape

Videotape is a polyester film on which television signals are stored magnetically. The most common format is VHS. Its main advantages are low cost and the widespread availability of video recorders for storing and reading material. Videotape is not normally used for storing technical documents, but it has a useful role where visual information is important, such as for product demonstrations and tours.

CD/DVD

CDs and DVDs are among the most popular storage media today. Information (text, images, film, music and documents) is encoded into a digital format and stored on the disc. The data can be read using a computer with suitable drive and software, or on a standalone device. The discs are low cost, and any information that exists as an electronic file can easily be transferred to a disc.

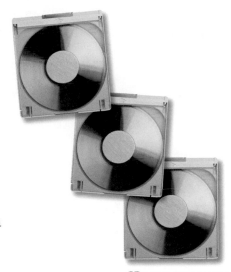

CDs can store a vast amount of information

Each type of information can be encoded using a number of different formats, each having certain advantages for different applications. Some of these are more common than others. To read the information, the reading device must be equipped with software to handle it.

Some typical file types (denoted by their normal file-name suffix) are:

Images	jpg, bmp, tif, pcx, gif, png
Movies	wmv, mpg
Text	txt
Documents	doc, pdf

Table 3.01 File types

Electronic file servers

The storage methods described so far need the reader to have a copy of the information, whether it is a book, CD or video; in effect, there is one copy per reader (although one copy may be shared among several readers, like a library book). The increasing use of electronic networking to connect computers together means that a single copy of some information can be stored and then accessed by many users at the same time. The information is held on a file server located anywhere in the world, provided that it is connected to the network. The Internet is the ultimate example of this. The information is stored in digital format, using the same formats as a CD or DVD. Access to the information can be limited to certain individuals or groups, or may be open to anyone.

USB Flash memory drive

USB Flash memory drives are small, rewriteable portable drives. Essentially they are a circuit board encased in a plastic or metal casing which makes them safe and secure for easy transportation. Their information can be accessed by plugging them into a USB port on a computer.

Information sharing

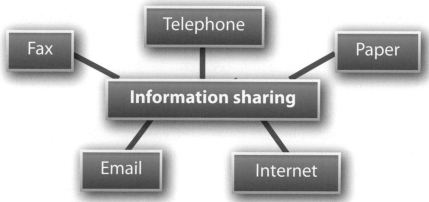

Figure 3.02 Methods of sharing information

Written material

This is one of the simplest ways of communicating, using printed material or by hand written notes and memos.

Telephone

Speaking directly to someone is sometimes the quickest way to obtain answers and information. Manufacturers nearly all have telephone helplines.

Facsimile (fax)

The fax machine enables written or printed material to be sent to another fax machine over a normal telephone line. Although email has greatly reduced the use of fax, it is still useful for sending handwritten documents and printed text without needing a computer, and it is quite fast. However, the end result is a sheet of paper, so it cannot be edited electronically. There are several different types of standalone fax machine, which vary as to how they print out the received document.

A fax machine is basically a scanner combined with a printer. Therefore a computer equipped with these two items, along with a fax modem and suitable software, can send and receive faxes without the need for a standalone fax machine.

Did you know?

Faxes have been around in one form or another for about 100 years

Printers

There are several types of printers in use with computers today. These include dot matrix, thermal paper, thermal ribbon and laser.

Dot matrix are rather old fashioned. They used patterns of pins pressing against an ink ribbon to create type.

Thermal paper printers use special paper that is coated with chemical to turn black when heated. They are cheap, have few moving parts, and do not need ink or ribbon supplies. However, they cannot work with ordinary paper and, over time, the image fades or disappears, particularly if it is left in a hot place.

Thermal ribbon printers have a heat-sensitive ribbon or film. As the paper passes over the ribbon (which also moves), the ink is transferred to the paper to make a mark. They do not need special paper, but the ribbon or film has to be replaced regularly.

Inkjet printers work with ordinary paper, but the ink cartridges need to be replaced. If the machine is not used for some time, the ink may dry and clog the cartridge, giving a poor image. They can produce good black-and-white and colour images.

Laser printers produce the best images. They are expensive to buy. They use toner cartridges but are more economical to run than inkjet printers, and use ordinary paper. They come in black and white, or colour models.

Email

Email is now one of the most popular methods of communicating in the business world and elsewhere. You can transmit simple text messages and complex files to almost any country in the world in seconds, and you can also attach other files, such as documents, images and movies. To send and receive email messages you need:

- a computer running a software application called a 'mail client' (e.g. Outlook Express or Pegasus)
- connection to the Internet (via modem or broadband)
- an account with an Internet service provider (ISP).

Did you know?

The Internet started as a defence project by the Advanced Research Projects Agency (ARPA) for the US government in 1969 and was first known as the ARPANET

The Internet

The Internet, sometimes called simply 'the net', is a worldwide system of computer networks.

In practice, the Internet is really a 'network of networks'. Today, the Internet is accessible to hundreds of millions of people across the world. It enables users to get information from other computers on the net, and is used to transmit voice, radio and video as well as data, particularly email. More recently, Internet telephony hardware and software has made it possible to use the Internet for normal voice conversations.

Using the World Wide Web (often abbreviated to 'www', or simply 'the web'), you have access to billions of pages of information. Web browsing is done using web-browser software, of which Microsoft Internet Explorer and Netscape Navigator are the most popular. Most companies now have websites that allow you to browse products or technical information.

Drawings and diagrams

On completion of this topic area the candidate will be able to identify types of drawing and diagrams used in the electro-technical sector.

Drawings, diagrams and symbols

A technical diagram is simply a means of conveying information more easily or clearly than can be expressed in words. In the electrical industry drawings and diagrams are used in different forms. Those most frequently used are:

- block diagrams
- circuit diagrams
- wiring diagrams
- schematic diagrams

- assembly drawings
- record (as fitted) drawings
- layout/location diagrams
- site plans.

Block diagrams

A block diagram can be used to relate information about a circuit without giving details of components or the manner in which they are connected. In block diagrams the various items are represented by a square or rectangle clearly labelled to indicate its purpose. This type of diagram shows the sequence of control for installations in its simplest form, as shown in Figure 3.03.

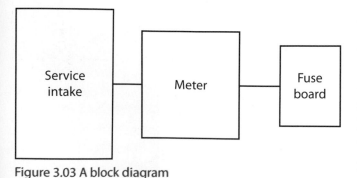

Figure 3.03 A block diagram

Circuit diagrams

A circuit diagram uses symbols to represent all circuit components and shows how these are connected. The circuit diagram should be as clear as possible and should follow a logical progression route from supply to output. In all other respects the circuit diagram cannot be regarded as a direct source of information.

For example, the shape of the diagram does not represent the physical outline of the circuit; it has no dimensions; and the symbols that are used need not bear the slightest resemblance to the components they represent. The symbols used are BS EN 60617 circuit diagram symbols. Some of the more common symbols are included at the end of this section. Figure 3.04 illustrates the use of a circuit diagram in relation to a rectification circuit.

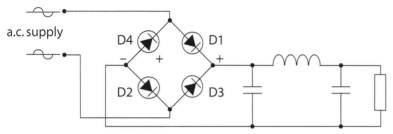

Figure 3.04 A circuit diagram

Wiring diagrams

In a wiring diagram (such as Figure 3.05) the physical layout is taken into consideration. The components and connections show a pictorial version of those found in the actual circuit. Wiring diagrams can be used to carry information of a specific nature regarding the wiring or connection of components. Wiring diagrams do not, as a rule, use circuit or location symbols, but there is no hard-and-fast rule about this. Since the object of the diagram is to relate information, this point should be borne in mind when drawing up the wiring or circuit diagram.

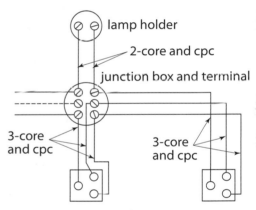

Figure 3.05 A wiring diagram

Schematic diagrams (working diagrams)

Schematic diagrams like Figure 3.06 are similar in concept to a circuit diagram. They do not show how to wire components but they do show how the circuit is intended to work. The diagram is a control circuit of an automatic star/delta starter. These types of diagram tend to be used for larger, more complicated electrical diagrams such as control systems for motor starters and heating systems.

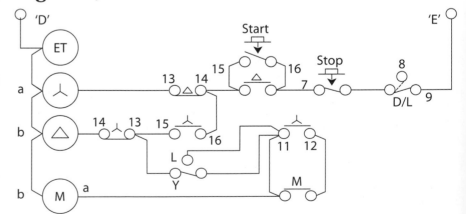

Figure 3.06 A schematic diagram

Assembly drawings

Assembly drawings (see Figure 3.07) should show how the individual parts (or modules) of a product fit together. They normally contain scale drawings of all the components shown in their correct position relative to each other, with some overall dimensions. Where there are internal components, these are shown by sectioning. Each component is listed and described on the drawing.

The example below is the assembly drawing for a push-button enclosure.

A   Enclosure base with built in contact block clips

B   Contact blocks/lamp holders

C   Locking ring

D   Enclosure lid

E   Legend plate

F   Captive screws (after screw in) loose in enclosure on delivery

G   Actuators and lens cap

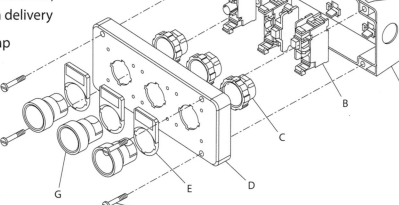

Figure 3.07 An assembly drawing

Record (as fitted) drawing

In an ideal world, every installation would follow exactly the consulting engineer's original plans. In practice, however, problems can come up which mean that conduit must be fixed in a different position or cables routed in another way. These changes must be recorded so that, in future, if maintenance or alterations are needed, the actual layout of the installation is known. This becomes very important when the installation is hidden in the walls, for example.

This is where record (as fitted) drawings come in. They are the installation drawings supplied by the consulting engineer with the final actual installed cable, conduit and trunking routes marked on them.

Location drawings

These are scale drawings prepared under the supervision of an electrical consultant or engineer responsible for a particular installation, and are based upon architect's drawings of the building in which the installation is to be installed. These drawings show the required position of all equipment, metering and control gear that is to be installed. They normally show the plan view of the installation, and standard BS EN 60617 location symbols are used. These diagrams are used to show the sequence of control of large installations.

Figure 3.08 shows a layout for a warehouse. Note the legend at the bottom of the page: this legend should always be included to help in understanding the diagram.

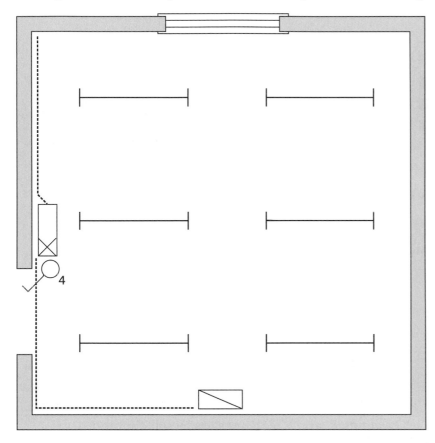

Figure 3.08 A layout diagram

Warehouse lighting (scale 1:50)
-------- 50 x 50mm trunking run
4-gang 1-way switch
lighting distribution board
main control
fluorescent luminaire

Manufacturers' data and service manuals

Almost all equipment will have the manufacturer's fitting instructions and other technical data or information sheets. These should be read and understood before fitting the item concerned.

Once installation of the item is complete, they should be kept safely in a central file so that they can be given to the customer in a hand-over manual when the project is completed. Sometimes you may need more details about an item. You can get this from the manufacturer's catalogue or datasheets, website, or by speaking directly to the manufacturer. It may be useful to talk to the contracts' engineer to make sure that he or she knows what is happening.

Using charts and reports

In addition to drawings and specifications, charts and reports are two other methods of showing and communicating technical information and data. For this Technical Certificate, you must be able to interpret the data contained in a chart.

Charts

Charts can often make information easier to understand and allow the user to see clearly what they need to know. The most popular chart used within construction work is the bar chart. When it shows activities against time, it is sometimes referred to as a **Gantt chart**, after its inventor, Robert Gantt.

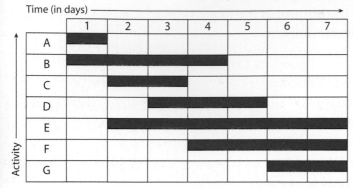

Figure 3.09 Bar chart 1

In Figure 3.09, the bar chart shows several activities and when they are due to happen. This helps the supervisor to keep an eye on how the contract is actually progressing when compared to the original plan. Main contractors often use this sort of bar chart to show when individual trades should be on site at any time during the contract.

Looking at the chart we can see that:

- Activity A should take one day
- Activity B starts on the same day, and lasts four days
- Activity C lasts two days, but doesn't start until Day 2 ... and so on.

Bar charts can be used to show additional information by adding colours, codes and symbols.

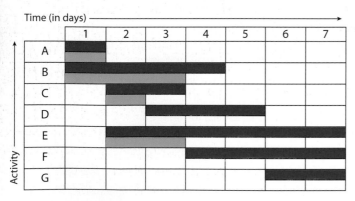

Figure 3.10 Bar chart 2

A further example of a bar chart is shown in Figure 3.10. The activities are the same as before but with the actual progress against each one shown by the shaded blue area beneath the original bars. The chart shows progress to the end of Day 3. It is easy to see which activities have been completed, and which ones are lagging behind.

Scale drawings

On completion of this topic area the candidate will be able to identify and describe how and why different scales are used to produce drawings, plans and diagrams, and state the necessity of using drawings and diagrams in conjunction with the specification.

Scale

Layout and assembly drawings give information about physical objects, such as the floor layout in a building, or a mechanical object. If we were to make the drawing the same size as the object, the drawings would often be far too big to handle.

To make the drawing a sensible size, we use **scaled drawings**. You may, for example, have built model aeroplanes from a kit. Quite often these are described as 1/32nd scale: in other words, every part in the model is 32 times smaller than the real thing. This type of scale is known as a ratio scale, and it makes the drawings easy to use. To find a measurement on the actual object, you measure the distance on the drawing and multiply it by the scale. It doesn't matter what unit of measurement you choose, because you are simply going to multiply it by a number (the scale).

For example, on most construction projects, the scale used to show the floor layout of a building is 1:100. So 10mm on the drawing represents something that is 100 times bigger in reality, (i.e. $10 \times 100 = 1000$mm (1 metre).

The drawing scale is chosen to make the drawing a reasonable size, according to its purpose. Although a scale of 1:100 may be fine for the layout of a building, it would be impractical for a road map, because you would only be able to get a few miles on each sheet. A scale of 1:500,000 (1cm = 5km) would be better.

In the same way, an assembly drawing for a wristwatch would be too small to read if we used 1:100; a better scale might be 20:1 (20mm on the drawing represents 1mm on the actual watch).

Specifications

All drawings and diagrams must be prepared and planned alongside the specifications for a project. The specifications demonstrate the final aims of the project and the intended end-result that the client is paying for. As such all preparation you carry out on site, must correspond with the contents of this document.

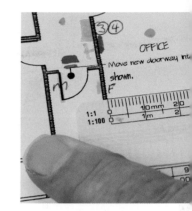

Using a scaled drawing

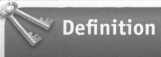

Definition

Scaled drawings – the size of everything on the drawing is drawn with a fixed ratio to the size of the actual object. This ratio is called the scale of the drawing

BS EN 60617 electrical symbols

On completion of this topic area the learner will be able to recognize and explain how BS EN 60617 electrical symbols are used.

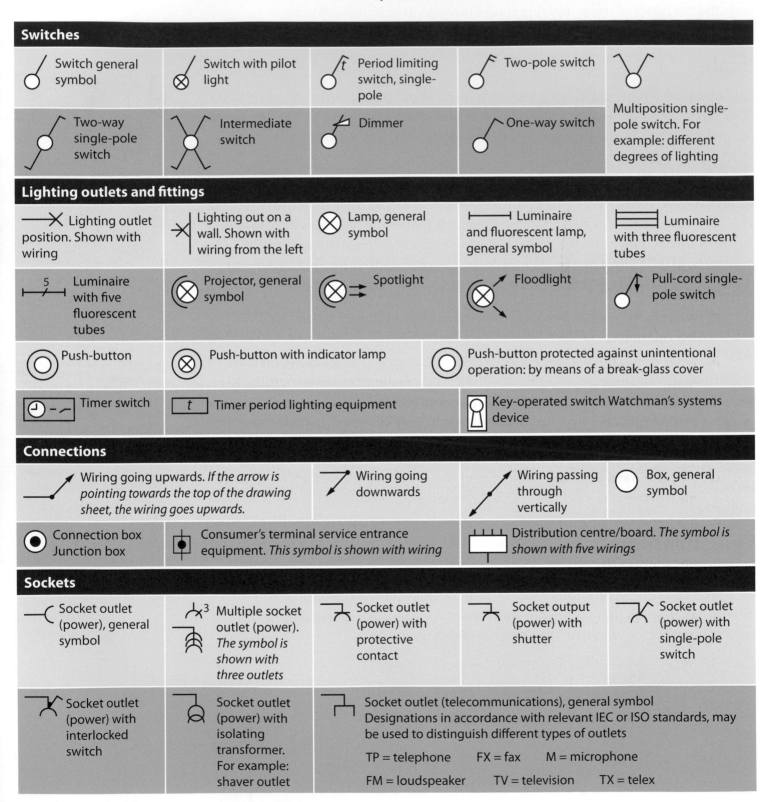

Figure 3.11 BS EN 60617 lists the standard symbols for use in installation drawings (continued)

Miscellaneous

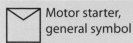 Motor starter, general symbol	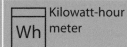 Direct-on-line starter with contactor for reversing the rotation of a motor	Star-delta starter	Fan. The symbol is shown with wiring
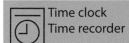 Time clock Time recorder	Hour meter h Hour counter	Wh Kilowatt-hour meter	

Figure 3.11 BS EN 60617 lists the standard symbols for use in installation drawings

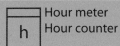

Activity

Produce a layout diagram for a single floor of a small installation (perhaps your house) using symbols for BS EN 60617.

FAQ

Q Why are the maintenance manuals still in book form and not on the computer?

A While computers are very good at storing and handling information, there is a snag – you need a computer to access the information! Sometimes information is best left in written form, particularly if large numbers of people need to access it or where a computer may not be available. That said the increasing power of portable devices mean on site computer use can only increase.

Q Why do I need to produce an 'as fitted' drawing when the designer has given me a drawing to work from in the first place?

A It is not always possible to install cables and equipment *exactly* where specified in the drawing because of variations in the building during construction. Any changes should be agreed with the designer and an 'as fitted' drawing is produced to reflect those changes.

Q My company has its own set of symbols for use on electrical drawings. Is this OK?

A No. According to BS 7671, any symbol used should conform to BS EN 60617.

Knowledge check

1. Describe how it would be best to share the following information.

   - A digital photograph between colleagues in different towns.

   - A message that you are unavoidably late for a meeting.

   - A non-urgent form for completion by a client.

   - Some instructions to be accessed by office based staff at different times.

   - Information about your company to be accessed by the public.

2. Draw a block diagram showing the sequence of control for a domestic lighting circuit.

3. With the aid of diagrams, describe a Gantt chart.

4. Explain why it is necessary to produce layout drawings to scale and which scales would be most suitable for a building.

chapter 4

Basic units used in electro-technology

Unit 2 Outcome 1

When carrying out installations, electricians need to be able to calculate the electrical features of the system. These include the current, resistance and efficiency. All these features have their own units, and it is these which form the foundation of electrical calculation. It is equally important to understand the relationships between these measurements and how they affect each other.

On completion of this chapter the candidate will be able to:

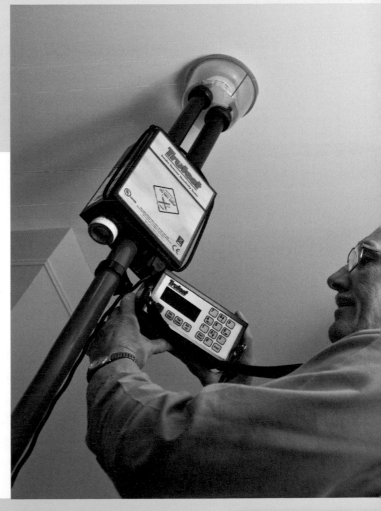

- state the relationships between force and movement; force, mass and acceleration; mass and weight; force times distance moved

- identify SI units

- describe how SI units relate to the principles of circuits, alternating current production, efficiency and lifting devices

- state the relationship of resistance, resistivity, length and area

- state the features and principles of a simple alternator and how it produces a sinusoidal waveform pattern

- state the efficiency of a machine in terms of output over input.

Mechanics

On completion of this topic area the candidate will be able to state the relationships between force and movement; force, mass and acceleration; mass and weight; force times distance moved.

The difference between mass and weight

Before we start talking about mechanics, we need to understand a very important concept – that is, the difference between weight (a force) and mass.

- Mass

  This is simply the amount of stuff or matter contained in an object. Assuming we do not cut or change the object, the mass of the object will stay the same wherever we are.

  The unit of mass is the kilogram (kg)

- Weight

  This is a force and depends on how gravity pulls on a mass. This can vary according to where we are (the higher above sea level you go, the less you weigh). The change in weight is tiny but can be measured with very sensitive and expensive scientific equipment.

  The unit of weight, and force in general, is the newton (N)

On Earth, if we disregard the effect of height above sea level, the weight acting on 1kg of mass is equal to 9.81 newtons (N). So, 1kg weighs 9.81 N. In many situations this can be rounded up to 10 N.

The human race is very inventive. We have devised many means of overcoming simple problems, such as lifting a heavy object and moving it from one place to another. In this section we will be looking at the following areas:

- simple machines for lifting and handling
- force, work, energy and power
- efficiency.

Force

Force is a push or pull that acts on an object. If the force is greater than the opposing force, the object will change motion or shape. Obvious examples of forces are gravity and the wind. Force is measured in **newtons**.

The presence of a force is measured by its effect on a body, e.g. a heavy wind can cause a stationary football to start rolling; or a car colliding with a wall causes the front of the car to deform and the occupants of the car to be forced forwards towards the windscreen (hence the use of seat belts).

Remember

It is vital you understand the difference between weight and mass

Remember

Mass and weight are **not** the same. Mass is the amount of material in an object. Weight is a force – e.g. a person who weighs a certain amount on Earth would weigh less on the Moon due to the decreased gravitational force, but they will still have the same mass

Equally, gravitational force will cause objects to fall towards the earth. Therefore, a spring will extend if we attach a weight to it, because gravity is acting on the weight.

As the force of gravity acts on any mass, such a mass tends to accelerate and exert a force that depends upon the mass and the acceleration due to gravity. This acceleration due to gravity is agreed worldwide as being 9.81 m/s^2 at sea level and therefore a mass of 1kg will exert a force of 9.81 N.

Expressed as a formula:

Force (N) = Mass × Acceleration

(Note: In calculations, it is often assumed for ease that the value of acceleration is taken as 10 m/s^2.)

Work

If an object is moved, then work is said to have been done. The unit of work done is the joule. Work done is the relationship between the effort (force) used to move an object and the distance that the object is moved. Expressed as a formula:

Work done (J) = Force (N) × Distance (m)

Example

A distribution board has a mass of 50kg. How much work is done when it is moved 10m?

Work = Force × Distance

$= (50 \times 9.81) \times 10$

$= 490.5 \times 10$

$= 4905$ J

Remember

Do not assume that the acceleration due to gravity is = 10 m/s^2. It is 9.81 m/s^2. Only take it as 10 m/s^2 if you are told to do so in a question, or if you are doing a rough calculation for your own purposes

Simple machines for lifting and handling

A simple machine is a device that helps us to perform our work more easily when a force is applied to it. A screw, wheel and axle and lever are all simple machines.

A machine also allows us to use a smaller force to overcome a larger force and can also help us change the direction of the force and work with a faster speed. The most common simple machines are as follows.

Levers

Levers let us use a small force to apply a larger force to an object. They are grouped into three classes, depending on the position of the fulcrum (the pivot).

Remember

To make any simple machine work for us, we need to apply a force on it

Class 1

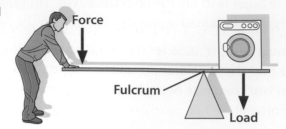

The fulcrum is between the force and the load, like a seesaw.

Class 2

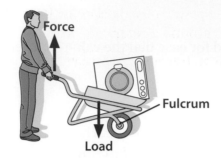

The fulcrum is at one end, the force at the other end, and the load is in the middle. A wheelbarrow is a good example.

Class 3

The fulcrum is at one end, the load at the other end and the force in the middle, like a human forearm.

Figure 4.01 Classes of levers

A small force at a long distance will produce a larger force close to the pivot. If we look at Figure 4.02:

$$10 \times 2 = F \times 0.5$$

$$F = \frac{10 \times 2}{0.5} = \frac{20}{0.5} = 40 \text{ N}$$

Figure 4.02 Large force close to pivot

Gears

Gears are wheels with teeth; the teeth of one gear fit snugly into those around it. You can use gears to slow things down or speed them up, to change direction, or to control several things at once. The gears, when placed together, have to be fitted into the teeth. Each gear in a series changes the direction of rotation of the previous gear. The smaller gear will always turn faster than the larger gear and in doing so, turns more times.

Figure 4.03 Gears

The inclined plane

The inclined plane is the simplest machine of all, as it is basically a ramp or sloping surface. The shortest distance between two points is generally taken as a straight line, but it is easier to move a heavy object to a higher point by using stairs or a ramp. If you think of the height of a mountain, the shortest distance is straight up from the bottom to the top. However, we always build a road on a mountain as a slowly winding inclined plane from bottom to top.

As an electrician, you will use the inclined plane most days in the form of a screw, which is simply an inclined plane wound around a central cylinder.

So the inclined plane works by saving effort, but you must move things a greater distance.

Pulleys

A pulley is made with a rope, belt or chain wrapped around a wheel and can be used to lift a heavy object (load). A pulley changes the direction of the force, making it easier to lift things. There are two main types of pulleys: the single fixed pulley and the movable pulley.

A **single fixed pulley** is the only pulley that uses more effort than the load to lift the load from the ground. The fixed pulley, when attached to an immovable object, e.g. a ceiling or wall, acts as a first class lever with the fulcrum being located at the axis but with a minor change – the bar becomes a rope. The advantage of the fixed pulley is that you do not have to pull or push the pulley up and down. The disadvantage is that you have to apply more effort than the load.

A **movable pulley** is one that moves with the load. The movable pulley allows the effort to be less than the weight of the load. The movable pulley also acts as a second class lever. The load is between the fulcrum and the effort.

There are many combinations of pulleys – the most common being the block and tackle – that use the two main types as their principle of operation.

Example

Let us look at examples of the two main types to understand their operating principles. Imagine that you have the arrangement of a 20 newtons (N) weight suspended from a rope, but actually resting on the ground as shown in diagram (Figure 4.04).

In this example, if we want to have the load suspended in the air above the ground, then we have to apply an upward force of 20 N to the rope in the direction of the arrow. If the rope was 3m long and we wanted to lift the weight up 3m above the ground, we would have to pull in 3m of rope to do it.

Now imagine that we add a single fixed pulley to the scenario, as shown in the Figure 4.04). We have not really changed anything in our favour. The only thing that has changed is the direction of the force we have to apply to lift the load. We would still have to apply 20 N of force to suspend the load above the ground, and would still have to reel

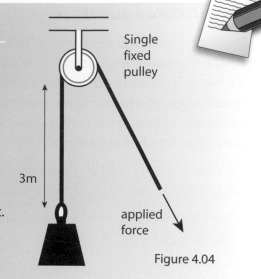

Figure 4.04

in 3m of rope in order to lift the weight 3m above the ground. This type of system gives us the convenience of pulling downwards instead of lifting.

Diagram (Figure 4.05) shows the arrangement if we add a second, movable pulley. This new arrangement now changes things in our favour because effectively the load is now suspended by two ropes rather than one. That means the weight is split equally between the two ropes, so each one holds only half the weight, or 10 N. That means that if you want to hold the weight suspended in the air, you only have to apply 10 N of force (the ceiling exerting the other 10 N of force on the other end of the rope). However, if you want to lift the weight 3m above the ground, then you have to reel in twice as much rope – i.e. 6m of rope must be pulled in.

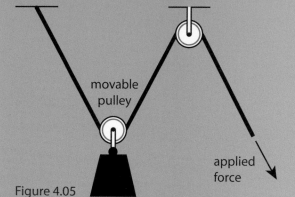

Figure 4.05

The more pulleys we have the easier it is to lift heavy objects. As rope is pulled from the top pulley wheel, the load and the bottom pulley wheel are lifted. If, in Figure 4.05, 2m of rope is pulled through, the load will only rise 1m (there are two ropes holding the load and both have to shorten by the same amount).

With pulley systems, to calculate the effort required to lift the load, we divide the load by the number of ropes (excluding the rope connected to the effort). Figure 4.06 shows a four pulley system, where the person lifting the 200kg mass or 2000 N load (remember: 1kg weighs 10 N) has to exert a pull equal to only 500 N (i.e. 2000 N divided by four ropes).

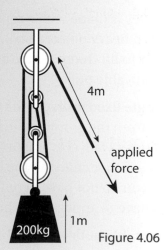

Figure 4.06

Mechanical advantage

The common theme behind all of these machines is that because of the machine, we can increase our ability and gain an advantage over nature. This is a relationship between the effort needed to lift something (input) and the load itself (output) and we call this ratio the **mechanical advantage**. Consequently, when a machine can put out more force than is put in, the machine is said to give a good mechanical advantage. Mechanical advantage can be calculated by dividing the load by the effort. **There are no units for mechanical advantage, it is just a number.**

Remember

Mechanical advantage is just a number. It has no units

$$\text{Mechanical Advantage (MA)} = \frac{\text{Load}}{\text{Effort}}$$

Example

Using Figure 4.06, what is the mechanical advantage of the pulley system?

$$\mathbf{MA} = \frac{\text{Load}}{\text{Effort}} = \frac{2000 \text{ N}}{500 \text{ N}} = 4$$

In a lever, an effort of 10N is used to move a load of 50 N. What is the mechanical advantage of the lever?

$$\mathbf{MA} = \frac{\text{Load}}{\text{Effort}} = \frac{50}{10} = 5$$

This effectively means that for this lever, any effort will move a load that is five times larger.

Here is a summary.

- Where MA is greater than 1: The machine is used to magnify the effort force (e.g. a class 1 lever)

- Where MA is equal to 1: The machine is normally used to change the direction of the effort force (e.g. a fixed pulley)

- Where MA is less than 1: The machine is used to increase the distance an object moves or the speed at which it moves (e.g. the siege machine).

Velocity ratio

Sometimes machines translate a small amount of movement into a larger amount (or *vice versa*). For example, in Figure 4.04, a small movement of the piston causes the load to move a much greater distance. This property is known as the velocity ratio, and is found by dividing the distance moved by the effort by the distance moved by the load in the same period of time. There are no units for velocity ratio, it is just a number.

$$\text{Velocity Ratio (VR)} = \frac{\text{Distance effort moves}}{\text{Distance load moves}}$$

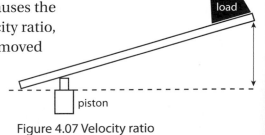

Figure 4.07 Velocity ratio

Example

In Figure 4.06, the piston moves 1m to move the load 5m. The velocity ratio is:

$$\text{Velocity Ratio (VR)} = \frac{\text{Distance effort moves}}{\text{Distance load moves}} = \frac{1}{5} = 0.2$$

Energy

Energy, measured in joules (J), is the ability to do work, or to cause something to move or the ability to cause change. Machines cannot work without energy and we are unable to get more work out of a machine than the energy we put into it. This is due mainly to friction. Friction occurs when two substances rub together. Try rubbing your hands together. Did you feel them get warmer?

Work produced (output) is usually less than the energy used (input). Energy can be transferred from one form to another, but energy cannot be created or destroyed. The loss of energy by friction usually ends up as heat.

There may be many forms of energy, but there are only two types:

- **Potential energy** (energy of position, or stored energy)

- **Kinetic energy** (energy due to the motion of an object).

Some forms of energy are: solar, electrical, heat, light, chemical, mechanical, wind, water, muscles and nuclear.

Potential energy

Anything may have stored energy, giving it the potential to cause change if certain conditions are met. The amount of potential energy something has depends on its position or condition. A brick on the top of scaffolding has potential energy because it could fall – due to gravity. The bow used to propel an arrow has no energy in its normal position, but draw the bow back and it now possesses a stored potential energy. A change in its condition (releasing it) can cause change (propelling the arrow).

Potential energy due to height above the Earth's surface is called gravitational potential energy, and the greater the height, the greater the potential energy.

There is a direct relation between gravitational potential energy and the mass of an object; more massive objects have greater gravitational potential energy. There is also a direct relation between gravitational potential energy and the height of an object. The higher an object is above the earth, the greater its gravitational potential energy. These relationships are expressed by the following equation:

PE_{grav} = mass of an object × gravitational acceleration × height

$$PE_{grav} = m \times g \times h$$

Another example of potential energy is the spring inside a watch. The wound spring transforms potential energy to kinetic energy of the wheels and cogs etc. as it unwinds.

Kinetic energy

Kinetic energy is energy in the form of motion. The greater the mass of a moving object, the more kinetic energy it has.

Power

When we do work in a mechanical system, the energy we put into the system does not appear instantaneously. It takes a certain time to move an object, lift a weight etc. The power that we put into a system must depend not only on the amount of work we do but also how fast we carry out the work.

To try to understand this, think of a 100 metre runner and a marathon runner. The sprinter has a burst of energy for maybe 10 seconds or so whereas the marathon runner may only use up slightly more energy but at a much slower pace. Let's face it, both events would leave you feeling shattered! But it is usual to say that the sprinter had greater power because he used his energy very quickly.

We usually say that:

Power = the rate of doing work

In terms of equations we can say that:

$$Power\ (P) = \frac{Work\ done\ (W)}{Time\ taken\ to\ do\ that\ work\ (t)}$$

or

$$Power\ (P) = \frac{Energy\ used\ (E)}{Time\ taken\ to\ do\ that\ work\ (t)}$$

In terms of units, energy or work is measured in joules (J), and time is measured in seconds (s). Power is measured in joules per second or J/s known as watts (W). So the units of power = W. Also 1000 watts (W) = 1 kilowatt (kW).

Remember

Be really careful, the shorthand for work is W and the units for power is W. Do not get them confused with each other

Example 1

A distribution board has a mass of 50kg and it is moved 10m.

Work = Force × Distance = (50 × 9.81) × 10 = 490.5 × 10 = 4905J

That is where we got to last time. Now what if it took 20s to move the distribution board by 10m. How much power did we put into moving it?

In this case:

$$P = \frac{Work\ done\ in\ moving\ the\ distribution\ board\ by\ 10m\ (W)}{The\ time\ it\ took\ to\ move\ the\ distribution\ board\ (t)}$$

Now, W = 4905 J, and t = 20 s. So:

$$P = \frac{4905}{20}$$

Therefore: **P = 245.25 W**

Example 2

In the diagram opposite, a trolley containing lighting fittings is pulled at constant speed along an inclined plane to the height shown. Assume that the value of the acceleration due to gravity is 10m/s^2. If the mass of the loaded cart is 3.0kg and the height shown is 0.45m, then what is the potential energy of the loaded cart at the height shown?

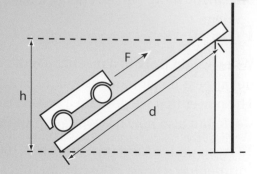

$PE = m \times g \times h$

$PE = 3 \times 10 \times 0.45$

PE = 13.5 J

If a force of 15.0 N was used to drag the trolley along the incline for a distance of 0.90m, then how much work was done on the loaded trolley?

$W = F \times d$

$W = 15 \times 0.9$

W = 13.5 J

SI units

On completion of this topic area the candidate will be able to identify the SI units, with multiple and submultiples in terms of quantities, as well as being able to describe how the SI unit applies to the electro-technical industry.

Why SI units?

Imagine if each country had a different idea of how long a metre is, or how much beer you get in a pint. What if a kilogram in Newcastle was different from one in Southampton? A common system for defining properties such as length, temperature and time is essential if people in different places are to work together.

In the UK and Europe an international system of units, known as SI units, is used for measuring different properties. There are seven base units – the main units from which all the other units are created.

Listed in Table 4.01 are the commonly used symbols and their units. The 'Definition' column defines the unit and relates it to the fundamental principles of electro-technology.

Quantity	Symbol	Unit name	Unit symbol	Definition
Electric current	I	Ampere	A	A quantity (Q) of electricity crossing a section of a conductor in a time (t) commonly referred to as current 'flow'.
Potential difference	V	Volt	V	The cause of movement of electric charge from one point to another, or the voltage applied to a circuit.
Resistance	R	Ohm	Ω	The property of a resistor to resist the flow of charge through it (all conductors resist the 'flow' of current).
Resistivity	ρ	Ohm-metre	Ωm	The part of the resistance of a conductor that is affected by the material it is made from.
Temperature	No symbol	Kelvin	K	For practical purposes, the degree Celsius scale is used. Both have identical intervals, 1K = 1°C, but the Kelvin scale begins at absolute zero –273°C. This means freezing point of water (0°C) is 273 K and boiling point (100°C) is 373 K.
Mass	No symbol	Kilogram	kg	This is a measure of the amount of material in the substance measured in kilograms (not to be confused with weight).
Force	F f	Newton	N	The cause of mechanical displacement or motion. Force may cause stationary objects to move or bring a moving object to rest, such as the force on a conductor imposed by a magnetic field. It is used in calculations for sizing electric motors and their efficiency.
Magnetic flux	Φ	Weber	Wb	This refers to the lines of force that appear around magnets and conductors carrying electric currents (electromagnets) and relates to the generation of alternating electric current.
Magnetic flux density	B	Tesla	T or Wb/m²	This refers to the intensity of the lines of force that appear around magnets and current carrying conductors and relates to the generation of alternating electric current.
Frequency	F	Hertz	Hz	The number of complete cycles that occur in one second in an a.c. waveform. The UK supply has a frequency of 50 Hz.
Power	P	Watt	W	The power or energy dissipated as a result of current flow through a load, found by the product of voltage and current. Used in calculations for supplies to loads, motors and efficiencies.
Energy	W	Joule	J	The capacity to do work over a period of time. Consumers pay by the amount of energy, in kW, used in 1 hour.
Time	t	Seconds	s	The unit of time, expressed in seconds.
Length	l	Metre	m	The unit of length measured in metres. Used for measurement of quantities of cables and equipment and setting out.
Area	A, a	Square metre	m²	The surface, enclosed by the sides of a two-dimensional shape, the area of a workshop.

Table 4.01 SI units

SI unit prefixes

We often deal in quantities that are much larger or smaller than the base units. If we have to stick with them, the numbers become clumsy and it is easy to make mistakes. For example, the diameter of a human hair is about 0.0009m, and the average distance of the Earth from the Sun is around 150,000,000,000m.

To make life easier, we can alter the symbols (and the quantities they represent) by adding another symbol in front of them (a prefix). These represent base units multiplied or divided by one thousand, one million etc.

Table 4.02 shows the most common prefixes.

Multiplier	Name	Symbol prefix	As a power of 10
1 000 000 000 000	Tera	T	1×10^{12}
1 000 000 000	Giga	G	1×10^{9}
1 000 000	Mega	M	1×10^{6}
1 000	kilo	k	1×10^{3}
1	unit		
0.001	milli	m	1×10^{-3}
0.000 001	micro	μ	1×10^{-6}
0.000 000 001	nano	n	1×10^{-9}
0.000 000 000 001	pico	p	1×10^{-12}

Table 4.02 Common prefixes

Here are some common examples of using the prefixes with the unit symbol:

- km (kilometre = one thousand metres)
- mm (millimetre = one thousandth of a metre)
- MW (megawatt = one million watts)
- μs (microsecond = one millionth of a second).

Laws of resistance

On completion of this topic area the candidate will be able to state the relationship of resistance, with regard to length, cross-sectional area and resistivity when applied to conductors and cables.

Resistance

An electric current passing through a material will produce heat. This is because of the opposition to the current flow. This property of a material is called its electrical resistance. There are four laws that determine the resistance of conductors:

- length
- area
- resistivity
- temperature.

We will now look at how the resistance is affected from the measurement of a single piece of material, when we change these factors.

Length

If two equal lengths of the same material, with the same cross-sectional area, are joined end to end and the resistance measured, then the total resistance would be found to have doubled. Add two more equal lengths of the same cross sectional area and the resistance then would be quadrupled.

This gives us law 1.

- The resistance of a material having a uniform cross-sectional area is **directly proportional** to its length.

Area

If two equal lengths of the same material, with the same cross-sectional area, are joined side by side and the resistance measured, the value will be found to be halved. Similarly, if four pieces of the same material of equal length, are all joined side-by-side the resistance, when measured, would be one quarter of the original but the cross-sectional area will be four times as large.

This gives us law 2.

- The resistance of the material of constant length is **inversely proportional** to its cross-sectional area.

If we combine law 1 and law 2 we can say:

$$\text{Resistance} \propto \frac{\text{length}}{\text{cross-sectional area}} \qquad \text{or} \qquad R \propto \frac{l}{a}$$

Resistivity

Resistivity is measured in **ohm metres (Ωm)**[3] and its symbol is the Greek letter ρ pronounced **rho**.

It provides us with law 3.

- The resistance of a material also depends upon what is made of, its atomic structure and the number of electrons within the structure.

If we now combine this with the previous statements regarding length and cross-sectional area, we can say:

$$\text{Resistance} = \frac{\text{Resistivity} \times \text{Length}}{\text{Cross-sectional area}} \qquad \text{or} \qquad R = \frac{\rho l}{a}$$

Resistivity tables are based on a piece of the material of unit length and unit cross sectional area (unit cube).

Material	Ωm	Material	Ωm
Silver	16.5×10^8	Glass	1×10^{13}
Copper	17.8×10^{-9}	Ebonite	2×10^{14}
Aluminium	28.4×10^{-9}	Porcelain	2×10^{14}
Brass	66.0×10^{-9}	Mica	9×10^{14}

Table 4.03 Resistivity table

As can be seen from Table 4.03, the resistivity of conductor material is very small, while the resistivity of insulator material is large.

Temperature

We know from applying laws 1, 2 and 3 that the resistance of a conductor is dependent upon its length, its cross-sectional area and its resistivity. However, this is only true if the temperature remains constant. Most conductors have a positive temperature coefficient. This means that their resistance **increases** as the temperature increases. The only exception to this is carbon, which has a negative temperature coefficient, resulting in its resistance **decreasing** as the temperature increases.

This gives us law 4.

- The resistance of conductors (except carbon) increases as their temperature increases.

Therefore the four laws of resistance are as follows:

1. The resistance of a material having a uniform cross-sectional area is **directly proportional** to its length.

2. The resistance of the material of constant length is **inversely proportional** to its cross-sectional area.

3. The resistance of a material also depends upon what is made of, its atomic structure and the number of electrons within the structure.

4. The resistance of conductors (except carbon) increases as their temperature increases.

Construction of a simple alternator

On completion of this topic area the candidate will be able to state the construction features and operating principles of a simple alternator and how the alternator produces a sinusoidal waveform output.

Conductors

When a conductor cuts through a magnetic field, at right angles to the magnetic flux, an **e.m.f.** (electromotive force) is induced in the conductor. The strength of the induced e.m.f. is determined by:

- the strength of the magnetic flux density, between the poll faces of the magnet
- the length of the conductor in the magnetic field
- the velocity or speed of the conductor through the magnetic field.

We calculate this with the following formula:

$e = Bl v$

where: e = induced e.m.f. in volts

B = Magnetic flux density in tesla (T)

l = Length of the conductor in metres (m)

v = velocity of the conductor in metres/sec (m/s)

We use this calculation about e.m.f. to construct a single loop alternator.

Single loop alternator

Figure 4.05 shows the construction of a simple single loop alternator.

In the simple single loop alternator shown, the conductor is formed into a loop and is positioned within the magnetic field of a permanent magnet. The ends of the wire loop are brought out on to slip rings. Carbon brushes, sprung loaded to maintain contact, connect the slip rings to an external circuit, in this case a lamp.

If the loop is now turned in the direction shown, the conductor cuts through the lines of magnetic flux, and an e.m.f. (voltage) is induced into the conductor. This completes the circuit, and the e.m.f. causes current to flow, making the lamp illuminate.

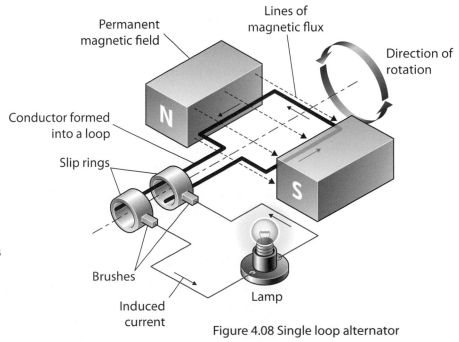

Figure 4.08 Single loop alternator

The voltage output of the alternator is in the form of a sinusoidal waveform, with an alternating current (a.c.). Figure 4.06 shows how the conductor rotates in the magnetic field and the resulting sinusoidal waveform produced.

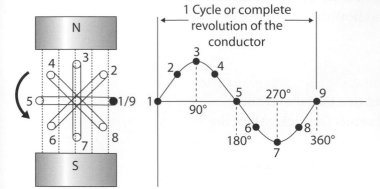

Figure 4.09 Waveform in a single loop conductor

When the conductor is in the position shown, between points 1 and 5, no lines of magnetic flux are being cut, and therefore the e.m.f. produced at this point is zero, as can be seen at point 1 on the waveform.

As the conductor rotates 45° anticlockwise, to the position shown between points 2 and 6, lines of magnetic flux are cut. An e.m.f. is now induced into the conductor, as can be seen at point 2 on the sinusoidal waveform. The conductor continues to rotate a further 45° to the position shown between points 3 and 7. At this point, the conductor is travelling at right angles to the lines of magnetic flux and is therefore cutting through the maximum number of lines of flux. As a result, the e.m.f. is at its maximum value, as can be seen at point 3 on the waveform diagram.

The conductor continues to rotate anticlockwise and at point 4 and 8 lines of magnetic flux are being cut, although not as many as at the previous point. The e.m.f. produced reduces, as can be seen at point 4 on the waveform diagram. The conductor moves on 45° further, to lie between points 5 and 1. At this point no lines of magnetic flux are being cut, and therefore the e.m.f. produced at this point is zero, as can be seen at point 5 on the waveform diagram. As the conductor continues to rotate it is now cutting the lines of magnetic flux in the opposite direction. Therefore the e.m.f. produced is also in the opposite direction.

When the conductor is between points 6 and 2 lines of magnetic flux are cut and an e.m.f. induced, as can be seen at point 6 on the waveform diagram. The conductor then continues to rotate a further 45° to lie between points 7 and 3. Again the conductor is travelling at right angles to the lines of magnetic flux, and the maximum number of lines of flux are being cut so the e.m.f. produced reaches its peak – as can be seen at position seven on the waveform diagram.

The conductor moves on, another 45° to the point between 8 and 4, where lines of magnetic flux are being cut but not as many as in the previous position, and therefore the e.m.f. produced reduces, as can be seen at point 8 on the waveform diagram. Finally, the conductor completes the 360° rotation to sit between points 9 and 5. No lines of magnetic flux are cut, and therefore no e.m.f. is induced. The waveform therefore returns to zero, as can be seen at point 9.

Every time, the conductor does a rotation, it produces a complete sine wave. This is one cycle and the number of cycles that occur in each second is the frequency. The unit of frequency is the Hertz (Hz). In the United Kingdom, the standard frequency of the supply is 50 Hz.

Efficiency

On completion of this topic area the candidate will be able to state the efficiency of a machine in terms of output and input expressed as a percentage.

Efficiency in machines

We often think of machines as having an input and an output. Instead, try of thinking of a machine as having two outputs: one that is wanted and one that is not and is therefore wasted. The greater the unwanted component, the less efficient the machine is.

In all machines, the power at the input is greater then the power output, because of losses that occur in the machine such as friction, heat or vibration. This difference, expressed as the ratio of output power over input power, is called the **efficiency** of the machine. The symbol sometimes used for efficiency is the Greek letter η (eta).

That is:

$$\text{Efficiency} = \frac{\text{Output power}}{\text{Input power}}$$

To give the efficiency as a percentage, which is usually more convenient and understandable, we can say that:

$$\% \text{ efficiency} = \frac{\text{Output power} \times 100}{\text{Input power}}$$

We will now run through a series of steps to end up with a final equation for efficiency. It is quite complicated, but try to follow the steps. The main thing, though, is to remember the final formula.

Now, for any machine:

Work done at the input = Effort × Distance the effort moves (force × distance)

And:

Work done at the output = Load × Distance the effort moves (force × distance)

Dividing these two equations gives us:

$$\frac{\text{Work at Output}}{\text{Work at Input}} = \frac{\text{Load} \times \text{Distance moved by load}}{\text{Effort} \times \text{Distance moved by effort}} = \text{Efficiency}$$

Which can be rewritten as:

$$\frac{\text{Work at Output}}{\text{Work at Input}} = \frac{\text{Load}}{\text{Effort}} \times \frac{\text{Distance moved by load}}{\text{Distance moved by effort}} = \text{Efficiency}$$

Now, you already know that:

$$\text{Mechanical Advantage (MA)} = \frac{\text{Load}}{\text{Effort}}$$

And that:

$$\text{Velocity Ratio (VR)} = \frac{\text{Distance effort moves}}{\text{Distance load moves}}$$

So:

$$\frac{1}{\text{VR}} = \frac{\text{Distance load moves}}{\text{Distance moved by effort}}$$

So:

$$\text{Efficiency} = \frac{\text{Work at Output}}{\text{Work at Input}} = \frac{\text{Load}}{\text{Effort}} \times \frac{\text{Distance moved by load}}{\text{Distance moved by effort}} = \text{MA} \times \frac{1}{\text{VR}}$$

Therefore:

$$\text{Efficiency} = \frac{\text{Mechanical Advantage}}{\text{Velocity Ratio}} = \frac{\text{MA}}{\text{VR}} \text{ or } \% \text{ efficiency} = \frac{\text{MA}}{\text{VR}} \times 100$$

That was quite tricky, but you must remember the final equation above.

If a machine has low efficiency, this does not mean it is of limited use. A car jack, for example, has to overcome a great deal of friction and therefore has a low efficiency, but it is still a very useful tool as a small effort allows us to lift the weight of a car when changing a tyre.

Let us look at three examples that will help us get to grips with some of these concepts.

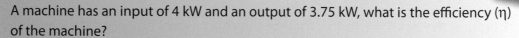

Example 1

A machine has an input of 4 kW and an output of 3.75 kW, what is the efficiency (η) of the machine?

$$\text{Efficiency} = \frac{\text{Output}}{\text{Input}} \qquad \eta = \frac{3750}{4000} = \mathbf{0.94}$$

$$\% \text{ Efficiency} = \frac{\text{Output}}{\text{Input}} \times 100 \qquad \% \text{ Efficiency} = \frac{3750}{4000} \times 100$$

Example 2

A motor with an output of 5kW is 85% efficient. What is the input to the motor?

$$\% \text{ efficiency} = \frac{\text{Output}}{\text{Input}} \times 100$$

Transpose to find the Input $\text{Input} = \frac{\text{Output}}{\% \text{ efficiency}} \times 100$

$$\text{Input} = \frac{5000}{84} \times 100 = \textbf{5952.4 watts} \quad \text{or} \quad \textbf{5.952 kW}$$

Example 3

A motor control panel arrives on site. It is removed from the transporter's lorry using a block and tackle that has five pulley wheels. Establish the percentage efficiency of this system given that the effort required to lift the load was 200 N, the panel has a mass of 80kg and acceleration due to gravity is 10 m/s².

The load = mass × acceleration due to gravity

$$= 80 \times 10$$

$$= 800 \text{ N}$$

$$\text{Mechanical Advantage} = \frac{\text{Load}}{\text{Effort}} = \frac{800 \text{ N}}{200 \text{ N}} = 4$$

Remembering what we said earlier

Velocity Ratio = the number of pulley wheels = 5

$$\text{Efficiency} = \frac{\text{Mechanical advantage}}{\text{Velocity ratio}} = \frac{4}{5} = 0.8$$

So, % efficiency = 0.8 × 100 = 80%

Therefore the system is 80 per cent efficient.

Activity

With the aid of some lengths of wire and an ohmmeter, find out what happens to the resistance of the wire when:

- the length is doubled

- the length is halved

- the cross-sectional area is doubled

- the cross-sectional area is halved.

FAQ

Q I don't understand how something can weigh less in space than on the ground.

A Weight and Mass are often confused because, in everyday life we often use weight as measure of the amount something. In fact, the amount of something should be referred to as its Mass. This remains the same wherever you are. An objects weight is variable depending on the amount of gravity there is to pull it downwards. In other words, to make it feel heavy. If you get far enough away from a gravitational pull, then the object would become weightless, although its mass would remain the same.

Q How does a pulley system make it easier to lift a load? I thought you couldn't get something for nothing?

A A pulley makes it easier to lift a load by reducing the amount of force required to move it. However, you have to pull the rope much further than you did before the pulley system was there, so the energy used to move the object remains approximately the same.

Knowledge check

1. Explain with the aid of diagrams the difference between class 1, 2 and 3 levers.

2. Explain the difference between potential energy and kinetic energy.

3. Convert the following numbers into whole numbers by removing the prefixes:

    - 2.7kg

    - 6.4 mWb

    - 1000 μA

    - 2200 nm

    - 2 GHz

    - 1 TW

    - 4700 pF

    - 100 MW

4. Sort the following resistivity figures into good conductors and good insulators:

    - 66×10^{-6}

    - 2×10^{15}

    - 5×10^{11}

    - 16.5×10^{-6}

    - 9×10^{15}

    - 28.4×10^{-6}

5. Explain with the aid of diagrams how a.c. is produced by rotating a coil of wire inside a magnetic field.

chapter 5

Basic scientific concepts in electro-technology

Unit 2 Outcome 2

In order to understand how to use the basic units found in electro-technology, an electrician needs to be aware of, and understand, the science and theory that lies behind them. The basis of all electrical theory lies in the particles that make up the atom, in particular the electron.

The scientific concepts that are covered in this chapter form the basis for all the work that electricians carry out in their day-to-day jobs. In order to complete installations safely and efficiently it is vital that you understand the scientific implications in order to correctly identify the practical implications.

On the completion of this chapter the candidate will be able to:

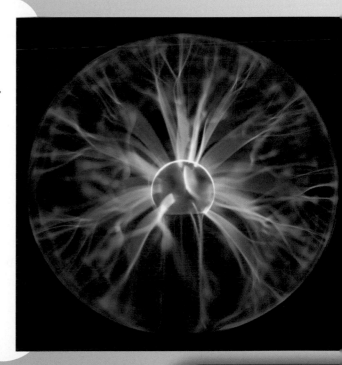

- state the reaction of electrons when charged and how an electromotive force may be produced by chemical, magnetic or thermal means

- list materials used as conductors and insulators and their properties and applications

- explain how current flow differs in series and parallel circuits using Ohm's law

- describe how voltmeters and ammeters are connected into circuits in order to quantify circuit voltages, current and resistance

- describe the magnetic fields and flux patterns set up by arrangements of permanent magnets, conductors and solenoids

- describe the construction of transformers and the benefits gained by their use.

Electron theory, conductors and insulators

On completion of this topic area the candidate will be able to state in simple terms; the reaction of electrons when charged, forming the concept of an electric current, list materials used as conductors and insulators and state the basic effects of an electric current.

Electron theory

This section is where we really start to look at electricity and electrical circuits in detail. You cannot even think of becoming an electrician unless you have a sound knowledge of the principles involved, starting with the atomic theory of matter and how this gives rise to an electric current.

In this section we will be looking at the following areas:

- molecules and atoms
- the electric circuit
- the causes of an electric current
- the effects of an electric current.

Molecules and atoms

Every substance is composed of molecules, which in turn are made up of atoms.

Atoms are not solid but consist of even smaller particles. At the centre of each atom is the nucleus, which is made up from particles known as protons and neutrons. Protons are said to possess a positive charge (+), and neutrons are electrically neutral. The neutrons act as a type of 'glue' which holds the nucleus together.

You probably know that:

- **like charges repel each other (+ and + or – and –)**
- **unlike charges attract each other (+ and –).**

So in a world without neutrons the positively charged **protons** would repel each other and the nucleus would fly apart. It is the job of the neutrons to hold the nucleus together. Since neutrons are electrically neutral they play no part in the electrical properties of atoms.

The remaining particles in an atom are known as **electrons**. These circle in orbits around the nucleus and are said to possess a negative charge (–).

All atoms possess equal numbers of protons and electrons. Thus the positive and negative charges are cancelled out, leaving the atom electrically neutral. In some cases it is possible to remove or add an electron to a neutral atom and leave it with a net positive or negative charge. Such atoms are then known as **ions**.

So what is the relationship between protons and electrons, and how do they form an atom?

Perhaps the simplest explanation is to look to our solar system – where we have a central star (the Sun) around which are the orbiting planets. In the atom the protons and neutrons form a central nucleus (sun) and the electrons are the orbiting particles (planets).

The three states of matter

Molecules are always in a state of rapid motion, but when they are densely packed together, this movement is restricted and the substance formed by these molecules is solid. When the molecules of a substance are less tightly bound, there is a great deal of free movement and the substance is a liquid. Finally when the molecule movement is almost unrestricted, the substance can expand and contract in any direction and is a gas.

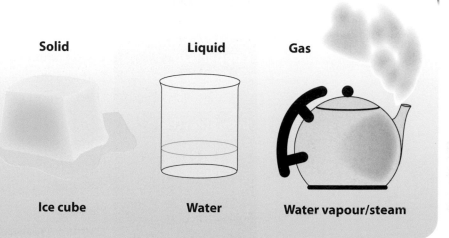

Solid   Liquid   Gas

Ice cube   Water   Water vapour/steam

The simplest atom is that of hydrogen which has one proton and one electron. Figure 5.01 shows the hydrogen atom.

Electrons are arranged in layers at varying distances from the nucleus; those nearest to the nucleus are more strongly held in place than those farthest away. These distant electrons are easily moved from their orbits and so are free to join those of another atom, whose own distant electrons may in turn leave to join another atom and so on.

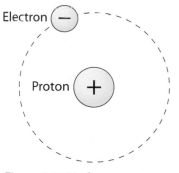

Electron ⊖

Proton ⊕

Figure 5.01 Hydrogen atom

It is these wandering or 'free' electrons moving about the molecular structure of a material that give rise to electricity.

We call a material that allows the movement of free electrons a **conductor**, and one which does not an insulator.

Conductors

Try to think of conductors as being materials that have their atoms packed together loosely. This, therefore, allows the free electrons to move through them.

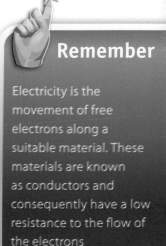

Remember

Electricity is the movement of free electrons along a suitable material. These materials are known as conductors and consequently have a low resistance to the flow of the electrons

Gold and silver are among the best conductors, but cost inhibits their use. Table 5.01 is a guide to the most common conductors and what they are used for.

Aluminium (Al)	Low cost and weight. Not very flexible. Used for large power cables.
Brass (alloy of Copper and Zinc)	Easily machined. Corrosion resistant. Used for terminals and plug pins.
Carbon (C)	Hard. Low friction in contact with other materials. Used for machine brushes.
Copper (Cu)	Good conductor. Soft and ductile. Used in most cables and busbar systems.
Iron/Steel (Fe)	Good conductor. Corrodes. Used for conduit, trunking and equipment enclosure.
Lead (Pb)	Flexible. Corrosion resistant. Used as an earth and as sheath of a cable.
Mercury (Hg)	Liquid at room temperature. Quickly vaporises. Used for contacts. Vapour used for lighting lamps.
Sodium (Na)	Quickly vaporises. Vapour used in lighting lamps.
Tungsten (W)	Extremely ductile. Used for filaments in light bulbs.

Table 5.01 Common conductors

Insulators

Something has to stop electricity from leaking everywhere, otherwise we would get an electric shock every time we used a piece of electrical equipment. The materials that we use to do this are called insulators.

Think of an insulator as being a material whose atoms are so tightly packed together, there is no room for the free electrons to move through them.

Surprisingly, one insulator that is used in cable manufacture is paper! Others are shown in Table 5.02.

Rubber/plastic	Very flexible. Easily affected by temperature. Used in cable insulation.
Impregnated paper	Stiff and **hygroscopic.** Unaffected by moderate temperature. Used in large cables.
Magnesium oxide	Powder, therefore requires a containing sheath. Very hygroscopic. Resistant to high temperature. Used in cables for alarms and emergency lighting.
Mica	Unaffected by high temperature. Used for kettle and toaster elements.
Porcelain	Hard and brittle. Easily cleaned. Used for carriers and overhead line insulators.
Rigid plastic	Less brittle and less costly than porcelain. Used in manufacture of switches and sockets.

Table 5.02 Common insulators

Measuring electricity

This would seem a simple task, apart from the fact that electricity is invisible. But what exactly shall we measure?

Electricity is simply the flow of free electrons along a conductor, so it would seem obvious to measure the number of electrons moving along the conductor. However, the electron is far too small to be of any practical use. So we group a number of electrons together and then measure the number of groups of electrons moving along. This grouping is known as a **coulomb**, and contains an unimaginable 6,240,000,000,000,000,000 electrons (give or take a couple).

A plumber will measure the amount of water flowing in litres not drops, as drops are too small a unit of measurement. If the plumber wishes to know how much water is being used at any one time, in other words 'the rate of flow' of the water, this would be measured in litres per minute.

The movement of the water is thought of as its current.

Similarly, the electrician may wish to know the amount of electrons flowing at any one time (rate of flow of electrons). In electricity, just as with water, this rate of flow of electrons is called the current and is defined as being one coulomb (an imaginary bucket full) of electrons passing by every second.

Remember

A good insulator has high resistance

Definition

Hygroscopic – the ability to absorb water

Remember

Think of a coulomb as an imaginary bucket of electrons

If one coulomb of electrons passes along the conductor every second, we say that the current flowing along is a current of one ampere. We use the symbol I to represent current.

In other words, one ampere equals one 'imaginary bucket full' of electrons passing by every second.

The electric circuit

We now know that electricity is the movement of electrons along a conductor and that the amount of electrons flowing along is called the current. So what causes the current to flow? What force causes the electrons to move?

Battery

In this circuit, the battery has an internal chemical reaction that provides what is known as an **electromotive force** (e.m.f. for short), which will push the electrons along the conducting wire and into the lamp.

The electrons will then pass through the lamp's filament, causing it to heat up and glow and then leave via the second conductor, returning to the battery and thus completing the circuit. If either of the two wires becomes broken or disconnected the flow of electricity will be interrupted and the lamp will go out.

It is this principle that we use to control electricity in a circuit. By inserting a switch into one of the wires connected to the lamp, we can physically 'break the circuit' with the switch and thus switch the lamp off and on.

Figure 5.02 A simple circuit

To summarise, for practical purposes a working circuit should:

- have a source of supply (such as the battery)
- have a device (fuse/MCB) to protect the circuit
- contain conductors through which current can flow
- be a complete circuit
- have a load (such as a lamp) that needs current to make it work
- have a switch to control the supply to the equipment (load).

Electron flow and conventional current flow

An electromotive force is needed to cause this flow of electrons. This has the quantity symbol **E** and the unit symbol **V** (volt). Any apparatus which produces an e.m.f. (such as a battery) is called a power source and it will require wires or cables to be attached to its terminals to form a basic circuit.

If we take two dissimilar metal plates and place them in a chemical solution (an electrolyte) a reaction will take place in which electrons from one plate travel through the electrolyte and collect on the other plate.

Remember

One ampere equals one coulomb of electrons passing by every second

One plate now has an excess of electrons, which will make it more negative than positive. The other plate will now have an excess of protons, which makes it more positive than negative. This process is the basis of how a simple battery or cell works.

Now select a piece of wire as a conductor (which we already know will have free outer electrons) and connect this wire to the ends of the plate as shown in Figure 5.03.

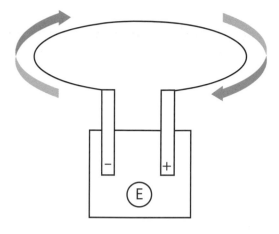

Figure 5.03 Electron flow through an electrolyte

Since unlike charges are attracted towards each other while like charges repel each other, you can see that the negative electrons will move from the negative plate, through the conductor towards the positive plate.

This drift of free electrons is what we know as electricity and this process will continue until the chemical action of the battery is exhausted and there is no longer a difference between the plates.

Note: The electron flow is actually negative to positive through the conductor. The conventional current flow is from positive to negative.

Potential difference

We now know that the force that pushes electrons along a conductor is the e.m.f. But how do we measure the e.m.f. or voltage? The electromotive force is measured in terms of the number of joules of work which are required to push one coulomb (our imaginary bucket full) of electrons along the circuit and is therefore measured with the unit joules/coulomb.

This unit is more commonly referred to as the volt. As an equation:

One volt = One joule/coulomb

The battery is acting as an energy conversion system, converting chemical energy into electric potential energy. This work increases the potential energy of the charge and thus its electric potential. The charge is moving from the 'low potential' terminal to a 'high potential' terminal inside the internal circuit of the battery. Once there it will then move through the external circuit (the conductor and equipment), before returning to the low potential terminal. The difference between our terminals is referred to as the **potential difference** and without it there can be no flow of charge.

As our charge moves through the external circuit, it can meet different types of component each of which acts as an energy conversion system, e.g. the lamp in Figure 5.02. Here the moving charge is doing work on the lamp to produce different forms of energy: heat and light. However, in doing so it is losing its electric potential energy and, therefore, on leaving the lamp it is less energized. Think of a marathon

Remember

Be careful, in circuit drawings we always show the current as flowing from the positive to the negative. You now know this is not the case

runner who starts the race fresh and full of energy. But as the race progresses, the runner uses up more and more energy, until at the end of the race the runner has no energy left.

At the start of the race the runner has the energy or the ability (the potential) to run the race. At the end, having used up all the energy, they no longer have the ability (potential) to run a race. The gap between the two is the potential difference.

We also need a 'reference' point that potential is measured against. In the case of electricity we use the general mass of earth, which has no potential at all, or 0 V. So when we say that a voltage is 230 V, we mean that its potential is 230 V above zero, the potential of the mass of earth.

Remember

A current cannot exist without a potential difference

Basic electric principles

Consider the diagram:

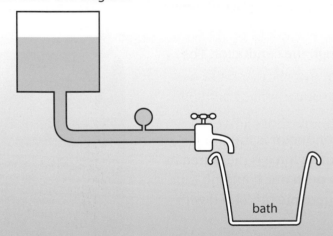

bath

Most water systems are known as gravity fed systems. In other words, the force that will drive the water through the system pipe work is gravity. However, for gravity to be able to make the water come out of the tap, there has to be a difference in level between the supply (the water tank) and the load (in this case the bath) and therefore the water tank is higher than the tap. This gives us the equivalent of potential difference.

The tap performs the same job as a switch in that it is our means of controlling whether the water reaches the load. The gauge that you see is fitted in the water line and will measure the amount of water flowing past it every second as dictated by the load. In other words, it is measuring the current. In electrical terms, we would call this an ammeter.

A cup only requires a small amount of water for it to perform its role and therefore only a small amount of water will be shown as passing through the gauge. The bath requires a large amount of water flowing into it and therefore the gauge will show a large amount of water passing by.

In electrical terms, a lamp only requires a small current and therefore only a small current would register on the ammeter, whereas an item such as a heater would require a larger current and the ammeter would show a larger current passing.

The causes of an electric current

As we saw in chapter 4 (pages 83–84) we need an electromotive force (e.m.f.) to drive electrons through a conductor. The principal sources of an e.m.f. can be classed as being:

- chemical
- thermal
- magnetic.

Chemical

When we take two electrodes of dissimilar metal and immerse them in an electrolyte, we have effectively created a battery. So, as we have seen earlier, the chemical reactions in the battery cause an electric current to flow.

Thermal

When a closed circuit consists of two junctions, each made between two different metals, a potential difference will occur if the two junctions are at different temperatures. This is known as the Seebeck effect, based upon Seebeck's discovery of this phenomenon in 1821.

If we now connect a voltmeter to one end (the cold end) and apply heat to the other, then our reading will depend upon the difference in temperature between the two ends. When we have two metals arranged in this pattern, we have a thermocouple.

We can apply this to the measurement of temperatures, with the 'hot end' being placed inside the equipment (such as an oven or hot water system) and the 'cold end' connected to a meter that has been located in a suitable remote position.

Magnetic

Under certain circumstances, a magnetic field can be responsible for the flow of an electric current. We call this situation **electromagnetic induction**. If a conductor is moved through a magnetic field, then an e.m.f. will be induced in it. Provided that a closed circuit exists, this e.m.f. will then cause an electric current.

The effects of an electric current

The effects are categorised in the exact same way as the causes, namely:

- chemical
- thermal
- magnetic.

In Figure 5.04, a d.c. supply enters a contactor. When we close the switch on the contactor the coil is energised and becomes an electromagnet, thus pulling anything containing iron in the magnetic field towards it. This allows the main supply to flow through to the distribution centre.

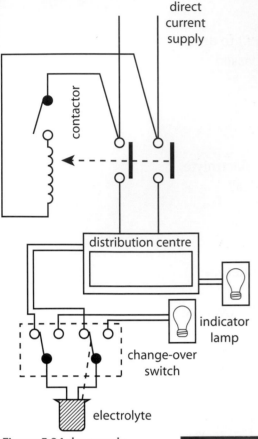

direct current supply

contactor

distribution centre

indicator lamp

change-over switch

electrolyte

Figure 5.04 d.c. supply

From the distribution centre a supply is taken to a change-over switch. In its current position, this switch allows current to flow into the electrolyte (dilute sulphuric acid and water) via one of the two lead plates. The current returns to the distribution centre via the other lead plate.

Also fed from the distribution centre is a filament lamp. We could equally have used an electric fire. This is because, when current flows through a conductor, heat is generated. The amount of heat varies according to circumstances, such as the conductor size. If sized correctly, we can make the conductor glow white-hot (a lamp) or red-hot (a fire).

If we run the system like this for a few minutes and then switch off the contactor, obviously we would see our filament lamp go out. However, if we now move our change-over switch into its other position, we would see that our indicator lamp would glow for a short while.

If we were to look at the lead plates, we would see that one of the plates has become discoloured. This is because the current has caused a chemical reaction, changing the lead into an oxide of lead. In this respect the plates acted as a form of re-chargeable battery, also known as a secondary cell.

Table 5.03 indicates how various equipment relates to these effects of current as their principle of operation.

Chemical effect	Heating effect	Magnetic effect
Cells	Filament lamps	Bells
Batteries	Heaters	Relays
Electro-plating	Cookers	Contactors
	Irons	Motors
	Fuses	Transformers
	Circuit breakers	Circuit breakers
	Kettles	

Table 5.03 Principles of operation

Series and parallel circuits

On completion of this topic area the candidate will be able to: explain how current flow differs in series and parallel circuits, transpose and apply basic formulae, and apply calculations related to Ohm's law.

Current flow

In order for current to flow, two conditions must be met. First, there has to be a potential difference (voltage) to force the current flow and, second, there must be a continuous circuit (circle) for the current to flow around.

When we start to look at circuits, we see that there are different combinations:

- series circuits
- parallel circuits
- parallel-series circuits.

In this section, we will be applying Ohm's law and looking at how current, resistance and potential difference are all related.

You will also find is useful to refer back to chapter 4 where basic calculations involving force, mass, energy, power and efficiency were covered in detail. More information on the relationship between a.c. and d.c. supply, and the benefits of using a.c. over d.c., can be found in chapters 6 and 10.

Ohm's law

So far we have established that **current** is the amount of electrons flowing past a point every second in a conductor and that a force known as the e.m.f. (or **voltage**) is pushing them. We now also know that the conductor will try to oppose the current, by offering a **resistance** to the flow of electrons.

Ohm's law, the means by which these three topics are linked together, is probably the most important electrical concept that you will need to understand and is stated as follows:

> *The current flowing in a circuit is directly proportional to the voltage applied to the circuit, and indirectly proportional to the resistance of the circuit, provided that the temperature affecting the circuit remains constant.*

Ohm's law was named after the nineteenth-century German physicist G.S. Ohm, who researched how current, potential difference and resistance are related to each other.

In simple language we could re-write Ohm's law as follows:

'*The amount of electrons passing by every second will depend on how hard we push them, and what obstacles are put in their way*'.

We can prove this is true, because if we increase the voltage (push harder), then we must increase the number of electrons that we can get out at the other end. Try flicking a coin along the desk. The harder you flick it, the further it travels along the desk. This is what we mean by directly proportional. If one thing goes up (voltage), then so will the other thing (current).

Equally we could prove that if we increase the resistance (put more obstacles in the way), then this will reduce the amount of electrons that we can get along the wire.

This time put an obstacle in front of the coin before you flick it. If flicked at the same strength, it will not go as far as it did before. This is what we mean by **indirectly proportional**. If one thing goes up (resistance), then the other thing will go down (current).

Ohm's law can be expressed by the following formula:

$$\text{Current (I)} = \frac{\text{Voltage (V)}}{\text{Resistance (R)}}$$

Remember

Remember Ohm's law,

$$I = \frac{V}{R}$$

Series circuits

If a number of resistors are connected together end to end and then connected to a battery, the current can only take one route through the circuit. This type of connection is called a series circuit.

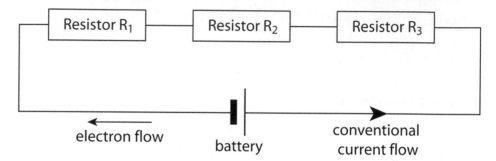

Figure 5.05
Series circuit

Features of a series circuit

- The total circuit resistance (R_t) is the sum of all the individual resistors. In our diagram, this means:

$$R_t = R_1 + R_2 + R_3$$

- The total circuit current (I) is the supply voltage divided by the total resistance. You'll recognise this as Ohm's law:

$$I = \frac{V}{R}$$

- The current will have the same value at every point in the circuit.

- The potential difference across each resistor is proportional to its resistance. If we think back to Ohm's law, we use voltage to push the electrons through a resistor. How much we use depends upon the size of the resistor. The bigger the resistor, the more we use. Therefore:

$$V_1 = I \times R_1 \qquad V_2 = I \times R_2 \qquad V_3 = I \times R_3$$

- The supply voltage (V) will be equal to the sum of the potential differences across each resistor. In other words, if we add up the p.d. across each resistor (the amount of volts 'dropped' across each resistor), it should come to the value of the supply voltage. We show this as:

$$V = V_1 + V_2 + V_3$$

- The total power in a series circuit is equal to the sum of the individual powers used by each resistor.

Calculation with a series circuit

Example

Two resistors of 6.2 Ω and 3.8 Ω are connected in series with a 12 V battery as shown.

We want to calculate:

(a) total resistance
(b) total current flowing
(c) the potential difference (p.d.) across each resistor.

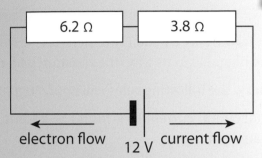

electron flow   12 V   current flow

Figure 5.06 Series circuit

(a) Total resistance

For series circuits, the total resistance is the sum of the individual resistors:

$$R_t = R_1 + R_2 = 6.2 + 3.8 = 10 \text{ ohms}$$

(b) Total current

Using Ohm's law:

$$I = \frac{\text{Voltage}}{\text{Resistance}} = \frac{12}{10} = 1.2 \text{ amperes}$$

(c) The p.d. across each resistor

$V = I \times R$, therefore:

Across R_1: $V_1 = I \times R_1 = 1.2 \times 6.2 = 7.44$ volts
Across R_2: $V_2 = I \times R_2 = 1.2 \times 3.8 = 4.56$ volts

Parallel circuits

If a number of resistors are connected together so that there are two or more routes for the current to flow, as shown in Figure 5.07, then they are said to be connected in parallel.

In this type of connection, the total current splits up and divides itself among the different branches of the circuit. However, note that the pressure pushing the electrons along (voltage), will be the same through each of the branches. Therefore any branch of a parallel circuit can be disconnected without affecting the other remaining branches.

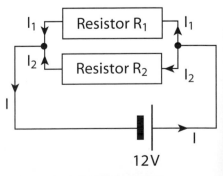

Figure 5.07 Parallel circuit

Explanation

The amount of electrons passing by (current) depends upon how hard we are pushing.

We have said that, in a parallel circuit, voltage is the same through each branch. Try to push two identical pencils in the same direction as the current flow towards a point on the circuit where the two branches split (shown as black circles on the drawing). When they reach that point, one pencil will travel towards R_1 and the other towards R_2. But look how the force pushing the pencils has stayed the same.

However, how easily a pencil can then pass through a branch will depend upon the size of the obstacle in its way (the resistance of a resistor).

In summary, the following rules will apply to a parallel circuit:

- the total circuit current (I) is found by adding together the current through each of the branches:

$$I = I_1 + I_2 + I_3$$

- the same potential difference will occur across each branch of the circuit:

$$V = V_1 = V_2 = V_3$$

- where resistors are connected in parallel and, for the purpose of calculation, it is easier if the group of resistors is replaced by one total resistor (R_t). So:

$$\frac{1}{R_t} = \frac{1}{R_1} + \frac{1}{R_2} + \frac{1}{R_2}$$

Calculations with a parallel circuit

Example

Three resistors of 16 Ω, 24 Ω and 48 Ω are connected across a 240V supply. There are two ways to find out the total circuit current.

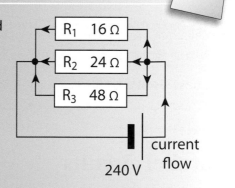

Figure 5.08

Method 1
Find the equivalent resistance, then use Ohm's law:

$$\frac{1}{R_t} = \frac{1}{R_1} + \frac{1}{R_2} + \frac{1}{R_2} = \frac{1}{16} + \frac{1}{24} + \frac{1}{48}$$

And therefore:

$$\frac{1}{R_t} = \frac{3 + 2 + 1}{48} = \frac{6}{48} \text{ or } R_t = \frac{48}{6}$$

and thus, $R_t = 8\,\Omega$

Now using the formula: $I = \dfrac{V}{R}$ we can say that: $I = \dfrac{240}{8}$

And therefore: **I = 30 A**

Method 2

Find the current through each resistor and then add them together.

Now for R_1: $I_1 = \dfrac{V}{R_1} = \dfrac{240}{16} = 15$ A

For R_2: $I_2 = \dfrac{V}{R_2} = \dfrac{240}{24} = 10$ A

For R_3: $I_3 = \dfrac{V}{R_3} = \dfrac{240}{48} = 5$ A

As: $I_T = I_1 + I_2 + I_3$

Then: $I_T = 15 + 10 + 5$

And therefore: **$I_T = 30$ A**

Series/parallel circuits

This type of circuit combines the series and parallel circuits as shown in Figures 5.05 and 5.07. To calculate the total resistance in a combined circuit, we must first calculate the resistance of the parallel group. Then, having found the equivalent value for the parallel group, we simply treat the circuit as being made up of series connected resistors and now add this value to any series resistors in the circuit, thus giving us the total resistance for the whole of the network.

Here is a worked example.

Example

Calculate the total resistance of this circuit and the current flowing through the circuit, when the applied voltage is 110 V.

Step 1: Find the equivalent resistance of the parallel group (R_p)

$$\frac{1}{R_p} = \frac{1}{R_1} + \frac{1}{R_2} + \frac{1}{R_3}$$

$$\frac{1}{R_p} = \frac{1}{10} + \frac{1}{20} + \frac{1}{30} = \frac{6 + 3 + 2}{20} = \frac{11}{60}$$

Therefore: $R_p = \mathbf{5.45\ \Omega}$

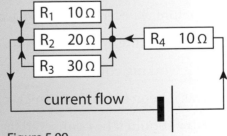

Figure 5.09

Step 2: Add the equivalent resistor to the series resistor R4

$R_t = R_p + R_4 = 5.45 + 10$

Therefore: $R_t = 15.45\ \Omega$

(continued over)

Step 3: Calculate the current

$$I = \frac{V}{R_t} = \frac{110}{15.45} = \textbf{7.12 amperes}$$

Instruments and measurements

On completion of this topic area the candidate will be able to describe how voltmeters and ammeters are connected into circuits, to quantify circuit voltages, current and resistance.

As electricians we are often responsible for the measurement of different electrical quantities. Some must be measured as part of the inspection and testing of an installation (e.g. insulation resistance). However, of the remainder, the most common quantities that we are likely to come across are shown in Table 5.04.

Property	Instrument
Current	Ammeter
Voltage	Voltmeter
Resistance	Ohmmeter
Power	Wattmeter

Table 5.04 Electrical quantities

In this section we will be looking at the following areas:

- measuring current (ammeter)
- measuring voltage (voltmeter)
- measuring resistance (ohmmeter)
- using an ammeter and voltmeter
- loading errors
- meter displays (digital and analogue).

But before we ever use a meter, we must always ask ourselves these questions?

- Have we chosen the correct instrument?
- Is it working correctly?
- Has it been set to the correct scale?
- How should it be connected?

Measuring current (ammeter)

An ammeter measures current by series connections. Although we know that we can use a multi-meter on site, the actual name for an instrument that measures current is an ammeter. Ammeters are connected in series so that the current to be measured passes through them. The circuit diagram in Figure 5.10 illustrates this. Consequently, they need to have a very low resistance or they would give a false reading.

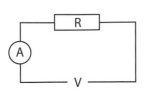

Figure 5.10 Ammeter in circuit

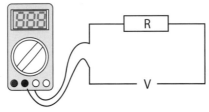

Figure 5.11 Multi-meter in circuit

If we now look at the same circuit, but use a correctly set multi-meter, the connections would look as in Figure 5.11.

Measuring voltage (voltmeter)

On site, or for general purposes, we can use a multi-meter to measure voltage, but the actual device used for measuring voltage is called a voltmeter. It measures the potential difference between two points (for instance, across the two connections of a resistor). The voltmeter must be connected in parallel across the load or circuit to be measured as shown in Figure 5.12.

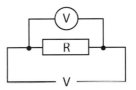

Figure 5.12 Voltmeter in circuit

If we now look at the same circuit, but use our correctly set multi-meter, the connections would look as in Figure 5.13.

The internal resistance of a voltmeter must be very high if we wish to get accurate readings.

Measuring resistance (ohmmeter)

There are many ways in which we can measure resistance. However, in the majority of cases we do so by executing a calculation based upon instrument readings (ammeter/voltmeter) or from other known resistance values. Once again we can use our multi-meter to perform the task, but we refer to the actual meter as being an ohmmeter.

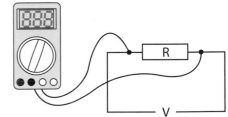

Figure 5.13 Multi-meter measuring circuit

The principle of operation involves the meter having its own internal supply (battery). The current, which then flows through the meter, must be dependent upon the value of the resistance under scrutiny.

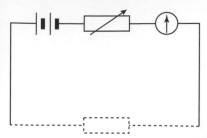

Figure 5.14 Ohmmeter

However, before we start our measurement, we must ensure that the supply is off and then connect both leads of the meter together and adjust the meter's variable resistor until full-scale deflection (zero) is reached.

The dotted line in Figure 5.14 indicates the circuit under test. All other components are internal to the meter.

Using an ammeter and voltmeter

If we consider the application of Ohm's law, we know that we can establish resistance by the following formula:

$$R = \frac{V \text{ (voltmeter)}}{I \text{ (ammeter)}}$$

If we therefore connect an ammeter and voltmeter in the circuit, then by using a known supply (battery) we can apply the above formula to the readings that we get.

In practice, we invariably find that we need a long 'wandering' test lead to perform such a test. Where this is the case, we must remember to deduct the value of the test lead from our circuit resistance.

Loading errors

Connecting a measuring instrument to a circuit will invariably have an effect on that circuit. In the case of the average electrical installation, this effect can largely be ignored. But if we were to measure electronic circuits, the effect may be drastic enough to actually destroy the circuit.

We normally refer to these errors introduced by the measuring instrument as loading errors, and below we will look at an example of this.

Let us assume in Figure 5.15(a) that we have been asked to measure the voltage across resistor B in the circuit, where both items have a resistance of 100 kΩ. Note that in this arrangement, the potential difference (p.d.) across each resistor will be 115 V.

Our instrument, a voltmeter, has an internal resistance of 100 kΩ and would therefore be connected as in the diagram.

(a)

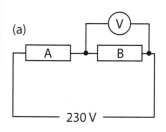

In effect, by connecting our meter in this way, we are introducing an extra resistance in parallel into the circuit. We could, therefore, now draw our circuit as in Figure 5.15(b).

(b)

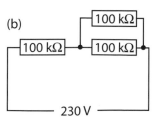

Figure 5.15 Loading errors

Now, if we find the equivalent resistance of the parallel branches:

$$\frac{1}{R_t} = \frac{1}{R_1} + \frac{1}{R_2} = \frac{1+1}{100} = \frac{2}{100}$$

Therefore: $R_t = \frac{100}{2} = 50 \text{ k}\Omega$

And then if we treat the parallel combination as being in series with 100 kΩ resistor A then total resistance:

$R = R_A + R_t = 100\ k\Omega + 50\ k\Omega = 150\ k\Omega$

If we apply Ohm's law again, we would find we now have a current of 0.00153 A (1.53 mA) flowing in the circuit but only half of this goes through resistor B, the other half goes through the meter. Hence, the p.d. across the resistor would be 76 V and not the 115 V we would expect to see. The same applies if we move the meter across resistor A.

The problem is caused by the amount of current that is flowing through the meter, and this is normally solved by using meters with a very high resistance (as resistance restricts current). So you now see why we said earlier that voltmeters must have a high internal resistance in order to be accurate.

Meter displays (digital and analogue)

There are two types of display, **analogue** and **digital**. Analogue meters have a needle moving around a calibrated scale, whereas digital tend to show results, as numeric values, via a liquid crystal or LED display.

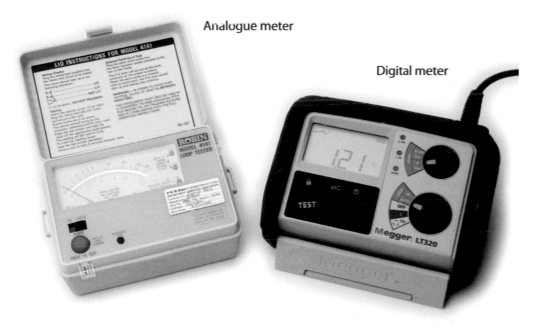

Analogue meter

Digital meter

For many years electricians used the analogue meter and indeed it is still common for continuity/insulation resistance testers to be of the analogue style. However, their use is slowly being replaced by the digital meter. Most modern digital meters are based on semiconductors and consequently have very high impedance, making them ideal for accurate readings and good for use with electronic circuits.

Generally speaking, as they have no moving parts they are also more suited to rugged site conditions than the analogue style.

What sort of meter will I need?

Although instruments exist to measure individual electrical quantities, most electricians will find that a cost-effective solution is to use a digital multi-meter similar to the one pictured on page 109. Meters like this are generally capable of measuring current, resistance and voltage, both in a.c. and d.c. circuits and across a wide range of values.

Magnetic fields and flux patterns

On completion of this topic area the candidate will be able to describe the magnetic fields and flux patterns of permanent magnets, current carrying conductors and solenoids.

Magnetism

The word magnetic originated with the ancient Greeks, who found natural rocks possessing this characteristic.

Magnetic rocks such as magnetite, an iron ore, occur naturally. The Chinese observed the effects of magnetism as early as 2600 BC when they saw that stones like magnetite, when freely suspended, had a tendency to assume a north and south direction. Because magnetic stones aligned themselves north–south, they were referred to as lodestones, or leading stones.

Magnetism is hard to define – we all know what its effects are: the attraction or repulsion of a material by another material. But why does this happen? And why do we only see it in some materials, notably metals and particularly iron? The physics behind this is too complex to go into here, but it is useful to remember that magnetism is a fundamental force (like gravity) and it arises due to the movement of electrical charge. Magnetism is seen whenever electrically charged particles are in motion.

Materials that are attracted by a magnet, such as iron, steel, nickel and cobalt, have the ability to become magnetised. These are called magnetic materials.

For the electrician, we say that a magnet is any device that produces an external magnetic field and, for the purpose of this book, we are only interested in two types of magnet:

- the permanent magnet
- the electromagnet (temporary magnet).

The permanent magnet

A permanent magnet is a material that, when inserted into a strong magnetic field, will not only begin to exhibit a magnetic field of its own, but also continue to exhibit a magnetic field once it has been removed from the original field.

This remaining field would allow the magnet to exert force (the ability to attract or repel) on other magnetic materials. This magnetic field would then be continuous without weakening, as long as the material is not subjected to a change in environment (temperature, demagnetising field etc.).

The ability to continue exhibiting a field while withstanding different environments helps to define the capabilities and types of applications in which a magnet can be successfully used.

Magnetic fields from permanent magnets arise from two atomic sources: the spin and orbital motions of electrons. Therefore, the magnetic characteristics of a material can change when alloyed with other elements.

For example, a non-magnetic material such as aluminium can become magnetic in materials such as alnico or manganese-aluminium-carbon.

When a ferromagnetic material (a material containing iron) is magnetised in one direction, it will not relax back to zero magnetisation when the imposed magnetising field is removed. The amount of magnetisation it retains is called its **remanence**.

It must be driven back to zero by a field in the opposite direction; the amount of reverse driving field required to demagnetise it is called its **coercivity**.

We have probably all experienced at some time the effect of a permanent magnet (although we cannot see the magnetic field with the naked eye), even if it was just to leave a message on the fridge door.

The magnetic field looks like a series of closed loops that start at one end (pole) of the magnet, arrive at the other and then pass through the magnet to the original start point.

Did you ever do the famous experiment at school where you took a magnet, placed it on a piece of paper and then sprinkled iron filings over it? If you did, you would see that it looks a bit like the diagram based on a bar magnet (Figure 5.16).

Each one of these lines is called a line of magnetic flux and has the following properties:

- they will never cross, but may become distorted
- they will always try to return to their original shape
- they will always form a closed loop
- outside the magnet they run north to south
- the higher the number of lines of magnetic flux, the stronger the magnet.

If we could count the lines, we could establish the magnetic flux (which we measure in webers), and we would find the more lines there were, the stronger the magnet would be. In other words the bigger the magnet, the bigger the flux produced.

The strength of the magnetic field at any point is calculated by counting the number of lines we have at that point and this is then called the **flux density** (measuring webers/square metre, which are given the unit title of a tesla).

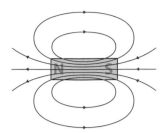

Figure 5.16 Bar magnet

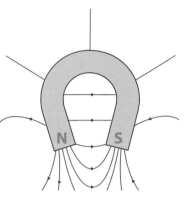

Figure 5.17 Horseshoe magnet

Remember

Like poles repel each other and unlike poles attract

We define this by saying that: '*If one weber of magnetic flux was spread evenly over a cross-sectional area of one square metre, then we have a flux density of one tesla (1T)*'. In other words the flux density depends upon the amount of magnetic flux lines and the area to which they are applied.

We use the following formula to express this:

$$\text{Flux density B (tesla)} = \frac{\text{(magnetic flux)}}{\text{(csa)}} = \frac{\Phi}{A} \text{ (webers/m}^2\text{)}$$

Here is an example to help you understand what is going on.

Example

The field pole of a motor has an area of 5cm² and carries a flux of 80 μWb. What will be the flux density?

Remembering the formula:

$$\text{Flux density B (tesla)} = \frac{\text{(magnetic flux)}}{\text{(csa)}} = \frac{\Phi}{A} \text{ (webers/m}^2\text{)}$$

We need to make allowance for the area being given in cm² and the flux in μWb.

Therefore:

$$\text{Flux density B (tesla)} = \frac{\text{(magnetic flux)}}{\text{(csa)}} = \frac{80 \times 10^{-6} \text{ Wb}}{5 \times 10^{-4} \text{ Wb}} = 0.16 \text{ T}$$

The electromagnet

An electromagnet is produced where there is an electric current flowing through a conductor, as a magnetic field is produced around the conductor. This magnetic field is proportional to the current being carried, since the larger the current, the greater the magnetic field.

An electromagnet is defined as being a temporary magnet because the magnetic field can only exist while there is a current flowing. If we have a typically shaped conductor such as a wire, the magnetic field looks like concentric circles and these are along the whole length of the conductor. However, the direction of the field depends on the direction of the current.

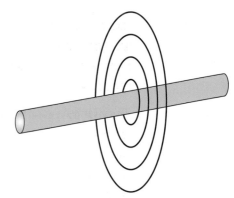

Figure 5.18 Lines of magnetic force set up around a conductor

The 'Screw Rule'

The direction of the magnetic field is traditionally determined using the 'Screw Rule'.

In the Screw Rule we think of a normal right-hand threaded screw. The movement of the tip represents the direction of the current through a straight conductor and the direction of rotation of the screw represents the corresponding direction of rotation of the magnetic field:

Rotation of screw = rotation of magnetic field

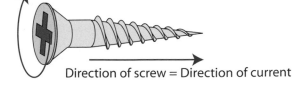

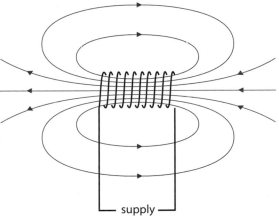

Figure 5.19 The 'Screw Rule'

The solenoid

A **solenoid** is a long hollow cylinder around which we wind a uniform coil of wire. When a current is sent through the wire, a magnetic field is created inside the cylinder.

The solenoid usually has a length that is several times its diameter. The wire is closely wound around the outside of a long cylinder in the form of a helix with a small pitch. The magnetic field created inside the cylinder is quite uniform, especially far from the ends of the solenoid. The larger the ratio of the length to the diameter, the more uniform the field near the middle.

Essentially, the magnetic field produced by a solenoid is similar to that of a bar magnet.

If an iron rod were placed partly inside a solenoid and the current turned on, the rod will be drawn into the solenoid by the resulting magnetic field. This motion can be used to move a lever or operate a latch to open a door, and is most commonly seen in use inside a door bell.

It is important to note that, with a solenoid, we can use an electric switch to energise the solenoid and therefore produce a mechanical action at a remote location, e.g. the doorbell where alternating current causes the rod to move in and out very fast.

Now, rather than having a current-carrying conductor placed in a magnetic field causing it to move, what if we took a conductor with no current flowing in it and instead moved the conductor through the magnetic field? Look back at chapter 4 page 83 where we described a simple alternator.

Figure 5.20 Magnetic field of solenoid

Construction of basic transformers

On completion of this topic area the candidate will be able to describe the construction of a basic transformer.

Transformer principles

A transformer is a device that transfers energy from one circuit to another through a magnetic field (magnetic coupling). It is used on a.c. supplies to transform one level of voltage to another. It is made from two or more coupled wirings around a magnetic core. The changing voltage moves from one winding to the other.

The transformer operates on the principle of mutual induction. This means that it contains a conductor lying within a magnetic field which constantly changes in magnitude and direction, inducing an e.m.f.

Although transformers come in many different sizes, and for a number of different uses, they all share these common operating principles.

Figure 5.21 shows the basic construction of a double wound transformer.

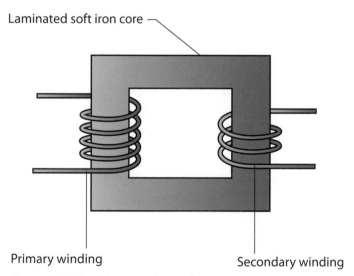

Laminated soft iron core

Primary winding

Secondary winding

Figure 5.21 Double wound transformer

Remember

Transformers can be both step-up and step-down

As the diagram shows, a double wound transformer has two windings mounted on a common magnetic core. One of these windings is the primary winding, with the other the secondary. In the diagram the transformer is a step-down transformer in the ratio of 2:1. This means that if an input voltage of 200 V was connected to the primary winding, the output voltage would be 100 V.

However, if an input voltage of 200 V was connected to the secondary winding, then the output voltage from the primary side would be 400 V.

As a transformer has no moving parts it is therefore seen as very efficient. However, there are two types of loss associated with transformers, which are:

- iron losses (caused in the transformer core)
- copper losses (caused in the windings).

Laminations

The windings of the transformer are placed upon a soft iron core. The core is used to improve the magnetic flux linkage between the two windings. However, **eddy currents** are induced into the core. These generate heat and produce a loss in power, unless action is taken. Eddy currents can result in heat which damages the insulation on the conductors.

The main method of reducing eddy currents is to construct the core from insulated laminated soft-iron segments. These laminations prevent the core from being simply a piece of steel, by separating out the layers. The thinner the lamination layers, the less energy will be lost to eddy currents.

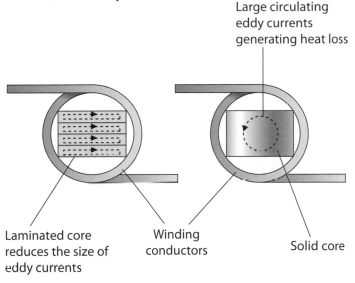

Large circulating eddy currents generating heat loss

Laminated core reduces the size of eddy currents

Winding conductors

Solid core

Figure 5.22 Plan view of the winding around a core

Enclosures

These losses can be seen in the form of heat – the greater the heat, the greater the energy loss. Hence, transformers are often contained within an enclosure designed to help them run cooler. This generally allows an air gap around the windings. Some smaller transformers used for 110 V power tools can be found to be totally encased in an epoxy resin material.

Large and high power transformers are usually encased within oil-filled steel enclosures. The oil is fire resistant and helps reduce the heat level within the windings. The insulating oil completely covers the core and the windings. When the transformer heats up, circulating convection currents are formed in the oil. The hot oil then travels to the top of the transformer, which forces the oil already at the

Remember

Transformers are often thought to be around 98% efficient

Definition

Eddy current
– caused when a moving magnetic field intersects a conductor; this causes a flow of electrons (current) in the conductor that oppose the magnetic field

top to move into the cooling tubes on the side of the enclosure. Air passing over the tubes causes the oil to cool and it drops to the bottom of the transformer, where the process can start again.

Saftey isolating transformer

BS 7671 requires extra precautions to be taken where there is an increased risk of an electric shock, due to a decrease in body resistance. Examples of these places may be damp or confined spaces, or areas where the body is in contact with large metallic surfaces such as pipelines, oil tanks or even the bodywork of a car.

In such situations, the method of protection required is to reduce the supply voltage to a level where it is impossible for the body to sustain damage. Tests done on humans' shows that the body can withstand indefinitely, without sustaining damage, a maximum voltage of 50 volts.

This is know as Separate Extra Low Voltage (SELV). SELV prevents the voltage from exceeding the limits of 50 V in a.c. and 120 V in d.c. under normal and fault conditions. It has a protective separation from all other circuits. This is normally assured by use of a transformer, as this guarantees minimum distances between conductors.

This transformer must be a safety isolating, double wound transformer complying with BS 3535. A safety isolating transformer is one in which there is no connection between the output winding and the body or the protective earth of the supply circuit. This ensures that the output of the transformer is totally isolated from the supply.

Transmission of energy

On completion of this topic area the candidate will be able to state the benefits gained by the use of transformers with the transmission and distribution of electrical energy (National Grid).

Transformers used in transmission

The production, transmission and distribution of energy will be looked at fully in chapter 11. In this section we will look at why transformers are used in the transmission and distribution of energy supplies.

Power can be calculated by the use of the following formula:

$$P = I^2R$$

Therefore, a reduction in the current will result in a reduction in the power lost in the lines. For example, if the current is halved, power loss is divided by four.

Most modern alternators used in the generation of electricity have an output of 25 kV. This has to be transformed up to 400 kV to be transmitted around the National Grid. Transmitting the voltage at this level minimises power loss and keeps cable sizes to a minimum.

From using this information we can see that the benefits from using transformers in the transmission and generation of electrical energy around the national grid are:

- smaller cable sizes
- smaller switch and control gear
- pylons can be lighter in construction as they have to bear less weight
- reduction in power losses in the power lines.

This all results in cost savings and creates a more cost-effective method of transmission. Which in turn is reflected in your energy bills!

Activity

Given a selection of resistors and an ohmmeter, measure the resistance of the individual resistors. Now calculate what would happen if the resistors were connected:

    (a) in series

    (b) in parallel.

Connect the resistors in series and parallel, and measure them to compare your results.

FAQ

Q How do electrons whizz around the nucleus in solid materials?

A Even solid materials are not actually solid; they are made up of atoms which are mostly empty space. If a nucleus was the size of a football, the nearest electron would be a mile away!

Q Where does electricity go when it flows to earth?

A It is a common misconception that electricity just flows into the ground and gets 'soaked-up', a bit like water. It doesn't. Electricity (conventional current flow) will only flow from one place to another if it can get back to where it came. Electricity that leaves a sub-station to earth will return to the sub-station through the ground.

Q Electricity must be able to do more than three things?

A No. Amazingly, every electrical device that you can think of uses one or more of the three effects of an electric current for its operation.

Q Why does the resistance go down when conductors are connected in parallel with each other?

A Think of the conductors as two water pipes run side by side. They would be able to carry twice as much flow as one pipe. Twice the flow means half the resistance, according to Ohms law. So, whenever you connect one conductor in parallel with another, the resistance always goes down.

Knowledge check

1. Atoms consist of several smaller particles. Which one of these particles possesses a negative charge?

2. State the three states of matter.

3. Define the terms voltage, current and resistance.

4. State the three effects of an electric current.

5. Two resistors of value 6.2 Ω and 3.8 Ω are connected in series with a 12 volt battery. What is the total resistance and total current flowing?

6. Explain, with the aid of a simple diagram, the operation of a transformer.

chapter 6

Basic electrical circuitry and applications

Unit 2 Outcome 3

The primary role of an electrician is the installation of electrical circuits and applications. In order to install any type of application, however, it is vital that from the start of your career you are able to put together basic circuitry. This requires knowledge of all parts of the electrical circuit, from the source of supply to the load.

This chapter will cover the various wiring systems, enclosures and equipment you will need to become familiar with in order to carry out your work as an electrician.

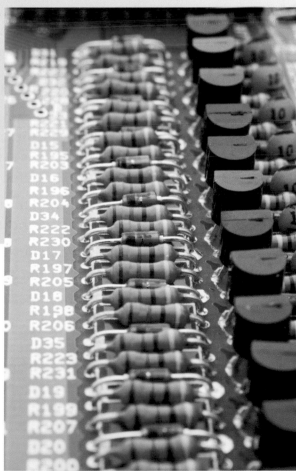

We have looked at a number of the terms covered in this chapter earlier in this book. Earthing and bonding, conduction and shock protection all have important roles to play in understanding basic circuitry, and it is these factors you will have to remember when assembling any piece of electrical equipment.

On completion of this chapter the candidate will be able to:

- state the component parts of an electrical circuit
- determine appropriate wiring systems
- state the component elements of electrical cables
- differentiate between earthing and bonding
- list possible conductive parts of structures
- state the purpose and function of earth protection
- list basic principles or shock protection, circuit overload and short-circuit protection.

The component parts of an electrical circuit

On completion of this topic area the candidate will be able to state the component parts of an electric circuit.

BS 7671 defines a circuit as 'an assembly of electrical equipment supplied from the same origin and protected against overcurrent by the same protective devices'. So what does make up the circuit?

Source of supply

Often a circuit is thought of as being the load and the conductors supplying the load. However in reality they are only part of the circuit. In order for current to flow two conditions have to be met.

1. There has to be a potential difference applied across the circuit (voltage).

2. There has to be a complete circuit (circle) for current to flow around.

The source of supply can be a.c. or d.c. If a d.c. supply is required, then this is derived from a battery. For an a.c. supply, either an a.c. generator or a d.c. generator with electrical components to rectify the supply, can be used.

An a.c. supply can be obtained directly from the mains. Single phase 230 V or three phase 400 V are those generally available in the UK.

The size and type of voltage required for the supply is determined by the load equipment to be used. All electrical equipment will have a plate attached to it indicating its safe working voltage.

Circuit conductors (cable)

The circuit conductors are those parts of the circuit which the current passes through. These are the cables. Cables have two components to them. One is the conductor itself. This is usually made of copper. The other part is insulation, usually made from PVC, which forms a sheath around the conductor. The insulation is required to:

● prevent the conductors touching together; this could short the circuit and preventing it from working

● prevent users of the circuit from coming into contact with the conductors and receiving an electric shock.

The type of insulation required is determined by the voltage which is to be applied to the cable.

Cables come as either single or multicore cables. Both have an overall sheath to keep all the associated cables together and to provide a minimal degree of mechanical protection. Appendix 4 of BS 7671 gives details on the sizes and types of cables available to us. We will be looking in more detail at types of cable later in this chapter.

Remember

If a d.c. supply requires a higher voltage, a series of batteries can be used rather than just one

Remember

A transformer can be used for an alternative voltage

Did you know?

The voltage will change depending on the intended use for the cable. A bell circuit which is to be supplied at 6 V will require considerably less insulation than a cable used to supply a heater at 230 V

The size of conductor used in a circuit is important and needs to be calculated accurately. We need to be sure that the conductor is large enough to carry the current produced by the load and to be sure that the load receives sufficient voltage for it to work safely.

Circuit protection

Every circuit requires protection if, in the event of a fault, damage is to be avoided. The inclusion into the circuit of a suitably rated fuse or protective device, such as a miniature circuit breaker (MCB) will protect both the load, and the cables supplying the load, from the heat damage associated with large fault currents.

Protective devices are used to provide a circuit with overload protection, short-circuit protection and shock protection. We will be looking at these in greater detail later in this chapter.

The protective device, or fuse, is deliberately designed to be a 'weak link' in a circuit. When current increases to a level where damage can be sustained, the fuse operates and automatically interrupts the supply to the circuit. As the fuse is a part of the circuit, once it is 'broken' it also breaks the flow within the circuit, ending the supply.

Circuits supplied at voltages below 50 V may be required to have circuit protection for functional reasons. Circuits supplied at 50 V and above are required to have circuit protection that complies with the requirements of BS 7671. For example, low voltage circuits (above 50 volts, but not exceeding 1000 volts) supplying socket outlets must automatically disconnect in the event of an earth fault within 0.4 seconds.

Protective device types include:

- Rewirable fuses to BS 3036
- Cartridge fuses to BS 1361
- Cartridge fuses to BS 88
- MCBs to BS EN 60898
- RCBOs to BS EN 61009.

Circuit control

It is vital that any circuit, no matter its level of complexity, can be controlled. This could be either:

- a simple switch, allowing us to turn the circuit on or off
- a time switch, which activates the circuit at certain times
- a float switch, which controls the water levels in a pump.

The last example is known as functional switching.

Isolation is the ability to remove the supply from all the live conductors, such as at the mains switch, to a complete installation. Emergency switching is another area of control that is required for some circuits. This allows us to press one button and turn off the main supply to a series of equipment, in the event of danger.

Remember

BS 7671 Appendix 4 has tables indicating the current carrying capacity and volt drop for given sizes and types of cable

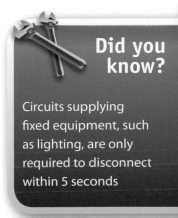

Did you know?

Circuits supplying fixed equipment, such as lighting, are only required to disconnect within 5 seconds

Did you know?

RCBO stands for residual current device (RCD) with overland protection, combining the capabilities of RCD and MCB

Load

The load refers to the item that requires the supply in order to function. For example, this might mean the lights in a circuit, a heater, a motor to drive a pump or any item of equipment (or combinations of these) that require an electrical supply in order for them to work.

Electrical loads are rated in watts or kilowatts (W and kW). The size of the load is generally stamped on the equipment or marked on a nameplate. This can contain the:

- rating in watts
- supply voltage
- frequency of the supply
- full load current.

If this information is not available on a nameplate, then it would be necessary to refer to the manufacturer's literature to establish the requirements of the load.

Determine appropriate wiring systems, enclosures and equipment

On completion of this topic area the candidate will be able to determine appropriate wiring systems, enclosures and equipment.

The role of BS 7671

Chapter 13 Part 1 of BS 7671 requires that all electrical installations shall be designed to provide for the protection of persons, livestock and property and the proper functioning of the electrical installation for the intended use.

In order to do this BS 7671 requires us to determine the characteristics of the available supply. This can be done by calculation, measurement, enquiry or inspection of existing supplies. The characteristics that need to be determined are:

- the nature of the current – either a.c. or d.c.
- the number of conductors
- voltage
- frequency
- maximum current allowed
- prospective short-circuit current
- earth loop impedance
- nature and size of the load
- number and type of circuits required
- the location and any special conditions that may apply. This should take into account the nature of the location and structure that will support the wiring system and the accessibility of the wiring to people and livestock.

Insulation colours

To identify cables the insulation is coloured in accordance with BS 7671, Table 51.

Table 51 Identification of conductors	
Function	**Colour**
Protective conductors	Green and yellow
Functional earthing conductor	Cream
a.c. power circuit[1]	
Phase of single-phase circuit	Brown
Neutral of single- or three-phase circuit	Blue
Phase 1 of three-phase a.c. circuit	Brown
Phase 2 of three-phase a.c. circuit	Black
Phase 3 of three-phase a.c. circuit	Grey
Two-wire unearthed d.c. power circuit	
Positive of two-wire circuit	Brown
Negative of two-wire circuit	Grey
Two-wire earthed d.c. power circuit	
Positive (of negative earthed) circuit	Brown
Negative (of negative earthed) circuit	Blue
Positive (of positive earthed) circuit	Blue
Negative (of positive earthed) circuit	Grey
Three-wire d.c. power circuit	
Outer positive of two-wire circuit derived from three-wire system	Brown
Outer negative of two-wire circuit derived from three-wire system	Grey
Positive of three-wire circuit	Brown
Mid-wire of three-wire circuit[2]	Blue
Negative of three-wire circuit	Grey
Control circuits, ELV and other applications Phase conductor	Brown, black, red, orange, yellow, violet, grey, white, pink or turquoise
Neutral or mid-wire[3]	Blue

NOTES

(1) Power circuits include lighting circuits.

(2) Only the middle wire of three-wire circuits may be earthed.

(3) An earthed PELV conductor is blue.

Table 6.01 To identify cable insulation in accordance with BS 7671

You will come across conductors with these colours of insulation for many years to come, but only in existing installations. From 1 April 2004, the colours of conductor insulation for new installations changed.

Conductor	Old colour
Phase	Red
Neutral	Black
Protective conductor	Green and yellow
Phase one	Red
Phase two	Yellow
Phase three	Blue
Neutral	Black
Protective conductor	Green and yellow

Table 6.02 Old conductor insulation colours

Single- or three-phase power supplies

In the UK the size of supply brought into most households is 100 amps single phase, 50 Hertz. Should the assumed current demand of an installation exceed 100 amps, then a three-phase supply is required. However the selection of a three-phase supply can also simply be down to the load, i.e. a requirement to install three-phase machines, motors, pumps etc.

Appendix 1 of the *On Site Guide* gives details of the requirements for calculating demand and diversity for installations. You will deal with single- and three-phase supplies in greater detail in chapter 11.

Switching of lighting circuits

There are numerous switching arrangements that make up a lighting circuit. This section will look at the most common ones. It would not be practical to look at all the possible combinations, as there are simply far too many.

Cabling	Switching arrangements
Wiring in conduit and/or trunking Wiring using multicore/composite cables.	Two-way switching Intermediate switching One-way switching

Table 6.03 Switching arrangements

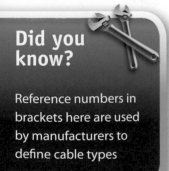

Wiring in conduit and trunking

Wiring in this type of installation is carried out using PVC single-core insulated cables (Ref. 6491 X). This code number is a manufacturer's code used to denote different types of cable. The line (live) conductor is taken directly to the first switch and looped from switch to switch for all the remaining lights connected to that particular circuit.

The neutral conductor is taken directly to the lighting outlet (luminaires) and looped between all the remaining luminaires on that circuit. The switch wire is run between the switch and the luminaire it controls.

Wiring using multicore/composite cables

This type of cable is normally a sheathed multicore twin and earth or three cores and earth (two-way and intermediate circuits only) (Ref. 6242Y and 6243Y respectively). A 'loop in' or 'joint box' method may be employed with this type of installation. In many instances a loop-in system is specified as there are no joint boxes installed and all terminations are readily accessible at the switches and ceiling roses.

With a joint box system normally only one cable is run to each wiring outlet. Where such joint boxes are installed beneath floors they should be accessible by leaving a screwed trap in the floorboard directly above the joint box. All conductors should be correctly colour identified.

On a composite cable installation, where the conductors other than brown are used as a line conductor, they should be fitted with a brown sleeve at their terminations.

All conductors must be contained within a non-combustible enclosure at wiring outlets (i.e. the sheathing of the cable must be taken into the wiring accessory). Throughout the lighting installation a circuit protective conductor (cpc) must be installed and terminated at a suitable earthing terminal in the accessory/box.

Where an earthing terminal may not be fitted in a PVC switch pattress, the cpc may be terminated in a connector. Where the sheathing is removed from a composite cable the cpc must be fitted with an insulating sleeve (green and yellow); this provides equivalent insulation to that provided by the insulation of a single-core non-sheathed cable of appropriate size complying with BS 6004 or BS 7211.

One-way switching

The most basic circuit possible is the one-way switch controlling one light, as shown in Figure 6.01. In this system, one terminal of the one-way switch receives the switch feed; the switch wire leaves from the other terminal and goes directly to the luminaire (a). Once operated, the switch contact is held in place mechanically and therefore the electricity is continually flowing through to the light (b).

In other words, we supply the switch feed terminal, point (A).

Operate the switch and it comes out at the switch wire terminal, point (B).

Figures 6.02 to 6.04 show the full circuit when wired using single-core cables, which would be run in either conduit or trunking.

Using the old cable colours again, Figure 6.02 now shows a second light point fed from the same switch wire. This means that the second light is now wired in parallel with the first light.

Remember

Red sleeving changes to brown sleeving

Remember

Cables supplying switches of any circuit and the cables feeding away from the switch to the light are identified. The cable from the distribution board to the switch is called the switch feed, and the cable from the switch to the light is called the switch wire

Find out

Find out what is pattress

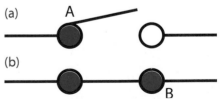

Figure 6.01 One-way switching

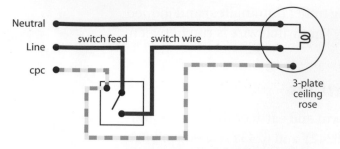

Figure 6.02 One-way switching for wiring with single-core cables (old cable colours)

Figure 6.03 One-way switching for wiring with single-core cables (new cable colours)

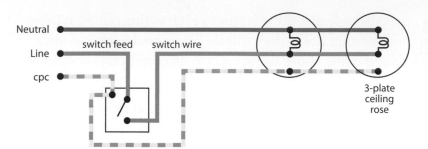

Figure 6.04 Extra lighting fed from the same switch, wired in parallel (new cable colours)

Two-way switching

Sometimes we need to switch a light on, or off, from more than one location, e.g. at opposite ends of a long corridor. When this is required, a different switching arrangement must be used, the most common being the two-way switch circuit. In this type of circuit, the switch feed is feeding one two-way switch, and the switch wire goes from the other two-way switch to the luminaire(s). Two wires known as 'strappers' then link the two switches together. In other words:

we supply the switch feed terminal, point (A).

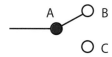

However, depending on the switch contact position, the electricity can come out on either terminal B or terminal C. In the following diagram it is shown energising terminal B.

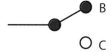

If we now operated the switch, the contact would move across to energise terminal C. Please note that actual switch terminals are not marked in this way.

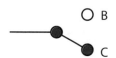

Switching of lighting circuits

By connecting together the two two-way switches we now have the ability at each switch to either energise the switch wire going to the light or to de-energise it (this is why this system is ideal for controlling lighting on corridors or staircases). In the first diagram below, the luminaire is off.

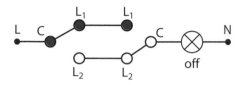

However, operate the second switch and we energise the common terminal (C) and the luminaire will now come on.

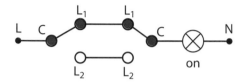

Figure 6.05 shows the full circuit when wired using single-core cable in conduit or trunking.

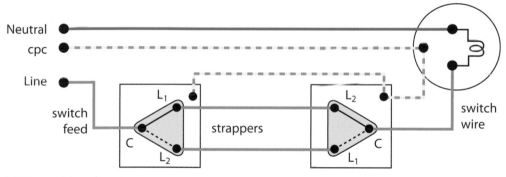

(a) New cable colours

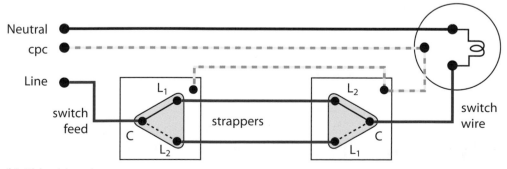

(b) Old cable colours

Figure 6.05 Full circuit wired with single-core cable. Old and new colours

We can also use two-way switches for other purposes. Figure 6.06 shows one two-way switch to control two indicator lamps. This sort of system is frequently used as an entry system outside offices or dark rooms, where the two lamps can be marked, for example as 'available' and 'busy'.

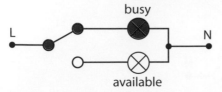

Figure 6.06 Two-way switch controlling two lamps

Intermediate switching

If more than two switch locations are required, e.g. in a long corridor with other corridors coming off it, then intermediate switches must be used. The intermediate switches are wired in the 'strappers' between the two-way switches.

The action of the intermediate switch is to cross-connect the 'strapping' wires. This gives us the ability to route a supply to any terminal depending upon the switch contact positions.

When we operate the switch into position two, the switch contacts cross over (Figure 6.07).

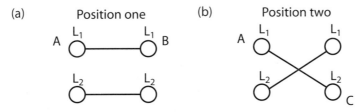

Figure 6.07

This means that a signal sent into terminal A can always be directed on to either terminal B or terminal C as required.

Ignoring terminal markings, Figure 6.08 shows an arrangement where the switch wire is de-energised and therefore the luminaire is off.

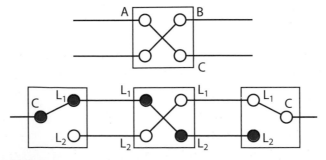

Figure 6.08

However, by operating the intermediate switch, we can route the switch feed along another section of the 'strappers' and energise the switch wire terminal, and therefore the luminaire will come on, as shown in Figure 6.09.

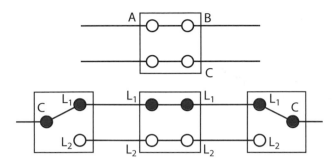

Figure 6.09

To use the example of a long hotel main corridor with other minor corridors coming off it, if we have an intermediate switch at each junction with the main corridor, anyone joining or leaving the main corridor now has the ability to switch luminaires on or off.

Figure 6.10 shows the full circuit in the new colours, when wired using single-core cable in conduit or trunking.

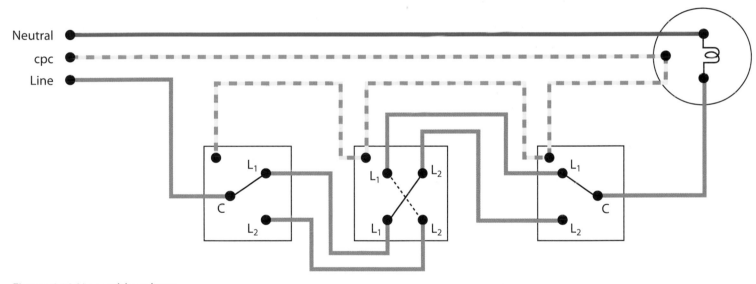

Figure 6.10 New cable colours

Note: in this diagram the light will be off.

Wiring with multicore cables

Multicore cables (commonly referred to as twin and earth) are basically a three-core cable consisting of a line, neutral and earth conductor. Both the line and neutral conductors consist of copper conductors insulated with coloured insulation (see Figure 6.11 and 6.12), the earth being a non-insulated bare copper conductor sandwiched in between the line and neutral. In the old colours, the correct use of this cable type would be to use two red conductors (switch feed and switch wire) from switch locations up to luminaires. However, for ease, it was common practice for many to use a cable containing red and black conductors.

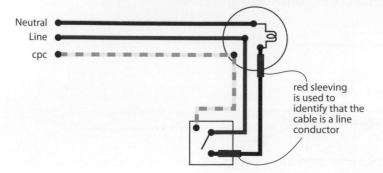

Figure 6.11 Old cable colours

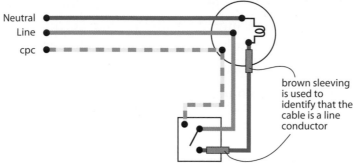

Figure 6.12 New cable colours

This means that the black conductor should be sleeved at both ends to indicate that it is a 'live' conductor. (Please be aware that many did not sleeve the conductors and therefore a luminaire connected across two black conductors is actually connected to a switch wire and a neutral.)

The outer white or grey covering is the sheath, and its main purpose is to prevent light mechanical damage of the insulation of the conductors.

This cable is often used for wiring domestic and commercial lighting circuits, and you would normally use 1.5mm² cable. Using this type of cable instead of singles is slightly more complicated because you are restricted as to where you plan your runs, i.e. in the old colours the cable will always consist of a red, black and earth conductor.

Two-way switching and conversion circuit

This method of switching using multicore cables requires the use of four-core cable. This is a cable that has three coloured and insulated conductors and a bare earth conductor. The three coloured conductors are in the old colours, coloured red, yellow, and blue. This type of cable is normally only stocked in 1.5mm² and used in a lighting circuit where its application is for switching that requires more than one switch position, i.e. two-way and intermediate switching.

It is also used for converting an existing one-way switching arrangement into a two-way switching arrangement. The first step in this process is to replace the existing one-way switch with a two-way switch. This would then be connected as shown in Figure 6.13.

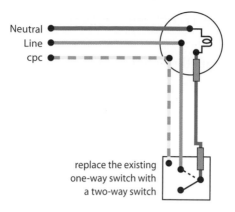

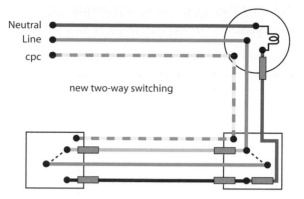

new two-way switching

replace the existing
one-way switch with
a two-way switch

Figure 6.13 Conversion of one-way
lighting into two-way switching

Figure 6.14 Conversion of one-way lighting into
two-way switching

The next step is to install the multicore cable between this location and the new
two-way switch location and complete the circuit connections as in Figure 6.14.

Note: In Figure 6.14 the light will be off.

Intermediate switching

Figure 6.15 illustrates an intermediate lighting circuit using multicore cables.

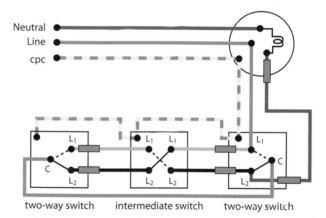

two-way switch   intermediate switch   two-way switch

Figure 6.15 Intermediate switching

Note: In Figure 6.15 the light will be on.

Wiring using a junction/joint box

This method of wiring is considered somewhat old-fashioned, as it has now been
superseded by the loop-in method. However, that is not to say that you will not come
across this method in the millions of households in the country that have a junction
box method installed. Care should be taken when wiring junction boxes, and the
following precautions should be observed:

● the protective outer sheath should be taken inside the junction box entries to a
  minimum of 10mm

● where terminations are made into a connector, only sufficient insulation should
  be removed to make the termination

- sufficient slack should be left inside the joint box to prevent excess tension on conductors
- cables should be inserted so that they are not crossing; they should be neat and fitted so that the lid fits without causing damage
- correct size joint boxes should be used
- joint boxes should be secured to a platform fitted between the floor and ceiling joists.

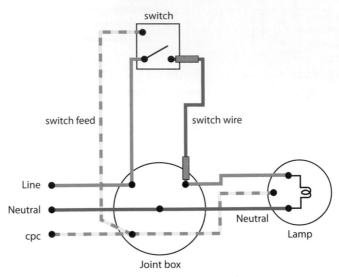

Figure 6.16 One-way switch using a joint box

Figure 6.16 illustrates the control of one light via a one-way switch using this method.

Rings, radials and spurs

Before the early 1950s different types and sizes of socket outlet and plug were in use in domestic premises, which was very annoying and inconvenient! In 1947, 5 A and 15 A socket outlets were deemed obsolete, and agreement was reached on a standard 13 A socket outlet with a fused plug, to BS 1363. In this section we will be looking at the standard power-circuit arrangements found in domestic premises. In particular the following will be investigated:

- ring circuits
- permanently connected equipment
- non-fused spurs
- fused spurs
- radial circuits.

Ring circuits

In this system the line, neutral and circuit protective conductors are connected to their respective terminals at the consumer unit, looped into each socket outlet in turn and then returned to their respective terminals in the consumer unit, thus forming a ring. Each socket outlet has two connections back to the mains supply.

The requirements for a standard domestic ring circuit are as shown in the bullet points below. An unlimited number of socket outlets may be installed provided the following points are taken into account.

- Each socket outlet of twin or multiple sockets is to be regarded as one socket outlet.

- The floor area served by a single 30 A or 32 A ring final circuit must not exceed 100m² in domestic installations.

- Consideration must be given to the loading of the ring main, especially in kitchens and utility rooms, which may require separate circuit(s).

- When more than one ring circuit is installed in the same premises, the socket outlets installed should be reasonably shared among the ring circuits so that the assessed load is balanced.

- Immersion heaters, storage vessels in excess of 15 litres capacity or permanently connected heating appliances forming part of a comprehensive space-heating installation, are supplied by their own separate circuit.

A typical domestic ring circuit is shown in Figure 6.17.

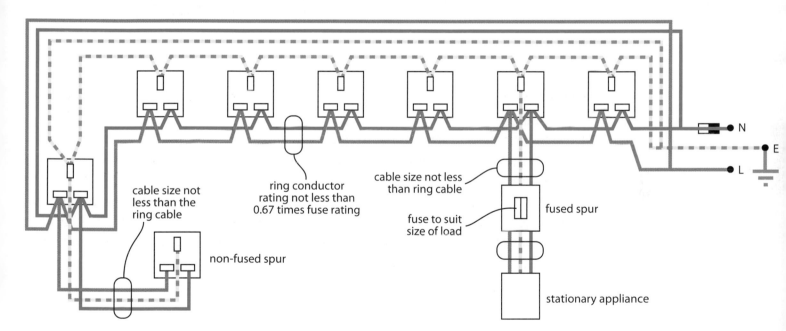

Figure 6.17 Domestic ring circuit with spurs

Permanently connected equipment

Permanently connected equipment should be locally protected by a fuse which does not exceed 13 A and be controlled by a switch complying with BS 7671 chapter 46, or be protected by a circuit breaker not exceeding 16 A rating.

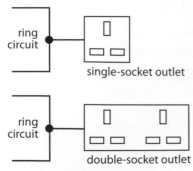

single-socket outlet

ring circuit

double-socket outlet

Figure 6.18 Non-fused spurs

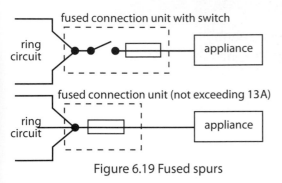

fused connection unit with switch

ring circuit

appliance

fused connection unit (not exceeding 13A)

ring circuit

appliance

Figure 6.19 Fused spurs

Non-fused spurs

Spurs may be installed on a ring circuit. These may be fused, but it is more common to install **non-fused spurs** connected to a circuit at the terminals of socket outlets, at junction boxes, or at the origin of the circuit in the distribution board. A non-fused spur may supply only one single or one twin-socket outlet or one item of permanently connected equipment. The total number of non-fused spurs should not exceed the total number of socket outlets and items of fixed equipment connected directly to the ring circuit. The size of conductor for a non-fused spur must be the same as the size of conductor used on the ring circuit.

Fused spurs

A **fused spur** is connected to a circuit through a fused connection unit. The fuse incorporated should be related to the current carrying capacity of the cable used for the spur but should not exceed 13 A. When sockets are wired from a fused spur the minimum size of the conductor is 1.5mm² for rubber- or PVC-insulated cables with copper conductors and 1mm² for mineral-insulated cables with copper conductors. The total number of fused spurs is unlimited.

Radial circuits

In a radial circuit the conductors do not form a loop but finish at the last outlet. As with the ring circuit, the number of outlets in any circuit is unlimited in a floor area up to the maximum allowed. In each case this will be determined by the estimated load and shock protection constraints.

A2 Radial circuit

The current rating of the cables is determined by the rating of the overcurrent protection device, i.e. 30 A or 32 A cartridge fuse or MCB. This means that copper-conductor PVC cables of not less than 4mm², or copper-conductor mineral-insulated cables of not less than 2.5mm², may be used, and the floor area served may not exceed 75m². See Figure 6.20 and Table 6.04.

Remember

The cable sizes shown are nominal, and if derating factors apply e.g. grouping or ambient temperature, these sizes may need to be increased. In practice this would mean that a 30 A or 32 A circuit breaker or fuse would be used as an overcurrent protection device and the circuit cable would be selected accordingly as required by Regulation 433-02-04

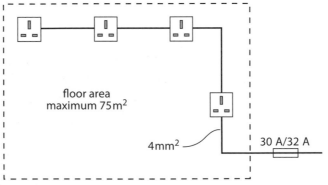

floor area maximum 75m²

4mm²

30 A/32 A

Figure 6.20 A2 radial circuit

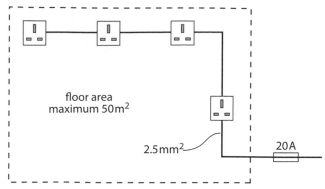

Figure 6.21 A3 radial circuit

A3 Radial circuit

In domestic premises this circuit is wired in 2.5mm² copper cable and protected by a 20A device. The floor area is limited to 50m²; in this area any number of socket outlets may be installed. See Figure 6.21 and Table 6.04.

Type of circuit	Overcurrent protection device		Copper conductor size		Maximum floor area served
1	2 Rating A	3 Rating A	4 (PVC) mm²	5 (MICC) mm²	6 m²
A1 Ring	30 or 32	Any	2.5	1.5	100
A2 Radial	30 or 32	Cartridge fuse or circuit breaker	4	2.5	75
A3 Radial	20	Any	2.5	1.5	50

Table 6.04 Overcurrent protection

Component elements of electrical cables

On completion of this topic area the candidate will be able to state the component elements of electrical cables.

Selecting a cable for an electrical installation is very important; consideration must be given to the following criteria in order to ensure the correct type of cable is chosen:

- conductor material
- conductor size
- insulation
- environmental conditions.

Conductor material

Copper and aluminium

The choice generally is between copper and aluminium. Copper has better conductivity for a given cross-sectional area and is preferable, but its cost has risen over the years. Aluminium conductors are now sometimes preferred for the medium and larger range of cables. All cables smaller than 16mm² cross-sectional area (csa) must have copper conductors.

Conductor	Advantages	Disadvantages
Copper	• Easier to joint and terminate • Smaller cross-sectional area for given current rating	• More costly • Heavier
Aluminium	• Cheaper • Lighter • Not recommended for use in hazardous areas	• Bulkier for given current rating

Table 6.05 Copper and aluminium conductors compared

Other conductor materials

● **Cadmium copper:** has a greater tensile strength for use with overhead lines.

● **Steel reinforced aluminium:** for very long spans on overhead lines.

● **Silver:** used where extremely good conductivity is required. However, it is extremely expensive.

● **Copperclad (copper-sheathed aluminium):** cables that have some of the advantages of both copper and aluminium but are difficult to terminate.

Whatever the choice of conductor material the conductors themselves will usually be either stranded or solid. Solid conductors are easier, and therefore cheaper, to manufacture but the installation of these cables is made more difficult by the fact that they are not very pliable. Stranded conductors are made up of individual strands that are brought together in set numbers. These provide a certain number of strands such as 3, 7, 19, 37 etc. and, with the exception of the 3-strand conductor, all have a central strand surrounded by the other strands within the conductor.

Conductor size

There are many factors that affect the choice of size of conductor.

● **Load and future development**
The current the cable is expected to carry can be found from the load, taking into account its possible future development, i.e. change in use of premises, extensions or additions.

● **Ambient temperature**
The hotter the surrounding area, the less current the cable is permitted to carry.

● **Grouping**
If a cable is run with other cables then its current carrying capacity must be reduced.

● **Type of protection**
Special factors must be used when BS 3036 (semi-enclosed) fuses are employed.

Did you know?

Solid conductors are not as common as stranded; their use has been restricted to either very small or very large conductor sizes

- **Whether placed in thermal insulation**
  If cables are placed in thermal insulation, de-rating factors must be applied.

- **Voltage drop**
  The length of circuit, the current it carries and the cross-sectional area of the conductor will affect the voltage drop. Regulation 525.1–4 states that the maximum voltage drop must not be more than 3 per cent for lighting and 5 per cent for other circuits (see Appendix 12) of the nominal voltage of the supply.

Insulation and sheathing

To insulate the conductors of a cable from each other and to insulate the conductors from any surrounding metalwork, materials with extremely good insulating properties must be used. This material around the cable is called the sheath. Cables can be installed in a variety of different situations, and you must take care that the type of insulation and sheath on the chosen cable is suitable for that particular situation.

Insulation types

Listed below are some of the working properties of the more common types of cable insulation:

- PVC
- synthetic rubbers
- silicon rubber

- magnesium oxide
- phenol-formaldehyde.

PVC

This is a good insulator: it is tough, flexible and cheap. It is easy to work with and easy to install. However, thermoplastic polymers such as PVC do not stand up to extremes of heat and cold, and BS 7671 recommends that ordinary PVC cables should not constantly be used in temperatures above 60°C or below 0°C. Care should be taken when burning off this type of insulation (to salvage the copper) because the fumes produced are toxic.

Synthetic rubbers

These insulators, such as Vulcanised Butyl Rubber, will withstand high temperatures much better than PVC and are therefore used for the connection of such things as immersion heaters, storage heaters and boiler-house equipment.

Silicon rubber

FP 200 cable using silicon rubber insulation and with an extruded aluminium over-sheath foil is becoming more popular for wiring such things as fire-alarm systems. This is due largely to the fact that silicon rubber retains its insulation properties after being heated up or burned and is somewhat cheaper than mineral-insulated metal-sheathed cables.

Magnesium oxide

This is the white powdered substance used as an insulator in mineral-insulated cables. This form of insulation is hygroscopic (absorbs moisture) and therefore must be protected from damp with special seals. Mineral-insulated cables are able to withstand very high temperatures and, being metal sheathed, are able to withstand a high degree of mechanical damage.

Phenol-formaldehyde

This is a thermosetting polymer used in the production of such things as socket outlets, plug tops, switches and consumer units. It is able to withstand temperatures in excess of 100°C.

Environmental conditions

Many factors affect cable selection. Some will be decided by factors previously mentioned and some by the following:

- risk of excessive ambient temperature
- effect of any surrounding moisture
- risk of electrolytic action
- proximity to corrosive substances
- risk of damage by animals
- effect of exposure to direct sunlight
- risk of mechanical stress
- risk of mechanical damage.

Ambient temperature

Current-carrying cables produce heat, and the rate at which the heat can be dissipated depends upon the temperature surrounding the cable. If the cable is in a cold situation, then the temperature difference is greater and there can be substantial heat loss. If the cable is in a hot situation, then the temperature difference of the cable and its surrounding environment will be small, and little, if any, of the heat will be dissipated.

Problem areas are boiler-houses and plant rooms, thermally insulated walls and roof spaces. PVC cables that have been stored in areas where the temperature has dropped to 0°C should be warmed slowly before being installed. However, if cables have been left out in the open and the temperature has been below 0°C (say, a heavy frost has attacked the cables), then you must report this situation to the person in charge of the installation. These low temperatures can damage PVC cables.

Find out

Appendix 4 of BS 7671 gives tables of different types of cable, and shows that the current rating changes in accordance with the conditions under which the cable is installed. Take a look at this and make a copy for your own reference

Moisture

Water and electricity do not mix, and care should be taken at all times to avoid the movement of moisture into any part of an electrical installation by using watertight enclosures where appropriate. Any cable with an outer PVC sheath will resist the penetration of moisture and will not be affected by rot. However, suitable watertight glands should be used for termination of these cables.

Electrolytic action

Two different metals together in the presence of moisture can be affected by electrolytic action, resulting in the deterioration of the metal. Care should be taken to prevent this. An example is where brass glands are used with galvanised steel boxes in the presence of moisture. Metal-sheathed cables can suffer when run across galvanised sheet-steel structures, and if aluminium cables are to be terminated on to copper bus bars then the bars should be tinned.

Corrosive substances

The metal sheaths, armour, glands and fixings of cables can also suffer from corrosion when exposed to certain substances. Examples include:

- magnesium chloride used in the construction of floors
- plaster undercoats containing corrosive salts
- unpainted walls of lime or cement
- oak and other types of acidic wood.

Metalwork should be plated or given a protective covering. In any environment where a corrosive atmosphere exists special materials may be required.

Damage by animals

Cables installed in situations where rodents are prevalent should be given additional protection or installed in conduit or trunking, as these animals will gnaw cables and leave them in a dangerous condition. Installations in farm buildings should receive similar consideration and should, if possible, be placed well out of reach of animals to prevent the effects of rubbing, gnawing and urine.

Did you know?

Many house fires are attributed to rodent damage to cabling

Direct sunlight

Cables sheathed in PVC should not be installed in positions where they are exposed to direct sunlight because this causes them to harden and crack: the ultraviolet rays leach out the plasticiser in the PVC, making it hard and brittle.

Earthing and bonding

On completion of this topic area the candidate will be able to differentiate between the terms earthing and bonding and give examples on the usage of each.

Earthing

BS 7671 defines **earthing** as a connection of any exposed conductive parts of an installation to its main earthing terminal. This is done to ensure that all the metalwork associated with the electrical installation is effectively connected to the general mass of the Earth.

This means that, in the event of an earth fault, the voltage developed to earth is restricted to 230 V. This is the case even on a 400 V three-phase supply, where automatic disconnection of the supply can be achieved.

Bonding

BS 7671 fails to define bonding, but does define a bonding conductor. This is a protective conductor providing equipotential bonding. This in turn is defined as an electrical connection maintaining various exposed conductive parts and extraneous conductive parts at substantially the same potential.

In order for current to flow, two conditions have to be met.

1. There has to be a circuit for current to pass along.

2. There has to be a potential difference (voltage).

By joining together all the metal work within the installation, it is now no longer possible for a difference of potential to exist between the various items of metalwork. Therefore there is no difference of potential, meaning no current flow. This prevents anyone, who comes into contact with a live piece of metal work, receiving an electric shock when they touch another piece of metal.

Protective equipotential bonding and automatic disconnection of the supply, sometimes referred to as **PEBADS**, is the main method of fault protection. This involves both **earthing** and **bonding**.

First, all the metal work within the installation is bonded together. This places all the metalwork at the same potential and prevents a difference of potential from existing between any of the items of metalwork. Having done this, the metalwork is then connected with the **general mass of earth**, via the **main earth terminal** and the **earthing conductor**. Now in the event of a fault to earth, automatic disconnection of the supply can be achieved.

Exposed and extraneous conductive parts

On completion of this topic area the candidate will be able to list exposed conductive parts and extraneous conductive parts of metallic structures and services.

Exposed conductive parts

These are the metallic parts of an electrical system, which can be touched and which are not normally live. However these can become live under fault conditions.

Examples of exposed conductive parts are:

- metal casings of appliances such as heaters
- kettles
- ovens
- wiring containment systems such as metal conduit, cable tray and trunking.

Extraneous conductive parts

These are metallic parts within a building, which do not form part of the electrical system, but can also become live under fault conditions. Examples are:

- water and gas pipes
- air conditioning
- boilers
- air ducting systems
- structural steel work of the building.

Earth protection

On completion of this topic area the candidate will be able to state the purpose of earthing and the function of earth protection.

Purpose of earthing

By connecting to earth all metalwork not intended to carry current, a path is provided for leakage current which can be detected and interrupted by fuses, circuit breakers and residual current devices. Figure 6.22 illustrates the earth return path from the consumer's earth to the supply earth.

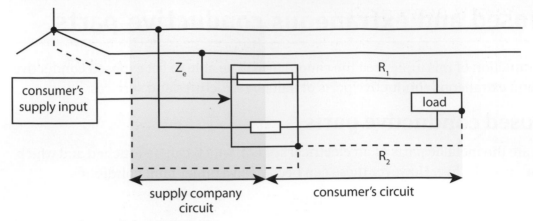

Figure 6.22 The earth-fault loop path

The connection at the consumer's earth can be by means of either an earth electrode at the building where the earth is required or may be in the form of a cable which runs back to the generator or transformer and is then connected to an earth point.

Because the transformer or generator at the point of supply always has an earth point, a circuit is formed when earth-fault currents are flowing. If these fault currents are large enough they will operate the protective device, thereby isolating the circuit. The star point of the secondary winding in a three-phase four-wire distribution transformer is connected to the earth to maintain the neutral at earth potential.

Results of an unearthed appliance

A person touching the appliance shown, which is live due to a fault, completes the earth circuit and receives an electric shock.

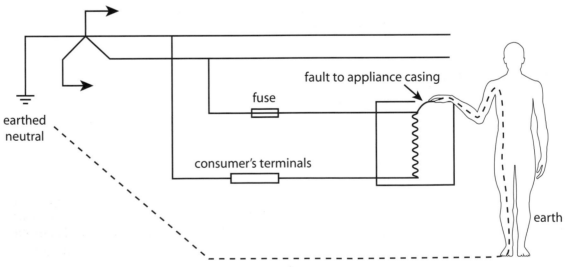

Figure 6.23 Electric shock

Results of a bad earth

A bad earth circuit – i.e. one with too large a resistance – can sometimes have more disastrous effects than having no earth at all. This is shown in the illustration, where the earth-fault circuit has a high resistance mainly due to a bad contact at point A.

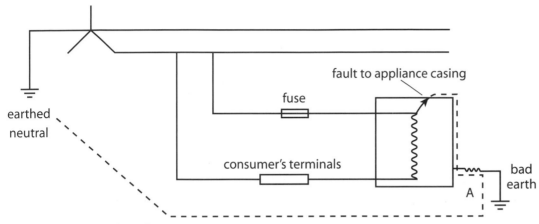

Figure 6.24 Bad earth path

The severity of shock will depend mainly upon the surroundings, the condition of the person receiving the shock and the type of supply. When the current starts to flow, the high resistance connection will heat up and this could be a fire hazard. Also, because the current flowing may not be high enough to blow the fuse or trip the circuit breaker, the appliance casing remains live.

Results of a good earth path

A good earth path, that is a low resistance one, will allow a high current to flow. This will cause the protective device to operate quickly, thereby isolating the circuit and giving protection against electric shock.

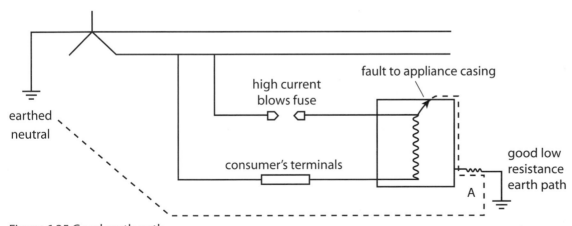

Figure 6.25 Good earth path

Earth-fault loop impedance

The path made or followed by the earth fault current is called the earth-fault loop or phase-earth loop. It is termed **impedance** because part of the circuit is the transformer or generator winding, which is inductive. This inductance, along with the resistance of the cables to and from the fault, makes up the impedance.

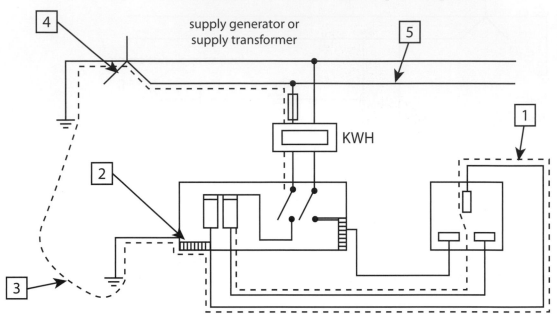

1 The circuit protective conductor (cpc) within the installation.
2 The consumer's earthing terminal and earthing conductor.
3 The earth return path, which can be either by means of an electrode or via the cable armouring.
4 The path through the earthed neutral point of the transformer and the transformer winding (or generator winding).
5 The phase conductor.

Figure 6.26 Earth-fault loop impedance

Electrical protection

On completion of this topic area and the candidate will be able to list basic principles of shock protection, circuit overload and short-circuit protection

Shock protection

BS 7671 classifies electric shock into two categories – shock resulting from either:

- failure of basic protection allowing direct contact with the electrical supply

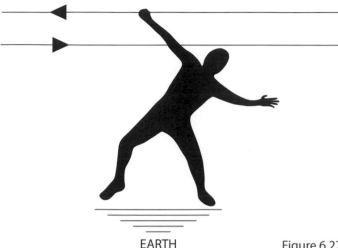

EARTH   Figure 6.27 Failure of basic protection

- failure of fault protection allowing indirect contact with the supply via exposed conductive parts or metalwork that have become live due to a fault.

EARTH   Figure 6.28 Failure of fault protection

BS 7671 requires protective measures to be taken to provide:

- **both** basic and fault protection
- or basic protection
- or fault protection

Requirements for basic and fault protection

Research has shown that the human body can withstand indefinitely, without sustaining damage, 50 volts or less. Logically then, if the voltage of an installation is reduced to 50 V or less, then it will not matter if anyone comes into contact either directly or indirectly with the supply, as they will not be hurt.

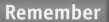

For this system to be effective the supply must be run through a safety isolating transformer. This device has no connection to earth, so the installation must meet the requirements of BS 7671 for a SELV (separate extra low voltage) supply.

This method of protection is used when there is an increased risk of an electric shock due to body resistance being reduced, such as in damp and wet conditions.

Basic protection

Basic measures to be taken to prevent electric shock include (Regulation 416):

(i) protection by insulating live parts

(ii) protection by barriers or enclosures

(iii) protection by obstacles, so preventing access

(iv) protection by placing out of reach

(v) IP codes/ratings are applied to electrical equipment to create an industry standard. These are covered below.

RCDs (residual contact devices) may be used as a supplementary means of shock prevention from direct contact, but only in addition to the measures listed in (i) to (v) above. We will look further at RCDs in chapter 12, pages 307–308.

Barriers or enclosures

All live electrical components should be within enclosures. In some cases they will also be protected by barriers, preventing easy access from the public.

In order to make sure that these requirements are followed correctly, an index of protection codes is applied to electrical equipment enclosures to create an industry standard. The first number in the code relates to the protection against direct contact in terms of the degree of accessibility, e.g. IP2X means that there is no contact with a probe of more than 12mm in diameter and less than 80mm long (approximately the length of a human finger). IPXXB is as above but for probes up to a maximum of 80mm. IP4X is used for protection against wires of 1 mm in diameter.

The horizontal top surface of the barrier or enclosure, which is readily accessible, should be protected to at least IP4X.

Fault protection

In this situation, conductive parts, such as metalwork, have become live due to fault conditions. The potential voltage on this metalwork rises above that of earth and electric shock results when a person touches the metalwork.

Measures taken to prevent electric shock arising from a fault include (Section 411, 412 & 413 in BS 7671) the following.

(i) Protective equipotential bonding and the automatic disconnection of the supply (PEBAD) formerly known as earthed equipotential bonding and the automatic disconnection of the supply (EEBADS).

(ii) Use of class II equipment and/or equivalent insulation. Class II usually relates to portable equipment or factory-built equipment. Equipment to this class must not have enclosing metalwork earthed. Furthermore for a class II installation, close supervision is required to ensure the installation or its equipment is not changed from class II. The symbol on the equipment that indicates that it is class II or double insulated is a square within a square.

(iii) Non-conducting location, i.e. physical separation from exposed conductive parts, no earth connection.

(iv) Earth free local equipotential bonding. Every part is bonded and therefore 'no voltage difference – no shock'.

(v) Electrical separation. A transformer fed secondary is not connected to earth or an isolated generator supply is used.

Methods (iii) to (v) have a limited application, as close supervision is required. The design must be specified by a qualified electrical engineer. The basic limitation of these methods is the need to maintain their design condition. There should be no incorrect modifications, additions or deterioration.

Overcurrent protection
Protection devices

In a healthy electrical system, insulation between line conductors and neutral conductors or between live conductors and earth is sound, with current flowing from the supply, through the load and returning via the neutral to its source.

Overcurrent protection is provided by means of a circuit breaker or fuse. These devices are designed to operate within specified limits, disconnecting the supply automatically in the event of an overload or fault current (short circuits or earth faults). Earth-fault protection is also provided by means of a circuit breaker or fuse, which operates under earth-fault conditions. If a high value current flows, these will safely disconnect the supply but may not be relied upon to detect and disconnect the supply in the event of low values of earth leakage current.

Residual current devices (RCDs) are designed to detect and automatically disconnect the supply in the event of earth-faults. Earth-faults can exist, for example, when a live conductor touches an equipment case or if you cut through a cable when cutting the grass. They are potentially dangerous and must be eliminated in order to prevent hazards resulting from electric shock or abnormal temperature rises and fires in the electrical installation.

Overcurrent protection device applies to fuses and miniature circuit breakers (MCBs). In this section we will be looking at the various means by which these devices provide protection to cables and circuits from damage caused by too much current flowing. The appropriate device must operate within 0.4 or 5.0 seconds.

Overload and short circuit protection

All circuit conductors must be sized correctly and must be of the correct current-carrying capacity for the load that they have to carry. All circuits on an installation must be protected against **overcurrent**. This is a current exceeding the rated value or current-carrying capacity of the cable conductor. Overcurrent conditions arise from either overloads or short circuits.

Overload current occurs in a circuit which is electrically sound, i.e. a circuit now carrying more current than it was designed to carry due to faulty equipment or too many pieces of equipment connected to the circuit. Overload is only a temporary fault condition and is easily corrected. Examples of how overload occurs include:

● several 13 A plugs connected to an adaptor and then plugged into one 13 amp socket outlet

● a stalled motor.

Such conditions cause the cable temperature to rise, resulting in an increased fire risk.

Fault current is the level of current flowing at a level up to and including the prospective short circuit current (the prospective short circuit current being the maximum current that the supply transformer is capable of delivering).

The prospective short circuit current must also be taken into account when selecting the type of overcurrent device to be installed.

This fault occurs at a point between live conductors of negligible impedance or between a live conductor and an exposed conductive part. It is more serious than an overload current and will usually not be as easily corrected.

Examples of faults are: a wiring fault (live connected to neutral or earth by mistake); a nail or screw being driven through an energised cable, making contact with the live conductors and either neutral or earth conductors.

A fuse is the weak link in a circuit which will melt when too much current flows, thus protecting the circuit conductors from damage.

Calculation of the cable size therefore automatically involves the correct selection of protective devices.

Activity

Have a look at the consumer unit in your own home and see how many circuits there are and how many outlets are supplied by each circuit. Do you think the way your house's circuits have been divided up is reasonable? If not, why not?

FAQ

Q What's the difference between an mcb and an RCD?

A An mcb is an overcurrent protective device. Like a fuse, it is designed to disconnect the circuit if the current flowing is excessive. An RCD measures the current flowing in the line and neutral conductors, and will trip out if the difference between them (the residual current) exceeds the rated value.

Q Why do I have to put coloured sleeving on switch wires and 'strappers'.

A Many accidents have occurred as a result of 'mistaken identity' of conductors, so marking them is essential. It is even more important since the 2004–2006 colour coding changes. It is possible to have a blue conductor that could be an old phase conductor or a new neutral. Similarly, black conductors could be old neutrals or new lines. Extreme caution is necessary!

Q Can I spur off the immersion heater circuit to supply the central heating system?

A No. Immersion heaters supplying vessels containing more than 15 litres of water must be permanently connected to their own separate circuit.

Q Why is a spur off a ring circuit called a 'non-fused spur'?

A Because a spur connected in this way breaks one of the 'golden rules' of circuit design, i.e. the fuse size (30/32 A) should always be smaller than the conductor current carrying capacity (27 A for 2.5mm², clipped direct), which it isn't! Therefore it is called unfused because the circuit fuse is too big to protect the spur cable.

Q What protects a non-fused spur cable from overload, then?

A Well, in a sense the spur cable is protected by the maximum load that can be connected to it. This 2 x 13 A socket outlets or 26 A, drawn through a 27 A cable. Now you see why 'spurs off spurs' are not allowed as, with two twin socket outlets, you could have 4 x 13 A – in other words 52 A flowing through a 27 A cable!

Q Why do properties have more than one ring circuit even though the floor area is less than 100m?

A Remember that the unlimited number of socket outlets per A1 ring circuit rule is always subject to loading. With the number of appliances increasing all the time, it would be very easy to overload a single ring circuit. Also the Regulations tell us to divide the circuits up to 'minimise inconvenience in the event of a fault' so it is not a good idea to have all outlets on one circuit. This is especially true of lighting circuits where loss of one circuit should not result in the loss of lighting on that floor of a building.

Knowledge check

1. List five types of circuit protection device and their BS numbers. Which of these are overcurrent protective devices and which are residual current devices?

2. Draw circuit and wiring diagrams for the following circuits.

   • One-way lighting circuit.

   • Two-way lighting circuit.

   • Two-way and intermediate lighting circuit.

3. How many socket outlets can be installed in a standard (A1) domestic ring circuit?

4. What is the maximum floor area standard domestic A2 and A3 radial circuits can serve?

5. What is the maximum floor area that a domestic ring circuit can serve when protected by a 30/32 A protective device?

6. What is the total number of fused spurs that could be fitted to a standard domestic ring circuit? What would be the limitation of this arrangement?

7. Define the terms exposed conductive parts and extraneous conductive parts.

8. Draw a labelled diagram showing the earth-fault loop path.

9. What is the minimum degree of protection allowed for protection by barriers or enclosures?

10. Explain the difference between an 'overload current' and a 'short circuit current'.

chapter 7

Tools, plants, equipment and materials

Unit 2 Outcome 4

To carry out day-to-day tasks as an electrician, a wide range of tools are required for use. These can range from hand tools, such as pliers and cutters, to electrically operated tools such as drills and saws. As with all things, these tools pose their own individual risks that need to be avoided.

In order to know which tool to use for which job, it is important to be familiar with the basic types of electro-technical systems and features, as well as the fitting and fixing activities involved in installations.

In order to keep tools in good working order, it is vital to practise safe storage procedures and good housekeeping.

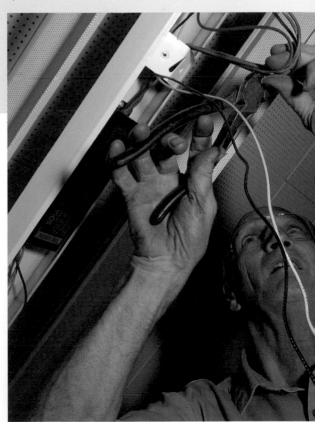

On completion of this chapter the candidate will be able to:

- state the application and safe use of hand and power tools
- state that electrically operated tools undergo inspection tests prior to and after use
- state the inherent risks of electric shock when using electrical equipment
- state the need for safe handling and storage of tools
- identify basic types of electro-technical systems and features
- identify fitting and fixing activities
- identify and deal with potential hazards when working at height or lifting and handling
- identify good housekeeping.

Hand and power tools

On completion of this topic area the candidate will be able to state the application and safe use of hand and power tools and their safe use and storage.

Tools and equipment

At one time or another, all of us have probably started on something and then been frustrated at not having the right tool for the job: 'If I only had...'. Used properly and safely, good-quality tools let you to work faster and more efficiently, so you need to look after them.

These days in the electrical trade we are using materials that can last for more than 50 years, but only so long as the system is well designed, complies with appropriate safety regulations and is properly installed. As part of an installation team, you could find yourself having to do jobs usually done by other trades, such as building and brickwork. To be a good electrician, you will need to get to grips with a number of different skills and learn how to use a wide range of tools.

It is not possible to list all the tools you might come across, so this section only describes only the most common ones.

Pliers and cutters

The main difference between electricians' pliers and any other sort is that, for obvious reasons, they have insulated handles. They have flat serrated jaws for gripping and bending and oval serrated jaws for gripping pipes and cylindrical objects.

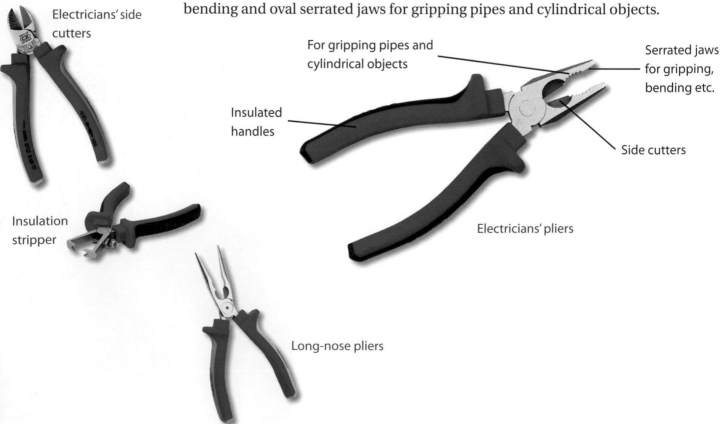

Electricians' side cutters

Insulation stripper

Long-nose pliers

For gripping pipes and cylindrical objects

Insulated handles

Serrated jaws for gripping, bending etc.

Side cutters

Electricians' pliers

Screwdrivers

The three main types of screwdriver tip are shown here. There are lots of other types for specialised jobs.

Flared slotted for
general use

Parallel slotted
head size same as shaft

Cross head (Philips, Pozidrive)
gives better 'purchase'

Hammers

Hammers are used to do three main jobs:

- drive a fixing (e.g. a nail)
- provide impact on another tool (e.g. a cold chisel)
- alter the shape of a work-piece (e.g. bend a piece of metal).

It's important to use the right hammer for the job, and to make sure that the shaft (handle) is firmly fitted into the head. Steel hammers are one-piece, so there's usually no problem. With wooden handles, which have wedges to tighten the shaft in the eye, always check that these aren't loose or missing and that the shaft is tight. Below are the three types of hammer you're most likely to use.

Ball-pein

Heavier ones are used with punches, cold chisels etc. The rounded end (ball) is used to shape metal and rivets. Its weight can be up to 2kg.

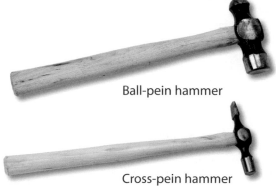

Ball-pein hammer

Cross-pein

These are often used by carpenters. The tapered end (pein) is used to start small nails held in fingers. Its weight can be up to ½kg.

Cross-pein hammer

Claw hammer

This is a general-purpose hammer. The claw end is used to lever out nails – but place a piece of hardboard between the work-piece and hammer to protect the work.

Saws

Electricians are most likely to use hacksaws and tenon saws in their work.

Hacksaw

This is used for basic metal cutting, e.g.:

- cutting tubes or sheets to length or size
- making thin cuts to help shape the metal.

Hacksaw

It consists of a frame, handle and blade. The blade is held in a handle, and tightened by a wing nut. In small 'junior' hacksaws the saw frame itself gives the tension and there is no wing nut.

Tenon saw

This is used mostly for cutting and making joints in timber. The metal strip along the top gives rigidity to the blade. To start a cut, angle the saw so that it cuts into an edge of the wood, then lower the angle for a straighter cut.

Tenon saw

Did you know?

The blade of a hacksaw has a correct way of being fitted. The teeth on the blade should face away from the handle

Flooring saw

The flooring saw is used for cutting the tongues of tongue-and-groove floorboards so they can be lifted. The saw has a curved blade that will slot into the gap between the boards.

Flooring saw

Pad saw.

Basically, this is a handle which can take differently sized narrow, tapered blades. The blade shape means you can saw in very tight spaces. It will also cut tight curves and closed shapes (by drilling a starter hole, then putting the blade through the hole)

Pad saw

Safety tip

All electrical power tools on site should be the reduced voltage type, operating at 110 volts

Drills

A drill makes a hole when the cutting edge at its tip is rotated with pressure applied. Drill bits must be chosen to match the hole size, the material being drilled and the tool used to rotate it. There are various manual and power tools for holding and turning drill bits.

Hand drill

This works by rotating the handle on the wheel.

Hand drill

Breast drill and carpenter's brace

Neither of these are used much these days now that cordless electric drills are available. The breast drill is like the 'big brother' to the hand drill, allowing you to apply pressure by using the weight of your body. The carpenter's brace has a cranked frame which you rotate while applying pressure to the domed head.

Power drills

The two most common are the hand-held electric drill and the drill press or bench-mounted drill. Both need to be handled carefully and safely to avoid accidents.

Power drills

Wrenches, spanners and supports

These are used for tightening nuts, bolts and setscrews. Spanners are usually made for one size of nut, marked on the spanner. (One exception is BSF spanners, which are designed to fit a nut one size larger than marked.) Double-ended spanners usually fit two different sizes.

The head of a ring or box spanner may be square, hexagonal (with 6 points) or bi-hexagonal (12 points). They grip all sides of the nut, reducing the possibility of damaging it. However they have to be placed over the nut, and so can't be used when you can't get to the end of its bolt or rod. Box spanners (which are cylindrical) can be used on deeply recessed nuts that are out of reach of normal spanners. The jaws of open-ended spanners close on four sides of a hexagonal nut (or three sides of a square one), giving easier access to nuts on a long bolt or rod, but providing a less secure grip.

Did you know?

Adjustable spanners or wrenches will fit a range of nut sizes

Adjustable spanner

Adjustable spanners or wrenches allow you to change the jaw opening to the size that is needed by adjusting the screw. They will fit a range of sizes of nut, but they do not give such good grip as a spanner and can slip, damaging the work piece. Use them only when you do not have the proper sized spanner, or need to grip something round.

Adjustable spanner

Footprint wrench

This has serrated jaws that adjust to the size of the nut.

Footprint wrench

Vice grip

Vice grip

This is adjusted with a screw to preset jaw width and released by a clip. The vice grip is good for holding work in progress securely. It is also called a mole wrench.

Stillson wrench

This is similar to the footprint wrench, but is larger and more powerful.

Stillson wrench

Files

Files have a rough face of hardened metal that is pushed across the surface of a work piece to remove particles of material. They can be used to make an object smoother or smaller, or to change its shape.

Files are classified according to length, cut, grade and shape. The main shapes are flat, square, half-round and round.

The main cuts are bastard, fine and medium. Files of this type must be used with a firmly fitted handle.

Surform files have a perforated blade (like a cheese grater) held in a frame. They can be used on wood, plastic and mild steel, and are good for removing material quickly without clogging.

Files

Chisels

As with most tools, there is a wide range available, with many used for special purposes. The ones you are more likely to need are shown below.

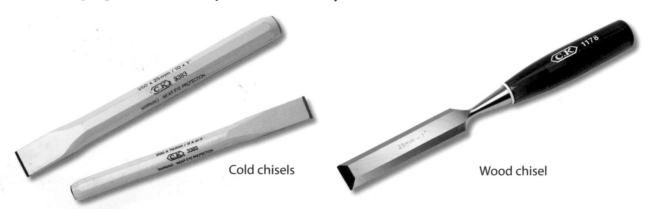

Cold chisels

Wood chisel

Crimping tools

Many electrical cables are terminated using metal lugs. These are fixed to the conductors either by soldering or crimping, where part of the lug is squeezed tightly on to the conductor using a crimping tool. These can be operated by hand or, for larger sizes, by hydraulics.

Crimping tool

Measuring tools

You will often need to check that something is level or measure lengths and distances. Available measuring tools range from a simple rule to a laser levelling device. Although measuring tools are robust, they should be handled carefully to maintain their accuracy. Here are some of the most common.

- Steel rulers are used for general measurement of trunking or tray plates.

- Steel tape measures are used for longer measurements, such as room sizes, trunking and timber.

- Spirit levels are used to check that an object is vertical or horizontal; a bubble in a glass tube shows when this is so.

- Laser rangefinders are used for measurements over longer distances; rooms, corridors etc. (typically up to 150m).

- Plumb bobs and chalk lines are used to find or check verticals. They consist of a weight on a piece of cord. When the weight stops moving, the string is a true vertical.

Steel ruler

Steel tape

Spirit level

Laser rangefinder

Chalk line

Power tools

So far, almost all the tools we have looked at are hand tools, driven by human effort alone. However, many tasks are done more quickly and easily with tools powered by electricity, batteries or compressed air.

Many of these tools are simply a powered version of hand tools, but some can do things that hand tools cannot. For example, electric drills can have a 'hammer' (or percussion) action that makes it much easier to drill into brick or concrete. Smaller electric tools used on site often operate at 110 V to reduce the risks.

Electric soldering iron

Bench grinder

The pillar/bench drill is used in a workshop, usually for repetitive, accurate drilling

Did you know?

Tools powered by rechargeable batteries (cordless) are now very common, and eliminate electrical risks

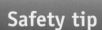

Safety awareness

You will use many different tools and equipment in your work. All of them can be dangerous if you do not respect them and use them properly and sensibly. For everyone's safety, you must continue to follow the warnings and advice you get all through your working life. See chapter 13 for more information on how to use tools and equipment safely; mean while, here is a quick checklist.

Cartridge tools, such as nail and staple guns, are widely used today. But they can also be very dangerous – they are not toys. Accidental 'firing' (or discharge) can result in serious injury. Even if the tool is not pointed directly at someone, there is a chance that the nail will bounce off another object and injure them seriously. If you need to use these tools in your work, you will be taught how to use them safely. Always follow the instructions carefully.

Basic rules for hand tools
✓ Always use the right tool for the job – don't make do with the nearest one.
✓ Keep tools clean and sharp – blunt tools are dangerous!
✓ Make sure handles are secure, tight and have no splinters.
✓ Never hit a wooden handle with a hammer.

Basic rules for electrical tools (drills, saws, sanders etc.)
✓ Check that the cable is not frayed or damaged.
✓ Check that the plug is not broken or that individual wires are showing.
✓ Check that mains tools (110 or 240 V) have been properly tested (PAT. tested).
✓ If in any doubt, don't use the tool and ask someone competent to check it out.

High-pressure air lines are frequently found in factories and workshops, and must be used with great care. Compressed air can be dangerous! It can cause explosions. **Never:**

- point it at yourself – or anyone else
- use it where tools or other items might be blown around
- use it to blow dust or dirt away.

Keep the working area tidy

Always try to keep your working area free from clutter and rubbish. Store away tools and materials that are not being used. Keep cables and air hoses tidy. Clear up spills, oily rags, paper etc. quickly. This will help you to work better and show that you are a skilled and efficient worker.

Basic types of electro-technical systems

On completion of this topic area the candidate will be able to identify basic types of electro-technical systems and features of power sources, wiring/cable systems, controls, electrical components, support and fixings.

Wiring and cable systems

PVC (Polyvinyl chloride)-insulated and sheathed cables are used extensively for lighting and power installations in domestic dwellings, being generally the most economical method of wiring for this class of work. This section will look at different types of cables and cords.

Types of cable and cords

Polyvinyl chloride is tough, cheap and easy to work with; it is also easy to install. It is not surprising, therefore, that PVC-insulated/PVC sheath cable is the most popular type of cable in current use. This form of insulation has its limitations, though, in conditions of excessive heat and cold. It can also be subject to mechanical damage unless you apply additional mechanical protection in certain situations. However, provided this and other factors mentioned later in this section are taken into consideration, then the PVC/PVC wiring system is probably the most versatile of all the wiring systems.

Single-core PVC-insulated unsheathed cable

Designed for drawing into trunking and conduit, its construction is that of PVC-insulated solid or stranded copper conductor (Ref 6491X), coloured brown or blue for single-phase systems. Other colours available include blue, green, grey, yellow, white and green/yellow stripes for use as a circuit-protective conductor.

Single-core PVC-insulated and sheathed cable

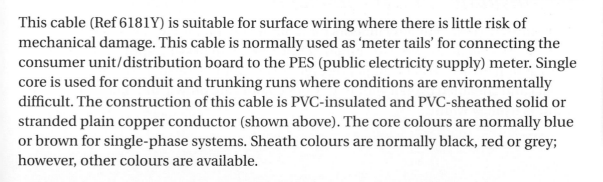

This cable (Ref 6181Y) is suitable for surface wiring where there is little risk of mechanical damage. This cable is normally used as 'meter tails' for connecting the consumer unit/distribution board to the PES (public electricity supply) meter. Single core is used for conduit and trunking runs where conditions are environmentally difficult. The construction of this cable is PVC-insulated and PVC-sheathed solid or stranded plain copper conductor (shown above). The core colours are normally blue or brown for single-phase systems. Sheath colours are normally black, red or grey; however, other colours are available.

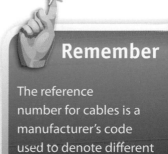

Remember

The reference number for cables is a manufacturer's code used to denote different types of cable

Remember

From 1 April 2004 the colours changed as follows: red changed to brown, black changed to blue; green/yellow remained the same

Single-core PVC-insulated and sheathed cable with a cpc

This cable (Ref 6241Y) is used for domestic and general wiring where a circuit-protective conductor is required for all circuits. Its construction is that of PVC-insulated copper conductor laid parallel with a plain copper circuit-protective conductor and PVC-sheathed overall. The core colours are brown or blue, the cpc is plain copper, and sheath colour is normally white or grey.

PVC-insulated and sheathed flat-wiring cables

This cable (Ref 6242Y and 6243Y) is used for domestic and industrial wiring. It is suitable for service wiring where there is little risk of mechanical damage. In its construction there are two or three core cables with two or three plain copper, solid or stranded, conductors insulated and sheathed overall with PVC. The colours for two cores are brown and blue for single-phase systems. For three cores (used for two-way switching) they are brown, black and grey. The sheath colours are normally grey or white. The construction of three-core cables is exactly as mentioned above with the inclusion of an uninsulated plain copper circuit protective conductor between the twin cables and between the black and grey cores of three core cables.

PVC-insulated and sheathed flexible cords

These flexible cords (Ref 3092Y and 3093Y) are suitable for use in ambient temperatures up to 85°C. They are not suitable for use with heating appliances. Their construction is plain copper flexible conductors insulated with heat-resisting (HR) PVC and (HR) PVC-sheathed. General purpose PVC flexible cords (3182Y, 3183Y etc.) are also available. The core colours are:

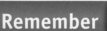

- twin-core: brown and blue
- three-core: brown, blue and green/yellow
- four-core: black, blue, brown and green/yellow
- five-core: black, black, blue, brown and green/yellow.

PVC-insulated and sheathed flat twin flexible cord

This is intended for light duty indoors for table lamps, radios and TV sets where the cable may lie on the floor. It should not be used with heating appliances. The construction is that of plain copper flexible conductor PVC-insulated two cores laid parallel and sheathed overall with PVC. Core colours are brown and blue; the sheath colour is usually white.

Remember

When installing PVC/PVC twin and cpc cables for lighting circuits, twin brown is readily available

PVC-insulated bell wire

This is used for wiring bells, alarms and other indicators that operate at extra low voltage. The construction is one single-core plain soft copper conductor insulated with PVC. Twin-core wire is produced from two single core wires laid parallel and insulated overall with PVC compound to form a figure 8 section. The standard colour of this wire is white; core identification is by a coloured stripe on the insulation of one-core.

Cables with thermosetting insulation

The thermosetting insulation is given the designation XLPE (cross-linked polyethylene). This means that the cable can be used in higher operating temperatures. The maximum continuous operating temperature for XLPE is 90°C compared with 70°C for PVC insulation. The increased temperature permits a reduction in conductor size if XLPE-insulated cables are used in preference to cables having PVC insulation. These cables are used mostly for mains distribution.

PILCSWA cables

These are Paper Insulated Lead Covered Steel Wire Armoured cables. They are found on systems at 3.3 kV and over. To work with these cables you need to be specially trained.

LSF cables

These are cables with a low smoke- and fume-giving capacity in the case of fires. Some local authorities may require the installation of this type of cable in special installations.

Installing PVC/PVC cables

Fixing with clips

Cables are fixed at intervals using plastic clips that incorporate a masonry-type nail. The maximum spacing of clips for cables run on the surface is specified in the 17th edition of the *Amicus Guide*, page 71, and states the maximum spacing for various types and sizes of PVC cable.

When bending PVC cable around corners the radius of the bend should be such that the cable or conductor does not suffer damage (Regulation 522.8.3). Guidance regarding the minimum internal radii of bends can be found on page 74 of the *Amicus Guide*. The image opposite illustrates some typical examples of clips used to fasten PVC cables.

Where PVC cables are installed on the surface the cable should be run directly into the electrical accessory, ensuring that the outer sheathing of the cable is taken inside the accessory to a minimum of 10 mm. If the cable is to be concealed, a flush box is usually provided at each control or outlet position.

Cable clips

Installing, clipping and terminating PVC/PVC cable

Installing and clipping

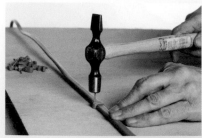

1. Fix the first clip using a small cross pein hammer.

2. To ensure a neat appearance PVC cable should be pressed flat against the surface between the cable clips.

3. The end clip should be fixed next.

4. Clips should be equally spaced between the first and end clip.

5. When a PVC cable is to be taken around a corner or changes direction the bend should be formed using the thumb and fingers as shown.

6. Final clip after insertion into enclosure.

Terminating

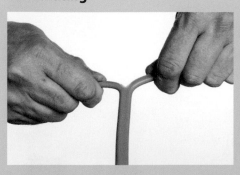

1. Nick the cable at the end with your knife and pull apart as shown.

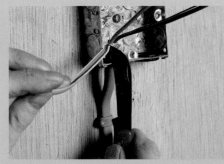

2. When the required length has being stripped, cut off the surplus sheathing with the knife as shown.

3. The insulation can be stripped from the conductors with the knife or with a pair of purpose-made strippers, as shown. Examine the conductor insulation for damage.

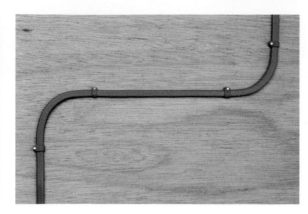

Care must be taken when bending cable

Care must be taken to ensure that the bend does not cause damage to the cable or conductors and that cable clips are spaced at appropriate intervals. The cable can be straightened by running the thumb over it before clipping. The palm of the hand can be used for bigger cables.

Cable suspension and catenary systems

Cables can be run outside between buildings by suspending them from a catenary wire. The catenary wire is usually a galvanised steel wire, which should be strained tight. The cables are fastened to this wire using tape or are suspended from it using hide hangers. In order to prevent travel of rainwater along the cable a drip loop should be formed at either end of the catenary wire. Figure 7.01 illustrates the catenary wire and the loop.

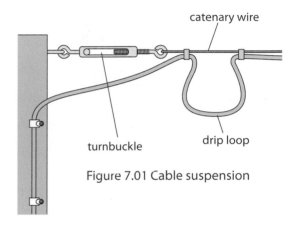

Figure 7.01 Cable suspension

Cable runs

Cable runs should be planned to avoid cables having to cross one another, which would result in an unsightly and unprofessional finish. When cables are to be installed in cement or plaster they should be protected against damage. They should be covered with metal or plastic channel or be installed in oval PVC conduit.

Care must be taken when installing PVC cables to ensure that they are not allowed to come into contact with:

- gas pipes
- water pipes
- any other non-earthed metal work.

PVC-sheathed cables should also not come into contact with polystyrene insulation, as a chemical reaction takes place between the PVC sheath and the polystyrene, resulting in the migration of polymers with the cable known as 'marring'.

Power sources and controls

These will be covered later in chapter 11. Please refer to this chapter for more information.

Fixings

The electrical industry uses a large variety of fixing and fastening methods. This can lead to confusion over the terminology used to refer to them and their associated devices. This section will look at the various types of fixings and fastenings and where they are used.

Wood screws

As the name suggests, they are primarily used when fixing items to wood. In the electrical field, however, they are more commonly employed in conjunction with rawlplugs where fastenings to masonry are required. When ordering screws it is important to give the correct description of the screw required. To do this, four pieces of information must be known:

- size of screw
- length of screw
- type of head
- type of metal finish used.

Size of screw

Screws are still available in imperial sizes, with the guage given a number between 2 and 24 based on the shank diameter as shown in Figure 7.02. Most screws are now metric with the thread the same diameter as the shank, apart from a taper near the tip for starting. Sizes range from 2mm to 6.5mm.

You can see opposite the difference between a clearance hole and a threaded hole. The clearance hole is designed to take the screw without the screw gripping its sides, while the pilot hole is made to give the thread of the screw a start. The pilot hole may be necessary if the screw is going to be used in thick or particularly hard wood. These holes are made either with a drill of the correct size or with a bradawl.

Length of screw

The length of a screw is measured from the tip to the part of the head flush with the material. Metric sizes range from 12mm to 150mm.

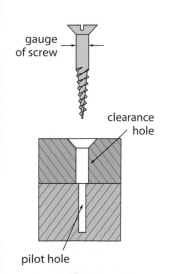

gauge of screw

clearance hole

pilot hole

Figure 7.02 Screwing two materials together

Type of head

The various types of screw head are appropriate in different situations, as shown in Figure 7.03.

Screwhead type	Usage
Countersunk (Pozidrive head) 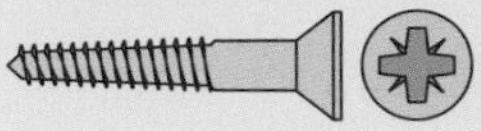	For general woodwork, fitting miscellaneous hardware, spacer bar saddles etc. This screw must be driven until the head is flush with the work surface or slightly below the surface. Used with special screwdrivers which will not slip from the cross-slots. Can be carried into confined spaces on the end of the screwdriver. They can be driven home in half the time of conventional screws.
Raised head 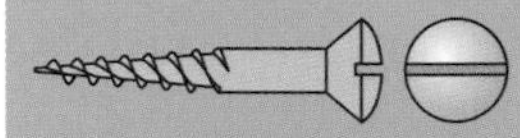	Used to fix door-handle plates and decorative hardware. They must be countersunk to the rim. They are usually nickel- or chrome-plated.
Round head 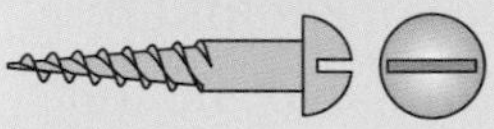	Used to fix surface work, fittings, accessory boxes etc. when countersunk screws are not required.
Dome head 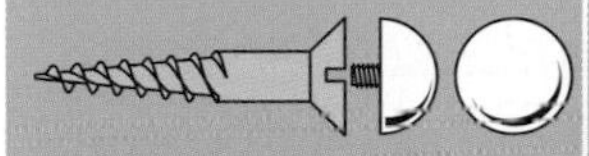	A concealed screw for fixing mirrors, bath panels and splash-backs where the head of the screw is covered by a chrome cap that is either screwed into the end of the screw or is push-fitted onto it.
Coach screw 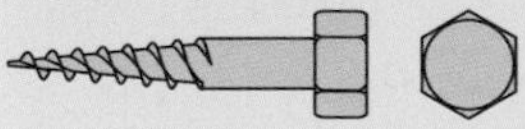	Provides strong fixing in heavy construction and framework. It is turned into the wood with a spanner.

Figure 7.03 Screw-head types and usage

Type of metal used

The material and finish of the screw must also be considered. The two main materials used are steel and brass. Steel screws may be left with a bare finish or may be sprayed black (black Japanned). They can also be cadmium-coated for rust prevention. Brass screws are either left bare or chrome-plated. The choice of material is dictated by strength requirement. If a load-bearing capability is required then steel screws should be used. The finish of screw is often just a matter of aesthetics.

Remember

Screws are available in metric and imperial units. You need to be familiar with both

Remember

If brass screws are tightened too much, the shaft of the screw is likely to break

Self-tapping hardened steel screws

Self-tapping screws are used primarily with sheet steel. A hole slightly smaller than the screw to be used is drilled through the steel, and when the screw is driven into steel it will cut its own thread and become fast. This is particularly useful when joining two pieces of steel together. Self-tapping screws and their head type are ordered as for wood.

They are used in particular on cookers, heaters and steel boxes.

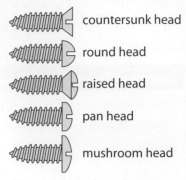

countersunk head

round head

raised head

pan head

mushroom head

Figure 7.04 Self-tapping screws

Machine screw

A machine screw, unlike a bolt, is threaded along its whole length. This avoids the need for long bolts when the parts to be joined are very thick, but requires the hole in one of the parts to be threaded, the other hole being a clearance hole.

Machine-screw heads vary in shape depending on their application. Some are designed to take a spanner. Others have a slot for a screwdriver or a socket for an Allen key. The heads may stand proud or may be sunk below the surface for neater appearance.

The machine screws that are particularly favoured in the electrical trade have, in the past, been the BA thread ranges. Metric conduit boxes now use 4mm screws, and socket outlet and switch covers use 3.5mm metric screws, which are also often found on panels and equipment.

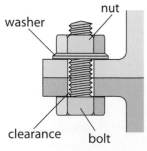

clearance

threaded

Figure 7.05 Machine-screw fixing

Stud, nut and washer

If a machine screw is frequently removed and replaced there is a tendency for the threads to wear and strip. This is overcome by the use of a stud, which is tightly fastened into the tapped hole and remains in this position when the nut is removed.

Bolts

The bolt in conjunction with nut and nut washer are used widely in all branches of engineering. The bolts passes through clearance holes in the parts to be joined. The clearance for general work is 1.5mm, e.g. for a 16mm bolt the hole would be drilled 17.5mm.

washer

nut

clearance

bolt

Figure 7.06 Bolt fixing showing clearance

Locking devices

Mechanical fastenings, e.g. bolts and studs, are used in order that parts may be removed for overhaul, replacement or to allow access to other parts. When these fastenings are subject to vibrations, such as on machines and engines, there is a tendency for them to work loose. This could result in serious damage and, to prevent this, locking devices are used. Some of the more common types are shown in the Table 7.01.

Locking device	Features and usage
Spring washer	• Similar to a coil spring. • When the nut is tightened, the washer is compressed, and because the ends of the washer are chisel-edged they dig into the nut and the component, thus preventing the nut from turning loose. • Spring washers may have either a single or double coil. • Depending on the condition, spring washers are used only once.
Locknut	• The bottom nut is tightened with a spanner. • The top nut is then tightened, and friction in the threads and between the nut faces prevents them from rotating. • Locknuts are always bevelled at the corners to ensure good setting of the faces.
Split pin	• This can be used with ordinary nuts or castle nuts (see below). • When used with ordinary nuts, ensure that the split pin is in contact with the nut when tightened. The split pin is opened out after insertion to prevent it falling out. • Split pins can be used only once. • The bolt is left two to three threads longer for drilling.
Castle nut	• The castle nut has a cylindrical extension with grooves. • The nut is tightened, the stud is drilled opposite a groove, then the split pin is passed through the nut and the stud prevents the nut from turning. The split pin is opened out after insertion to prevent it falling out. • Split pins can be used only once.
Simmonds locknut	• The Simmonds nut has a nylon insert. • When the nut is screwed down the threads on the end of the stud bite into the nylon. • Friction keeps the nut tightened. • This nut can be used only once.
Serrated washer	• When the nut is tightened, the serration is flattened out, causing increased friction between the faces thus preventing rotation. • This type of washer can be used only once.
Tab washer	• This is a more positive type of locking device. • When the nut is tightened, one tab is bent up on to the flat side of the nut and the other tab is bent over the edge of the component. • Tab washers can be used only once – they tend to fracture when straightened out and re-bent.

Table 7.01 Locking devices

Fixing devices

Whenever it is necessary to fix a piece of apparatus to a wall, ceiling or partition, a fixing device is required to ensure a good hold without causing damage. The drill, screw and device should all be of the correct size with respect to each other. This will depend on the material from which the fixing surface is made and on the weight to be supported.

Fixing holes should be made with an appropriate drill. A plain drill bit should be used for timber while a tungsten carbide-tipped drill should be used for masonry. Select a slow speed for drilling masonry with an electric drill. Hammer or percussion drills are recommended for concrete.

Light fixing devices

As their name suggests, these are used for relatively light fixing jobs and for partition walls and thin sheet materials.

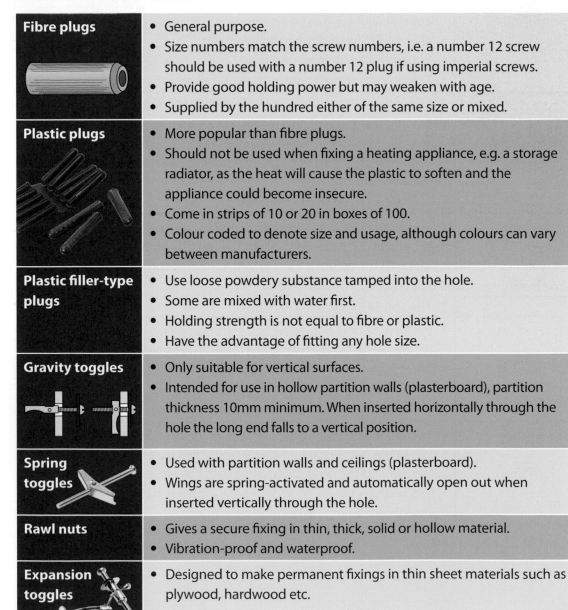

Fibre plugs	• General purpose. • Size numbers match the screw numbers, i.e. a number 12 screw should be used with a number 12 plug if using imperial screws. • Provide good holding power but may weaken with age. • Supplied by the hundred either of the same size or mixed.
Plastic plugs	• More popular than fibre plugs. • Should not be used when fixing a heating appliance, e.g. a storage radiator, as the heat will cause the plastic to soften and the appliance could become insecure. • Come in strips of 10 or 20 in boxes of 100. • Colour coded to denote size and usage, although colours can vary between manufacturers.
Plastic filler-type plugs	• Use loose powdery substance tamped into the hole. • Some are mixed with water first. • Holding strength is not equal to fibre or plastic. • Have the advantage of fitting any hole size.
Gravity toggles	• Only suitable for vertical surfaces. • Intended for use in hollow partition walls (plasterboard), partition thickness 10mm minimum. When inserted horizontally through the hole the long end falls to a vertical position.
Spring toggles	• Used with partition walls and ceilings (plasterboard). • Wings are spring-activated and automatically open out when inserted vertically through the hole.
Rawl nuts	• Gives a secure fixing in thin, thick, solid or hollow material. • Vibration-proof and waterproof.
Expansion toggles	• Designed to make permanent fixings in thin sheet materials such as plywood, hardwood etc.

Table 7.02 Light fixing devices

Heavy fixing devices

These are used for heavier jobs such as fixing a large fuseboard or securing a motor to a concrete plinth. Because of the possible dangers associated with these heavier fixings, the following should be seen as only a rough guide to different methods, and further information should be obtained prior to use.

Rawlbolt 	• Used for fixing materials to walls, floors etc. Two types:
	  • bolt end protruding from the body on to which the washer and nut are placed
	  • bolt threaded separately.

Self-drill anchor 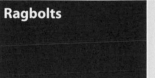	• Expensive but faster to use. • Self-drilling bolt, which is fastened in the chuck of the drill. • The bolt is then removed and a tapered plug inserted. • The bolt is then reinserted and, with the drill set to hammer, knocked into place. • The end of the bolt is then snapped off, leaving an inserted shaft ready to accept a bolt.
Ragbolts	• Bolts with a fluted end for use in floors. • A hole is drilled in the ground larger than the bolt and the whole thing is cemented in. • It is then left to dry before fixing the piece of equipment.

Table 7.03 Heavy fixing devices

Miscellaneous fixings

- Roundhead nail – used for general woodwork.

- Oval nail – used for general woodwork; prevents splitting of timber, especially thin or heavily grained timber.

- Brad – used as floorboard fixing; difficult to remove.

- Galvanised clout nail – handy for fixing channelling over cables prior to plastering.

- Panel pin – small pin for fastening hardboard or woodsheets, used with buckle clips.

- Masonry nail – hard nail for use with plastic clips (PVC-sheathed cable).

- Rivet – a device for joining together two or more pieces of metal. They should be of the same material as the metal being joined; if this is not possible the rivets should be of a softer metal than the sheets being joined.

The Gripple

The Gripple is a new system for supporting false ceilings, cable basket or other similar loads. It uses a principle of mini-tirfor jacks and can be easily tensioned into the correct position. The Gripple can be released using the small key provided.

Safety tip

Any holes through elements of building structures that are intended to take wiring systems must be sealed around the cables in order to prevent the spread of fire

Gripple supporting cable tray

Fitting and fixing

On completion of this topic area the candidate will be able to identify activities in relation to isolation procedures, specifications and determining appropriate fixing and fitting methods.

Nearly all modern domestic wiring is recessed into the walls, as this leaves a flat surface and a neater appearance. However, most rewiring of houses involves buildings that are at least 25 years old. Before 1956 wiring was usually installed in vulcanised rubber (VRI) insulated with a tough rubber sheath (TRS) or a lead sheath. This type of wiring is greatly affected by temperature changes and eventually the rubber becomes brittle; any interference with the cable usually results in the insulation breaking off. Modern PVC cables have a far greater lifespan. Sometimes wiring was installed in slip-gauge conduit (light-duty conduit with an open seam).

This type of conduit is not to be used as an earth return under any circumstances.

Both of these types of systems are now due for rewiring.

This section will cover some of the techniques required for the installation of rewiring an existing building and the regulations that apply for the protection of installed cables. Areas that will be covered include the following:

- rewiring an existing building
- floorboards
- cables run into walls
- chasing
- wiring in partition walls
- ceiling fittings
- protection of cables
- protection against heat damage and spread of fire
- cable support and bends
- miscellaneous.

Rewiring an existing building

Floorboards

Quite often it is necessary to lift floorboards to be able to install new cables. With some older buildings the boards are of the butt-type finish. This makes lifting the boards comparatively easy. Modern houses tend to have tongue-and-groove floorboards as shown in Figure 7.07.

Figure 7.07 Tongue-and-groove

Most present-day domestic properties have high-density chipboard panels, which interlock with each other and will require a different approach should you be required to install additional points in the future.

Floorboards are normally fitted starting from one wall, i.e. board number 1 placed in position and then board number 2 slotted into board number 1, then number 3 is slotted into number 2 etc. until the whole floor is covered. Then they are nailed down. Lifting the boards with minimum damage entails lifting the last board laid and reversing the laying procedure. However, this is a time-consuming operation, especially when only one or two boards need to be lifted in the area you wish to work in. Lifting a middle board means that the two tongues holding the board need to be removed. Ideally a circular saw with a narrow blade or a flooring saw can be used.

Before attempting to lift the floorboard the nails should be punched down to enable the board to be lifted more easily. Very great care must be taken to ensure that any other service pipes, e.g. water or central heating, are not cut or damaged during the work involved with lifting floorboards or chipboards.

(a)

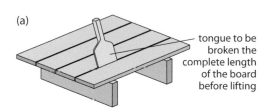

tongue to be broken the complete length of the board before lifting

(b)

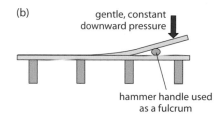

gentle, constant downward pressure

hammer handle used as a fulcrum

Figure 7.08 (a) and (b) Lifting floorboards

Once the end of the board has been prised free you should endeavour to get the rest up without excess damage.

Sometimes it is necessary to cut a floorboard because it disappears under the skirting board or a built-in wardrobe. The place to cut is where the board crosses a joist; this is usually where the board is nailed. If you are unable to lift the floorboard enough to cut it, then the board must be cut at the side of the joist, using a padsaw.

When refitting a board, all that is required is for a small piece of wood known as a fillet to be screwed to the side of the joist. The board can be laid on it and nailed down.

(a)

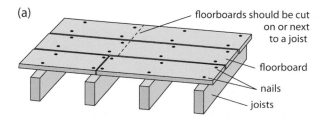

floorboards should be cut on or next to a joist

floorboard

nails

joists

(b)

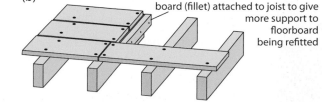

board (fillet) attached to joist to give more support to floorboard being refitted

Figure 7.09 (a) and (b) Cutting and refitting floorboards

Cables run into walls

Cables that are run in walls to feed socket outlets, lighting switches and other wall-mounted accessories, where the cable is at a depth of less than 50mm from any surface, must be suitably protected to prevent penetration by nails, screws etc. in accordance with Regulation 522.6.6. They may be installed either horizontally within 150mm of the top of the wall or partition or vertically within 150mm of the angle formed by two walls, or run horizontally or vertically to any accessory or consumer unit. Where this is not possible, an rcd of 30mA or less shall provide additional protection. (Reg 522.6.7)

Figure 7.10 Permitted cable routes

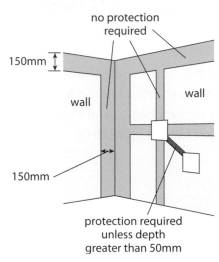

Chasing

Chasing is the name given to cutting slots into a wall for conduit or cable. This job is made extremely easy by using a chasing tool or a chasing tool attachment (available on most electric drills); all that is required is a line on the wall as a guide to work to. An alternative to this is a bolster chisel and hammer, which takes longer but does the same job. The back boxes for wall-mounted 'flush' fittings must be mounted in a hole cut into the brickwork. The hole is made slightly bigger than the box to allow plaster or filler to be applied around the edge and back of the hole. Once this has been completed the box can be fitted in and secured. The front edge of the box must not protrude from the surface of the wall. The purpose of the filler is to ensure that the box remains firm in the wall.

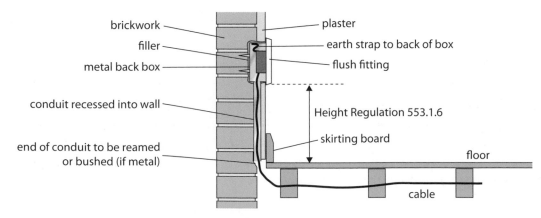

Figure 7.11 Flush fitting fed through the wall via the floor

Did you know?

In new buildings electrical wiring is usually fixed before the plasterers render the walls. In this way chasing is kept to a minimum. The accessories are then usually fixed when the plastering and decorations are completed

In older houses the bricks and mortar were not made to today's standards and they tend to be very powdery. Take, for example, a socket outlet recessed into a wall. The continual action of inserting and withdrawing a 3-pin plug puts strain on the back box, and the box may become loose and eventually unsafe.

Wiring in partitions

Wiring must be done before the lining (surface) is fixed in position, and if the cables are sheathed they must pass through the centre of the wooden framework and must be a minimum of 50mm from the finished surface.

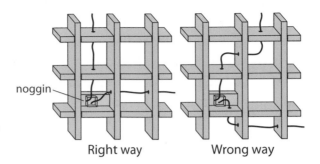

Figure 7.12 Wiring in a partition

However, if the cables are to be run in metal conduit then the conduit can fit into slots in the outer edges of the wooden framework. Regulations for conduit still apply to this form of installation.

In either case the back boxes for fitting must be securely fixed, preferably to **noggins** in the structure of the partition. Dry lining boxes can be used if the partition lining is of sufficient strength and thickness.

Ceiling fittings

Boxes for ceiling outlets must be securely fixed either to a joist or to a noggin between the joists.

Figure 7.13 Dry-lining box

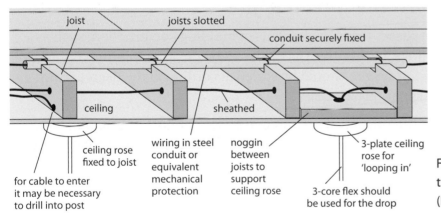

Figure 7.14 Cables through joists (Regulation 522.6.5)

Protection of cables

Where cables are installed under floors or above ceilings they must be run in such positions that they are not liable to be damaged by contact with the floor or ceiling or their fixings. Unarmoured cables passing through a joist shall be at least 50mm from the top or bottom as appropriate, or enclosed in steel conduit. Alternatively the cables can be provided with mechanical protection sufficient to prevent penetration of the cable by nails, screws etc.

> **Definition**
>
> **Noggin** – piece of wood fixed to joists to enable ceiling roses to be fixed

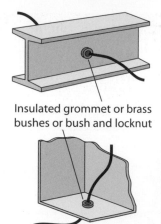

Insulated grommet or brass bushes or bush and locknut

Figure 7.15

Where cables pass through holes in metalwork, precautions should be taken to prevent abrasion of the cables on any sharp edges (see Regulation 522.6.1) – for instance by the use of insulated grommets (see Figure 7.15).

Non-sheathed cables for fixed wiring should be mechanically protected if less than 50mm above or below the surface (Regulation 522.6.6).

Where significant solar radiation or ultraviolet radiation is experienced or expected, a wiring system suitable for the conditions should be selected or adequate shielding provided. Special precautions may need to be taken for equipment subject to ionising radiation (Regulation 522.11.1).

Protection against heat damage

The insulation and sheathing of cables and flexes inside a fitting or appliance must be suitable for the heat likely to be generated inside the fitting under normal use. This may be achieved by fitting heat-resistant sleeves or beads on to the conductor or over the insulation. To prevent the effects of external heat sources (including solar gain) one or more of the following methods should be employed (Regulation 522.2.1):

- shielding
- placing sufficiently far from the heat source
- selecting a system with regard to temperature rises that may occur
- local reinforcement or substitution of insulating material.

Protection against the spread of fire

Where a wiring system passes through elements of building construction, such as floors, walls, roofs, ceilings, partitions or cavity barriers, the openings remaining after passage of the wiring system should be sealed according to the degree of fire resistance required (Regulation 527.2.1).

Where a conduit, cable ducting or cable trunking, busbar or busbar trunking penetrates elements of building construction having specific fire resistance it should be internally sealed so as to maintain the degree of fire resistance of the respective element as well as being externally sealed to maintain the fire resistance required – for example, by using cement (see Regulation 527.2.4).

Cable support and bends

When a conductor or cable is not continuously supported it should be supported by suitable means at appropriate intervals so that it doesn't suffer damage under its own weight (Regulation 522.8.4). The following tables list suitable spacing of supports for cables in accessible positions.

Overall diameter of cable	Non-armoured rubber, PVC or lead-sheathed cables				Armoured cables		MICC or aluminium cables	
	Generally		In caravans					
	Horizontal mm	*Vertical mm*	*Horizontal mm*	*Vertical mm*	*Horizontal mm*	*Vertical mm*	*Horizontal mm*	*Vertical mm*
up to 9mm	250	400	250	400	–	–	600	800
9mm–15mm	300	400	250	400	350	450	900	1200
16mm–20mm	350	450	250	400	400	550	1500	2000
21mm–40mm	400	550	250	400	450	600	–	–

Table 7.04 Cable support requirements (Table 4A, *On site Guide*)

Type of system	Maximum length of span (metres)	Maximum height of span above ground (metres)		
		At road crossings	*In positions accessible to vehicular traffic, other than crossings*	*Positions inaccessible to vehicular traffic*
Cables sheathed with PVC or having an oil-resisting and flame-retardant or HOFR sheath, without intermediate support	3			3.5
Cables sheathed with PVC or having an oil-resisting and flame-retardant or HOFR sheath in heavy-gauge steel conduit of diameter not less than 20mm and not jointed in its span	3			3
Bare or PVC-covered overhead lines on insulators without intermediate support	30			5.2
Cables sheathed with PVC or having an oil-resisting and flame-retardant or HOFR sheath, supported by a catenary wire	No limit			3.5
Aerial cables incorporating a catenary wire	See manufacturers' details			3.5
Bare or PVC-covered overhead lines installed in accordance with the overhead line Regulations	No limit	5.8 for all types	5.8 for all types	5.2

Table 7.05 Maximum length of span and minimum heights above ground for overhead wiring between buildings

| Normal size of conduit (mm) | Maximum distance between supports (metres) | | | | | |
| | Rigid metal | | Rigid insulating | | Pliable | |
	Horizontal	Vertical	Horizontal	Vertical	Horizontal	Vertical
Not exceeding 16	0.75	1	0.75	1	0.3	0.5
Exceeding 16 but not exceeding 25	1.75	2	1.5	1.75	0.4	0.6
Exceeding 25 but not exceeding 40	2	2.25	1.75	2	0.6	0.8
Exceeding 40	2.25	2.5	2	2	0.8	1

Table 7.06 Spacing of supports for conduits

Insulation	Finish	Overall diameter (a)	Factor to be supplied to overall diameter of cable to determine minimum internal radius of bend
XLPE, PVC or rubber (circular or circular stranded copper or aluminium conductors)	Non-armoured	Not exceeding 10mm Exceeding 10mm but not exceeding 25mm Exceeding 25mm	3 (20)* 4 (3)*
	Armoured	Any	6
XLPE, PVC or rubber (solid aluminium or shaped copper conductors)	Armoured or non-armoured	Any	8
Mineral	Copper sheath with or without covering	Any	6#

Table 7.07 Internal radii of bends in cable for fixed wiring

(a) For flat cables the diameter refers to the major axis

*   The figure in brackets relates to single-core circular conductors of stranded construction installed in conduit, ducting or trunking

Mineral insulated cables may be bent to a radius not less than three times the cable diameter over the copper sheath, provided that the bend is not reworked, i.e. straightened and re-bent

| Cross-section area of trunking (mm²) | Maximum distance between supports (metres) | | | |
| | Metal | | Insulating | |
	Horizontal	Vertical	Horizontal	Vertical
Exceeding 300 but not exceeding 700	0.75	1	0.5	0.5
Exceeding 700 but not exceeding 1500	1.25	2	1.25	1.25
Exceeding 1500 but not exceeding 2500	1.75	2	1.25	1.25
Exceeding 2500 but not exceeding 5000	3	3	1.5	2
Exceeding 5000	3	3	1.75	2

Table 7.08 Spacing for supports for cable trunking

Mechanical stress

PVC cables, when used for overhead wiring between buildings, can be subjected to mechanical stress if a catenary wire is not used to support them along the way. Figure 7.16 illustrates suspension heights for cables above road crossings. Flexible cables used to suspend heavy luminaries that exceed the recommended weights given in BS 7671 will feel the effects of mechanical stress. See Table 7.09.

1. Road crossing accessible to vehicles

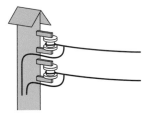

2. Accessible to vehicles but not a road crossing

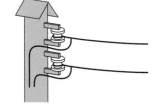

3. Inaccessible to vehicles

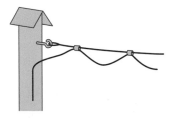

All methods of suspension 5.8m minimum above ground

All methods of suspension 5.2m minimum above ground

PVC cables supported by a catenary wire. 3.5m minimum above ground

Figure 7.16 Suspension heights for cables

Table 4F3A Flexible cords weight support			
Conductor (cross-sectional area mm²)	Current carrying capacity		Maximum mass
0.5	3 A	3 A	2kg
0.75	6 A	6 A	3kg
1	10 A	10 A	5kg
1.25	13 A	10 A	5kg
1.5	16 A	16 A	5kg
2.5	25 A	20 A	5kg
4	32 A	25 A	5kg

Table 7.09 Maximum loads on flexible cables

These cables can also suffer from stress when subjected to excessive vibration, causing breakdown of the insulation.

Mechanical damage

The main function of the cable sheath is to protect the cable from mechanical damage. All conductors and cables should also have additional protection where they pass through floors and walls or are installed in exposed positions where damage could occur. Cables to be installed underground should have a sheath or armouring resistant to any mechanical damage likely to occur. When cables pass through holes drilled in wooden joists these should be 50mm from the top or bottom of the joist

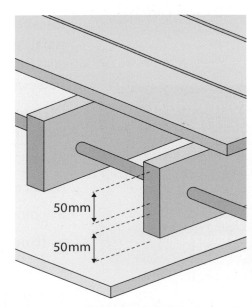

50mm

50mm

Figure 7.17 Clearance for cables in joists

measured vertically (see Figure 7.17). Numerous other examples of situations where cables can be subjected to mechanical damage can be found in chapter 52 of BS 7671.

Miscellaneous

Other factors that affect the choice of a wiring system and its associated enclosures and equipment include:

- ambient temperature (this affects the insulation properties)
- moisture (can the enclosures safely resist the ingress of water? IP ratings to be checked)
- corrosive substances (chemicals can destroy insulation, enclosures and equipment)
- UV rays (sunlight affects cable insulation)
- damage by animals (by chewing cables or the corrosive effect of animal urine)
- mechanical stress and vibration (use flexible conduits and helix loops at connections to motors to absorb vibration and make sure the cable is adequately supported throughout its length)
- aesthetic conditions (what looks right and appropriate for the building and customer).

Remember that when you are installing cables, someone has to repair the building fabric once the job is finished. While carrying out electrical installation work it will invariably be necessary to work closely with other trades (to co-ordinate the process). This will ensure a satisfactory completion of each stage of the work process and help reduce damage to a building's structure and fabric.

Patching up afterwards is more than likely going to be the job of the main contractor (builder). However, on smaller jobs you may have this responsibility. Therefore the less damage caused, the easier the repairs are likely to be.

The repair needed will obviously depend on the work done, and certainly in domestic installations you may only have to consider using substances such as Polyfilla to repair damage around ceiling outlets or switch boxes. However, in a domestic rewire also think about flooring, and replace floorboards with care, screwing them back in place after you work to prevent them from 'squeaking'.

On larger sites, you may have to think about replacing drop-in grid ceiling tiles or replacing fire barriers.

Always clean up after yourself and dispose of waste materials in the correct manner.

A successful installation will involve good relationships with all concerned, plus:

- compliance with requirements, specifications and drawings
- fixing of wiring systems and components
- ensuring electrical continuity and system integrity
- avoiding damage to property, components and systems.

Do these things and you will have a happy client.

Remember

Try to keep all damage to property to a minimum. If a hole is needed in a wall for a cable to pass through, use a drilling machine with an appropriately sized masonry bit; don't use a hand grenade!

Remember

Take care with repair work. The more care and attention taken in restoring the work area to its proper condition, the more likely you and your company are to be recommended for further work

Remember

If you are removing old fluorescent lamps, have them removed and disposed of in special 'crushers' or local authority facilities

Safe isolation of electrical supplies

We covered the safe isolation of electrical supplies in chapter 1 page 16–23.

Potential hazards

On completion of this topic area the candidate will be able to identify and deal appropriately with the potential hazards of working at height and lifting and handling.

Working at height

Working at height is the single biggest causes of workplace deaths and one of the main causes of injuries. In 2003/04 there were 67 fatal falls from height and 4000 major injuries connected to working at height.

In 2005, the government introduced the **Working at Height Regulations 2005**. These replaced all earlier regulations and consolidated previous legislation. The regulations implemented European Council Directive 2001/45/EC, which concerned the minimum health and safety requirements for working at height.

The regulations define working at height as follows:

- Regulation 2 – A place is 'at height' if (unless these Regulations are followed) a person could be injured falling from it, **even if it is at or below ground level**.

All industry sectors are exposed to the risks presented by this hazard, although the level of incidence varies considerably.

Ladders

Ladders, stepladders and trestles are perhaps the most commonly used access equipment on sites and they are also perhaps the most misused. It is essential that safe working practices should be followed if accidents are to be avoided.

Before you use a ladder or stepladder, check it – make sure it is safe to use. Look for:

- cracks in the rungs or the stiles (the sides of the ladder)
- missing, broken or weakened rungs
- rungs depending solely on nails or spikes for support
- mud, grease or oil either on the rungs or the stiles
- obvious signs of permanent bending in the rungs or stiles
- items stuck in the feet such as swarf or stones, or grease or dirt, that prevent the feet from making a direct contact with the ground
- missing, damaged or worn anti-slip feet on metal and fibreglass ladders and stepladders; with ladders, check those at the top and bottom
- missing or loose screws or rivets

Did you know?

According to the HSE, falls from height are the most common cause of fatal injury and the second most common cause of major injury to employees, accounting for 15 per cent of all such injuries

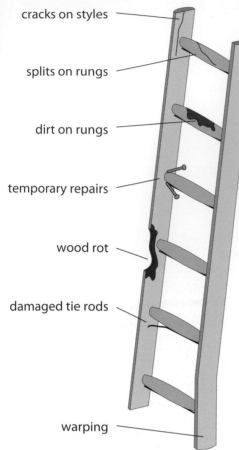

cracks on styles

splits on rungs

dirt on rungs

temporary repairs

wood rot

damaged tie rods

warping

Figure 7.18 Ladder with defects

- cracked or damaged welds on metal ladders or stepladders
- rot, woodworm or tie rods that are either missing or damaged
- painted wooden ladders or stepladders: these should not be used as the paint can hide defects.

Short ladders can be carried by one person, on the shoulder, in either the horizontal or vertical position. Longer ladders should be carried by two people, horizontally on the shoulders, one at either end holding the upper stile. When carrying ladders take care in rounding corners or passing between or under obstacles.

There are certain rules for erecting ladders, which must be followed to ensure safe working.

- The ladder should be placed on firm, level ground. Bricks or blocks should not be used to 'pack up' under the stiles to compensate for uneven ground.
- If using extension ladders they should be erected in the closed position and extended one section at a time. When extended there must be at least four or five rungs' overlap on each extension.
- Ladders should be placed clear of any excavation, and in such a position that they are not causing a hazard, or placed anywhere where they may be struck or dislodged. If the ladder is placed in an exposed position it should be guarded with barriers.
- The angle of the ladder to the building should be in the proportion of 4 up to 1 out, or 75°.
- The ladder should be secured at the top and as necessary at the bottom to prevent unwanted movement. Alternatively the ladder may be steadied by someone holding the stiles and placing one foot on the bottom rung. This is commonly known as 'footing' the ladder. This person must not under any circumstances move away from footing the ladder while someone is up it.
- When the ladder provides access to a roof or working platform, the ladder must extend at least 1m, or 5 rungs, above the landing place.
- Make sure that the ladder is not resting against any fragile surface or against fittings such as gutters or drainpipes, as these may give way, resulting in an accident.
- If a ladder has to be secured, never secure it by the rungs; always use the stiles.

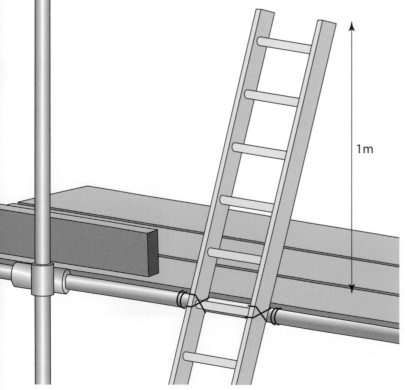

1m

Figure 7.19 Ladder attached to scaffold

- When climbing up ladders you must use both hands to grip the rungs. This will give you better protection if you should slip.
- The working position should be not less than five rungs from the top of the ladder.

All ladders and stepladders should be tested and examined on an annual basis. The results of the tests should be recorded. Ideally the item tested should be marked to show it has been tested. A competent person must carry out this test.

Stepladders

A lot of the rules that apply to ladders also apply to stepladders. However, specifically bear in mind the following.

- All four legs of a stepladder should rest firmly and squarely on the ground. They will do this provided that the floor or ground on which they stand is level and the steps themselves are not worn or damaged.
- Ensure the legs are fully opened.
- Check that the hinge is in good condition.
- Check that the rope or hinged bracket is of equal length and not frayed.
- When using the steps ensure that your knees are below the top of the steps.
- The top of the steps should not be used unless it is constructed as a platform.

Trestles

Some jobs cannot be done safely from a ladder or stepladder. In such cases, a working platform known as a 'trestle scaffold' should be used. This consists of two pairs of trestles or 'A' frames spanned by scaffolding boards, which then provide a simple working platform.

When erecting trestle scaffolds the following rules should be observed.

- As with ladders they should be erected on a firm, level base with the trestles fully opened.
- The platform must be at least 600mm wide.
- The platform should be no higher than two-thirds of the way up the trestle; this ensures that there is at least one-third of the trestle above the working surface.
- The scaffolding boards must be of equal length and thickness, and should not overhang the trestle by more than four times their own thickness, e.g. a 40mm board must not overhang by more than 160mm.

Fibreglass stepladder

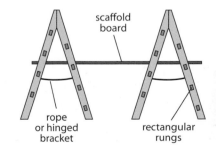

Figure 7.20 Trestles

- The trestles should be spaced at the following distances apart: 1 metre for 32mm thick boards; 1.5 metres for 38mm thick boards; and 2.5 metres for 50mm thick boards.

- If the platform is more than 2m above the ground, toe boards and guardrails must be fitted and a separate ladder provided for access.

- Trestles must not be used where anyone can fall more than 4.5m.

- Trestles over 3.5m tall should ideally be 'tied' to the structure of the building.

Mobile scaffold towers

Tower scaffolds are widely used and sadly are involved in numerous accidents each year. These usually happen because the tower has not been properly erected or used. Aluminium towers are light and can easily overturn. Towers rely on all the parts being in place to ensure adequate strength and can easily collapse if sections are left out.

Normally made from light aluminium tube, the tower is built by slotting the sections together until the required height is reached. Towers may be either mobile ones fitted with wheels or static ones fitted with plates. Consequently, they are useful for installing long runs – in, say, a factory lighting installation.

Mobile scaffold tower

When working with a mobile/static tower scaffold, the following points must be checked and followed.

- The person erecting the tower should be competent.

- Make sure the tower is resting on firm, level ground with the wheels or feet properly supported. Do not use bricks or building blocks to take the weight of any part of the tower.

- The taller the tower, the more likely it is to become unstable. As a guide, if towers are used in exposed conditions or outside, the height of the working platform should be no more than three times the minimum base dimension. If the tower is to be used inside, on firm level ground, the ratio may be extended to 3.5. Using this guide, if the base of the tower is 2m by 3m the maximum height would be 6m for use outside and 7m for inside.

- Before using the tower, always check that the scaffold is vertical and, in the case of a mobile tower, that all wheel brakes are on.

- There must be a safe way to get to and from the work platform. It is not safe to climb up the end frames of the tower except where the frame has an appropriately designed built-in ladder or a purpose-made ladder can be attached safely on the inside.

- Provide suitable edge protection on platforms where a person could fall more than 2m. Guard rails should be at least 910mm high and toe boards at least 150mm high. An intermediate guard rail or suitable alternative should be provided so that the unprotected gap does not exceed 470mm.

- When moving a tower, check that there are no power lines or other overhead obstructions; check that the ground is firm and level; push or pull only from the base; never move it while there are people or materials on the upper platforms; and never move it in windy conditions.

- Outriggers can increase stability by effectively increasing the area of the base, but if used they must be fitted diagonally across all four corners of the tower and not on one side only. When outriggers are used they should be clearly marked (e.g. with hazard marking tape) to indicate that a trip hazard is present.

- When towers are used in public places, extra precautions may be needed such as minimising the storage of materials and equipment on the working platform, erecting barriers at ground level to prevent people from walking into the tower or work area, and removing or boarding over access ladders to prevent unauthorised access if the tower is to remain in position unattended.

- Before you use a tower on a pavement, check whether you need a licence from the local authority.

- Tower scaffolds must be inspected by a 'competent person' before first use and following substantial alteration or any event likely to have affected their stability. If a tower remains erected in the same place for more than seven days, it should also be inspected at regular intervals (not exceeding seven days) and a written report should be made. Any faults found should be put right.

Safety tip

It is not safe to climb up the end frames of a mobile scaffold tower

Scissor and boom lifts

This category of access equipment is sometimes referred to as 'mobile elevating platforms'.

From a safety perspective, as well as with regard to cost, scissor lifts can offer quick and efficient access solutions for a wide range of installation and maintenance tasks. Compact dimensions and tight turning circles give these machines great versatility.

There is a wide range of machines available for flat-slab and rough-terrain applications, thus enabling them to be used in circumstances ranging from installations within large factories to repairs of external lighting.

However, scissor lifts can only extend upwards. Sometimes it is necessary to 'reach' over objects to be able to carry out the work, for example repairing a street-lighting column. In these circumstances a telescopic boom platform is likely to be more appropriate.

In both types of lift, it is essential that the workers wear a safety harness. This must be attached to the lift and never to the structure being worked on.

Particular attention must also be paid to whether overhead power supplies are present.

Scissor lift

Boom lift

Manual handling

The Manual Handling Operations Regulations 1992, as amended in 2002, apply to a wide range of manual handling activities, including lifting, lowering, pushing, pulling or carrying. The load may be either inanimate, such as a box or a trolley, or animate, such as a person or an animal.

Within the context of the electrical industry, manual handling can involve items such as scaffolding, tools, equipment, switchgear and motors.

Employer requirements	Employee responsibilities
Reduce the need for hazardous manual handling, so far as is reasonably practicable	Follow appropriate systems of work laid down for employee safety
Assess the risk of injury from any hazardous manual handling that cannot be avoided	Make proper use of equipment provided for employee safety
Reduce the risk of injury from hazardous manual handling, so far as is reasonably practicable	Co-operate with the employer on health and safety matters
	Inform the employer if you identify hazardous handling activities
	Ensure that your activities do not put others at risk

Table 7.10 Employer duty and employee responsibility regarding manual handling

Where possible, avoid manual handling. Check whether you need to move the item at all. For example, does a large work piece really need to be moved, or can the activity be done safely where the item already is? Think about using handling aids such as a conveyor, a pallet truck, an electric or hand-powered hoist or fork-lift truck.

However, beware of new hazards from automation or mechanisation. For example, automated plant still needs cleaning, maintaining etc., and fork-lift trucks must be suited to the work and have properly trained and certified operators.

The movement of loads requires careful planning to identify potential hazards before they cause injuries. This planning involves carrying out a risk assessment, which looks at a number of areas.

The task	Does the task involve: • holding loads away from the body? • twisting, stooping or reaching upwards? • large vertical movement? • long carrying distances? • strenuous pushing or pulling? • repetitive handling? • insufficient rest or recovery time? • a work rate imposed by a process?
The load	Is the load: • heavy? • bulky, unwieldy or difficult to grasp? • unstable or are the contents likely to shift? • sharp, hot or otherwise potentially damaging?
The working environment	Does the working environment have: • space constraints? • floors that are slippery or unstable? • poor lighting? • hot, cold or humid conditions?
Individual capability	Does the individual: • have a reach problem, that restricts their physical capability? • have knowledge of and training in manual handling?
Handling aids and equipment	• Is the device the correct type for the job? • Is it well maintained? • Are the wheels on the device suited to the floor surface? • Do the wheels run freely? • Is the handle height between the waist and shoulders? • Are the handle grips in good order and comfortable? • Are there any brakes? If so, do they work?

Table 7.11 Risk assessment of a manual handling operation

Lifting and carrying

Here are some practical tips for safe manual handling.

Think before lifting/handling, plan the lift

- Can handling aids be used?

- Where is the load going to be placed?

- Will help be needed with the load?

- Remove obstructions such as discarded wrapping materials.

- For a long lift, consider resting the load midway on a table or bench to change grip.

Keep the load close to the waist

- Keep the load close to the body for as long as possible while lifting.

- Keep the heaviest side of the load next to the body.

- If a close approach to the load is not possible, try to slide it towards the body before attempting to lift it.

Adopt a stable position

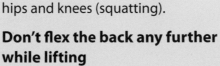

- The feet should be apart with one leg slightly forward to maintain balance (alongside the load, if it is on the ground).

- Test the weight of the load by pushing it with the foot.

- The worker should be prepared to move his or her feet during the lift to maintain stability.

- Avoid tight clothing or unsuitable footwear, which may make this difficult.

Get a good hold

Where possible the load should be hugged as close as possible to the body. This may be better than gripping it tightly with hands only.

Start in a good posture

At the start of the lift, slight bending of the back, hips and knees is preferable to fully flexing the back (stooping) or fully flexing the hips and knees (squatting).

Don't flex the back any further while lifting

- This can happen if the legs begin to straighten before starting to raise the load.

- Avoid twisting the back or leaning sideways, especially while the back is bent.

- Shoulders should be kept level and facing in the same direction as the hips.

- Turning by moving the feet is better than twisting and lifting at the same time.

- Keep the head up when handling.

- Look ahead, not down at the load, once it has been held securely.

Move smoothly

The load should not be jerked or snatched as this can make it harder to keep control and can increase the risk of injury.

Don't lift or handle more than can easily be managed

There is a difference between what people can lift and what they can *safely* lift. If in doubt, seek advice or get help.

Put down, then adjust

If precise positioning of the load is necessary, put it down first, then slide it into the desired position.

Pushing and pulling

Here are some practical points to remember when pushing and pulling loads.

Handling devices

- Aids such as trolleys should have handle heights that are between the shoulder and waist.

- Devices should be well-maintained with wheels that run smoothly (the law requires that equipment is maintained).

- When purchasing new trolleys etc. ensure they are of good quality with large-diameter wheels made of suitable material and with castors, bearings etc. that will last with minimum maintenance.

- Consultation with your employees and safety representatives will help, as they know what works and what doesn't.

Force

As a rough guide, the amount of force that needs to be applied to move a load over a flat, level surface using a well-maintained handling aid is at least 2 per cent of the load weight. You should try to push rather than pull when moving a load, provided you can see over it and control steering and stopping.

Slopes

You should enlist help whenever necessary if you have to negotiate a slope or ramp, as pushing and pulling forces can be very high. For example, if a load of 400kg is moved up a slope of 1 in 12 (about 5°), the required force is over 30kg even in ideal conditions – with good wheels and a smooth slope. This is above the guideline weight for men and well above the guideline weight for women.

Uneven surfaces

On an uneven surface, the force needed to start the load moving could increase to 10 per cent of the load weight, although this might be offset to some extent by using larger wheels. Soft ground may be even worse.

Stance and pace

To make it easier to push or pull, you should keep your feet well away from the load and go no faster than walking speed. This will stop you becoming too tired too quickly.

General risk assessment guidelines

There is no such thing as a completely 'safe' manual handling operation. But working within the guidelines shown in Figure 7.21 below will cut the risk.

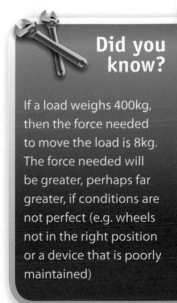

Did you know?

If a load weighs 400kg, then the force needed to move the load is 8kg. The force needed will be greater, perhaps far greater, if conditions are not perfect (e.g. wheels not in the right position or a device that is poorly maintained)

Remember

Moving an object over soft or uneven surfaces requires higher forces

Use the drawing to make a quick and easy assessment. Each box contains a guideline weight for lifting and lowering in that zone. (As you can see, the guideline weights are reduced if handling is done with arms extended, or at high or low levels, as that is when injuries are most likely to occur.)

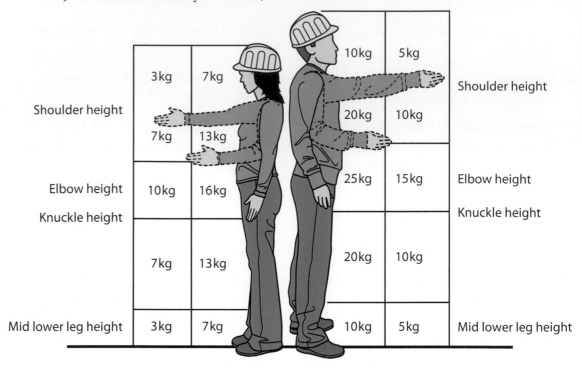

Figure 7.21 Lifting chart for men and women

Observe the work activity you are assessing and compare it to the diagram. First, decide which box or boxes the lifter's hands pass through when moving the load. Then, assess the maximum weight being handled. If it is less than the figure given in the box, the operation is within the guidelines.

If the lifter's hands enter more than one box during the operation, use the smallest weight. Use an in-between weight if the hands are close to a boundary between boxes.

The guideline weights assume that the load is readily grasped with both hands and that the operation takes place in reasonable working conditions, with the lifter in a stable body position.

Good housekeeping

On completion of this topic area the candidate will be able to identify the need for good housekeeping during and at the end of the installation.

Clean and tidy work area

Good housekeeping is essential if a project is to be completed cost effectively, safely, on time and within budget. Materials and equipment left lying around are liable to cause or lead to:

- damage
- theft
- accidents
- fire
- delay to the project
- vermin attraction.

Materials that are damaged will need to be replaced. As the customer will only pay once for the materials, this means that any and all damage after the initial purchase of material will need to be paid for by the contractor. This increased expenditure naturally reduces the profit level for the job.

Any stolen materials or equipment will lead to insurance claims being made in order to recover the value of the lost items. Frequent claims by a firm or individual could lead to difficulty in maintaining effective insurance cover.

Trip hazards can be caused by any items left 'lying around' a site. This could result in injury to anyone in the area. If you leave equipment in a dangerous place, both you and the company are liable for prosecution, both in the civil court and the criminal court. In order to make the site secure outside of working hours, it is necessary to lock away any potentially dangerous tools and equipment – including items like mobile scaffolds and step ladders – to avoid any hazards.

Similarly, if packaging is not disposed of correctly, it can also become a fire hazard. This is especially the case when it is added to the rest of the waste material that can be generated on site, such as waste paper, oily rags and cardboard boxes.

Vermin, such as rats and mice, are always attracted to food. To avoid having an infestation of these animals, be careful when disgarding food or, even better, do not eat on site at all.

All of the above can lead to a delay in the project. Any delay may see the use of a penalty clause by the client, which in turn can lead to the withdrawal of the contractor from preferential contractor lists and a poor reputation, affecting the level of work the contractor is commissioned to do.

So remember the following.

- Always keep the work area clear and tidy – any tools or equipment not being used at the time should either be kept to one side or within a safe lockable area.

- At the end of the day, or when going for meal breaks, always ensure that tools and equipment are safely locked away.

Remember

If a contractor finds it difficult to secure insurance, it may delay the start time for any work

Did you know?

The civil court will award compensation. The criminal court will prosecute any offences under the Health and Safety at Work Act

- Always dispose of waste materials in the skips or bins provided on site, or make sure that all waste materials are removed to the company van or any other such provision made for their removal.

- Never consume food in the work area.

Environmental issues

Although the majority of electrical installation has little direct environmental impact, major construction works can include the installation of equipment such as transformers and capacitors that contain cooling oils.

Storage and handling of materials

In England, if you store oil (such as petrol, diesel, vegetable, synthetic or mineral oil) in a container with a storage capacity of more than 200 litres (44 gallons), then you may need to comply with the Control of Pollution (Oil Storage) (England) Regulations 2001.

A **Safety Data Sheet** must accompany any material supplied to you that has potentially hazardous properties. The Safety Data Sheet gives information on how chemicals should be handled, stored and disposed of. If a Safety Data Sheet does not accompany the delivery, contact the supplier and ask for it. Suppliers who fail to provide adequate information for the safe use of their products are in breach of the law.

Always clean up spills immediately. Ensure that you have absorbent materials suitable for the type and quantity of fuel, oil and chemicals that you will store and use on site. Keep absorbent materials close to the storage area in a labelled weatherproof box or plastic bin to keep them in good condition. Depending on the type of spill involved, there are specialist spill kits available.

Water

If you discharge any effluents into a public sewer you must have either prior written authorisation from the Statutory Sewerage Undertaker in the form of a trade effluent consent or have entered into a **Trade Effluent Agreement** with them. In most cases the Statutory Sewerage Undertaker will be your local water company (in England and Wales), or Scottish Water (in Scotland) and the Water Service (in Northern Ireland).

If you discharge any sewage, effluent or contaminated surface water to surface waters or groundwater you must have prior written authorisation from your Environmental Regulator in the form of a **discharge consent**.

Air

The Clean Air Act 1993 applies to all small businesses that burn fuels in furnaces or boilers, or that burn material in the open, including farms, building sites and demolition sites.

You must prevent the emission of dark smoke from any chimney on your premises. This includes chimneys serving furnaces, fixed boilers or industrial plant whether they are attached to buildings or not. There are some instances where dark smoke may be emitted without an offence being committed, such as during start-up conditions, if all practicable steps have been taken to prevent or minimise the emissions.

You must prevent the emission of dark smoke from any industrial or trade premises, e.g. burning tyres and cables. This applies to burning materials on a site that you own or a site where you are working such as a building or demolition site or land used for commercial agriculture or horticulture.

In England, Wales and Scotland it is not necessary for local authorities to have witnessed the emissions of dark smoke for them to take action against you: evidence of the burning of materials that potentially give rise to dark smoke is sufficient (in this way the law aims to stop people creating dark smoke at night and using the lack of visual evidence as a defence!). This does not apply in Northern Ireland.

Burning tyres is against the law

You must inform your local authority before installing a furnace or a fixed boiler. Any new furnaces or boilers must be able to operate continuously without emitting smoke when burning the type of fuel for which they have been designed. Obtaining a planning permission/building warrant from the local authority for the construction of the chimney or plant is not sufficient: you need the local authority's specific approval under the Clean Air Act or the Clean Air (Northern Ireland) Order 1981.

Under the Sulphur Content of Liquid Fuels Regulations, you must not use gas oils with a sulphur content higher than 0.2 per cent by weight (this will be reduced further to 0.1 per cent from January 2008). For heavy fuel oils, you must not use heavy fuel oils with a sulphur content higher than 1 per cent sulphur by weight (unless you hold an exemption).

The burning of waste will almost always create smoke (dark or otherwise) and can as a consequence cause nuisance to people in the locality. Combustion under uncontrolled circumstances at low temperature (in comparison to the temperatures in a waste incineration plant) may lead to the release of noxious gases and/or dust and grit in the area.

Residues of harmful chemicals (such as lead paints, tars and oils) that are left in the ashes can be washed into the ground by rain, leading to lasting contamination of the soil or groundwater. Contaminants may also be washed into surface water.

Did you know?

In England, Scotland and Wales, if your plant emits excessive (i.e. amounting to a nuisance) levels of grit and dust, local authorities can place limits on these emissions; exceeding the limits may be an offence. This is likely to apply to old plant with a history of complaints or plant run with fuel or procedures it was not designed for

Remember

In general the burning of waste in the open is an environmentally unsound practice, and you should use less damaging options of waste disposal

Noise and statutory nuisance

Part III of the Environmental Protection Act 1990 (as amended) contains the main legislation relating to statutory nuisance. It applies in England, Wales and Scotland and is enforced by the local authorities; under Part III of the Pollution Control and Local Government (Northern Ireland) Order 1978, district councils have powers to deal with noise nuisance. Your local authority also operates a system of Consents and Prohibition Notices, which could affect your site.

In England, Scotland and Wales this system is set out in the Control of Pollution Act 1974: s.60 on Prohibition Notices and s.61 on Consents – A Code of Practice. The Noise Act 1996 also applies to domestic dwellings and therefore has an impact on installations, as does the Noise and Statutory Nuisance Act 1993.

Although there is no legal definition of a statutory nuisance, for action to be taken the nuisance complained of must be, or be likely to be, prejudicial to people's health or interfere with a person's legitimate use and enjoyment of land. This particularly applies to nuisance to neighbours in their homes and gardens.

Good practices is as follows.

- Establish whether your business might cause a nuisance to neighbours by checking noise, odours and other emissions near the boundary of your site during different operating conditions and at different times of the day. Take all reasonable steps to prevent or minimise a nuisance or a potential nuisance.

- Even if a complaint does not amount to a statutory nuisance you should consider if there are simple, practical things that you can do to keep the peace.

- Try to establish a good relationship with your neighbours, particularly in relation to transient events likely to affect them. Advise neighbours in advance if you believe that a particular operation, such as building work or an installation process for new plant, could cause an adverse effect. If neighbours are kept informed they will perceive the business as more considerate and are less likely to make a complaint.

- Make sure there is a good level of 'housekeeping' on your site and that your site manager and staff are aware of the need to avoid nuisances. Regularly check your site for any waste accumulations, evidence of vermin, noise or smell as applicable.

- Avoid or minimise noisy activities, especially at night; pay particular attention to traffic movements, reversing sirens, deliveries, external public address systems and radios.

- Where practical, schedule or restrict noisy activities to the normal working day (for example 0800 to 1800, Monday to Friday and 0800 to 1300 on Saturday).

- Consider where noisy operations are undertaken in relation to site boundaries and relocate them if you can, perhaps further away, or make use of existing buildings/ stockpiles/topography as noise barriers.

Remember

A statutory nuisance could arise from the poor state of your premises or any noise, smoke, fumes, gases, dust, steam, smell, effluvia, the keeping of animals, deposits and accumulations of refuse and/or other material

- Use good practice to minimise noise escaping from your buildings by, for example, keeping doors and windows closed.

- At noise-sensitive locations, undertake monitoring of background noise before your works begin.

- At noise-sensitive locations, note any actions you can take to reduce noise levels.

- Reduce noise levels outside your buildings by increasing insulation to the building fabric and keeping doors and windows closed.

- Ensure that any burglar alarms on your premises have a maintenance contract and a callout agreement.

- Consider replacing any noisy equipment and take account of noise emissions when buying new or replacement equipment. Maintain fans and refrigeration equipment.

- Do not have any bonfires; find other ways to re-use or recover wastes (see the *Clean Air and Waste Management Guidelines*).

- Keep abatement equipment, such as filters and cyclones, in good working order. Ensure boilers, especially oil or solid-fuel units, are operating efficiently and do not emit dark smoke during start up.

Pollution prevention and control

Within the UK, the new Pollution Prevention and Control (PPC) regime implements the EU's Integrated Pollution Prevention and Control (IPPC) Directive. The UK implementing legislation for IPPC (collectively referred to as the PPC Regulations) is the Pollution Prevention and Control (England and Wales) Regulations 2000.

PPC Part A (A(1) and A(2) in England and Wales) includes new issues not previously covered by IPPC such as:

- energy efficiency
- waste minimisation
- vibration
- noise.

The new regime also requires an effective system of management to be implemented to ensure that all appropriate pollution prevention and control measures are taken. Special emphasis is placed on the application of Best Available Techniques (BAT) to reduce the environmental impact of the process.

The Dangerous Substances and Preparations (Safety) and Chemicals (Hazard Information and Packaging for Supply) Regulations 2002 concern the marketing and use of substances and preparations that contain carcinogens, **mutagens** and substances toxic to reproduction (CMRs), and the subsequent restriction of their supply to the general public.

Remember

Where appropriate use mains-generated electricity in preference to diesel generators. This will help you to reduce noise levels and to reduce the risk of pollution through fuel spillage

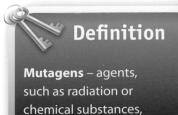

Definition

Mutagens – agents, such as radiation or chemical substances, which cause genetic mutation

Marking and labelling

On the completion of any installation, BS 7671 requires that a legible diagram, chart, table or equivalent form of information shall be provided from the contractor to the client. This document should contain the:

- type of circuit, for example, lighting or socket outlets
- location of the circuit
- number of points served
- size of the conductors used
- size of CPC used
- type of wiring used
- type of protective device used
- size of the protective device used
- method of protection used against indirect contact
- location of the distribution board.

Other notices and labels required by the Regulations are as follows.

1. At the origin of every installation as shown in Figure 7.22.

IMPORTANT
This installation should be periodically inspected and tested and a report on its condition obtained as prescribed in IEE Wiring Regulations BS 7671 Requirements for Electrical Installations.

Date of last inspection : ....

Recommended date of next inspection : ....

Figure 7.22 Label for origin of installation

2. Where different voltages are present in:

- equipment or enclosures within which a voltage exceeding 250 volts exists but where such a voltage would not be expected
- terminals between which a voltage exceeding 250 volts exists, which although contained in separate enclosures are within arm's reach of the same person
- the means of access to all live parts of switchgear or other live parts where different nominal voltages exist.

3. Earthing and bonding connections: a label with the words shown in Figure 7.23 shall be permanently fixed in a visible position at or near the following points:

- the point of connection of every earthing conductor to an earth electrode

> Safety Electrical Connection
> – Do Not Remove

Figure 7.23 Label for earthing and bonding connections

- the point of connection of every bonding conductor to an extraneous conductive part

- the main earth terminal of the installation where separate from the main switchgear.

4. **Residual Current Devices**: where RCDs are fitted within an installation, a suitable permanent durable notice marked in legible type no smaller than the example shown in BS 7671 shall be permanently fixed in a prominent position at or near the main distribution board. The notice shall contain the following words:

This installation, or part of it, is protected by a device that automatically switches off the supply if an earth-fault develops. Test quarterly by pressing the button marked 'T' or 'Test'. The device should switch off the supply and should then be switched on to restore the supply. If the device does not switch off the supply when the button is pressed, seek expert advice.

Caravan installations: all touring (mobile) caravans shall have a notice fixed near the main switch giving instructions on the connection and disconnection of the caravan installation to the electricity supply. Details of the wording required on the notice are given in full in BS 7671.

5. A warning notice should be supplied if alterations or additions have been made to an installation which now results in the installation having a mix of both the old colours and new colours of cables.

Caution

This installation has wiring colours to two versions of BS 7671. Great care should be taken before undertaking extension, alteration or repair that all conductors are correctly indentified.

Figure 7.24 Warning notice where wiring colours are mixed

This notice should be fixed at or near the appropriate distribution board.

Record finished work

Finally, on completion of the electrical installation an electrical installation certificate needs to be issued to the customer (the person ordering the work). This document confirms that the installation has been designed, installed and tested in compliance with BS 7671, and that it is safe for the installation to be used. The electrical installation certificate needs to be accompanied by both the schedule of inspections and the schedule of test results.

In addition to this, the customer will need to be provided with distribution charts as fitted drawings. These will indicate exactly what has been installed and where. You will also need to supply any manufacturer's literature and operation manuals that may be appropriate.

All of these documents should be contained within the O&M (operation and maintenance) manual and should be passed over to the customer at the handover.

The contractor must accompany the client around the installation to show that the installation works and that it complies with the original job specification. They will need to answer any questions that the customer may have regarding the functioning of the installation.

Activity

Assemble your hand and power tools and check them all to see if they are in good condition, fit for the purpose and properly maintained.

FAQ

Q   I've got insulated tools. Does this mean I don't have to isolate the supply?

A   No it doesn't. Using an insulated tool on a live supply may protect you against electric shock but it won't stop arcing and heat being generated. People have sustained very serious (sometimes fatal) burns as a result of electrical arcing. Don't risk it – always isolate the supply before working.

Q   Does metal capping provide mechanical protection for an installed plaster-depth cable?

A   No. Metal capping assists installation of cables and provides limited protection during installation. Plaster-depth cables should be run horizontally or vertically from an accessory. If not, they should be in earthed metallic conduit or buried to a depth of at least 50mm below the surface, or protected by an rcd ≤ 30mA.

Q   Does the 'horizontal and vertical rule' apply to cooker control units and their outlets?

A   Yes. Just because it's a cooker cable doesn't mean you can run it unprotected plaster-depth diagonally down the wall.

Knowledge check

1. Which type of pliers are most suitable for the following tasks.

   - Cutting cable to length.

   - Stripping-off insulation from conductors.

   - Tightening a conduit lock ring.

   - Forming a loop in a small conductor.

2. State the most likely type of cable to be installed using the following methods.

   - Plaster-depth in a domestic wall.

   - Installed in conduit or trunking.

   - Used for meter tails.

   - To connect a pump to its outlet.

3. Explain with the aid of a diagram, the terms 'countersink', 'pilot hole' and 'clearance hole' in relation to using fixings.

4. You have only a box of nuts and bolts to mount a motor to its base. Explain a method of providing a 'vibration-proof' fixing.

5. State the recommended horizontal and vertical clipping distances for an armoured cable of 17mm overall diameter.

6. What are the checks you should make before using a ladder?

7. Why would you choose to use trestles instead of a ladder?

8. Describe the process of lifting a heavy object.

Safe systems of working

chapter 8

Unit 3 Outcome 1

Health and safety in the workplace relies on having competent employees who are able to apply the principles of safe working to their place of work. It is the responsibility of employers to provide instructions, information and control measures to ensure safe systems of working. In turn, it is the responsibility of employees to follow any guidelines laid down by employers for safe working.

We looked at the principles of safe working earlier in this book. In this chapter we will see how technical and health and safety knowledge can be applied to the workplace, through the use of best practice.

On completion of this chapter the candidate will be able to:

- state the health and safety risks, precautions and procedures associated with tasks in the workplace
- list the five main stages of a risk assessment
- list the common categories of risk
- list the common options for risk control
- state the necessity to develop positive personal attitudes to safety
- list safety procedures to prevent accident and injury
- state methods of establishing safe access and exit from a site
- state the need for keeping tools and equipment secure
- identify emergency switches, isolators, alarms and emergency equipment
- state the need to ensure the workplace is cleaned and tidied on the completion of work
- state the rules for manual handling and lifting
- state correct procedures for disposing of waste.

Risks, precautions and procedures

On completion of this topic area the candidate will be able to state the health and safety risks, precautions and procedures associated with tasks in the workplace.

Risks

How do we prevent accidents? Understanding where accidents come from is an important step to accident prevention. There are two main causes of accidents: human and environmental.

Human causes

Accidents are covered in chapters 1 and 7. Here we will define which of these forms of accident are 'human' and which are 'environmental'. Human causes are accidents linked to the behaviour and actions of individuals. It would be nice to think that we all have a high level of common sense, but some very avoidable accidents still happen. Examples of human causes of accidents are:

- carelessness
- bad and foolish behaviour
- improper dress
- lack of training
- lack of experience
- poor supervision
- fatigue
- use of alcohol or drugs.

Environmental causes

Environmental causes relate to the environment that you are working in. Such causes can be unguarded machinery, defective tools and equipment, poor ventilation, excessive noise, poorly lit workplaces, overcrowded workplaces and untidy or dirty workplaces.

Precautions

The best way to reduce the risk of accidents is to try and remove the cause. If substances, such as oil, grease, cutting compounds, paints and solvents are spilled on the floor and not dealt with immediately, they can cause hazards. Even things like food, off-cuts from conduit, cables and tools are dangerous, if left underfoot. Clearly employees have some responsibility, both to themselves and others, to keep the workplace hazard free.

Remember

People cause accidents!

Safety tip

Hazards that cannot be removed should be guarded against: safety guards and fences can be put on or around machines, safe systems of work introduced, safety goggles, helmets and shoes issued. Your employer should provide any additional PPE required, such as ear defenders, respirators, eye protection and overalls

Where necessary the workplace should be tidy, with clearly defined passageways, good lighting, ventilation, reduced noise levels and non-slip floorings. Locking dangerous substances away into approved areas will remove another potential hazard, so that people will not be tempted to interfere with them, causing danger to themselves and others. Conduit, metal trunking and cable tray should be stored horizontally to minimise the hazard of their falling.

Health and safety in the workplace is something we each need to take personal responsibility for and, by thinking about what we are doing, or are about to do, we can avoid most potentially dangerous situations.

Equally, we can demonstrate a responsible approach by reporting potentially dangerous situations, hazards or activity to the correct people. Even if the hazard is something you can easily fix yourself still report it to your supervisor.

The dangers from health and safety risks can be prevented by following certain precautions:

- regularly carrying out risk assessments, and following their recommendations
- by conducting yourself with positive personal attitudes, acting and working responsibly to minimise risks.

Procedures

You may also have a role to play in other circumstances. For example, at present premises are inspected by local Fire Authorities. If you are aware of something unsafe, such as a fire exit that has been blocked, then tell them about it. To combat hazardous situations it is important to conduct thorough risk assessments of the workplace, and to follow the recommendations and advice that these give. Safety in the workplace generally falls into three main categories:

- personal protection, through PPE
- safety from electricity and electric shock
- emergency procedures for dealing with accidents – such as the ABC of recovery.

We covered the principles of these in chapter 1.

The secret is to be aware of all possible danger in the workplace and have a positive attitude towards health and safety. Follow safe and approved procedures where they exist, and always act in a responsible way to protect yourself and others. That way the construction site can be a happy and safe working environment for many years to come.

Remember

The first person to alert about a health and safety issue should be your Site Supervisor or Safety Officer

Safety tip

If the hazard is something that can be easily fixed, and as long as it is not dangerous to yourself, then you should fix it

Risk assessment

On completion of this topic area the candidate will know how to carry out and prepare a report identifying potential health hazards.

Why risk assessments?

Accidents and ill health can ruin lives, and affect business too if output is lost, machinery is damaged, insurance costs increase or employers have to go to court. Employers are, therefore, legally required to assess the risks in their workplace.

A risk assessment is nothing more than a careful examination of what, during working activities, could cause harm to people, so that an employer can weigh up whether it has taken enough precautions or should do more to prevent harm. The aim is to make sure that no one gets hurt or becomes ill.

The important things to ask yourself are:

- what are the hazards?

- are they significant?

- are hazards covered by satisfactory precautions so that the risk is small?

Employers need to check this when assessing risks. For example, we know that electricity is a hazard that can kill, but the risk of it doing so in a tidy, well-run office environment is remote, provided that the installation is sound, 'live' components are insulated and metal casings are properly earthed.

What is a hazard and what is a risk?

A hazard means anything that can cause harm (e.g. chemicals, asbestos, electricity, working from ladders or scaffolding etc.).

A risk is the chance, high or low, that somebody will be harmed by the hazard.

How to assess risks in the workplace

Some years ago the HSE produced guidance for employers to help with the process. Known as the *5 Steps to Risk Assessment*, copies of which can be accessed through the HSE website, it is an invaluable tool for grasping the essentials of risk assessment. It comprises the following.

- **Step 1:** Look for the hazard.

- **Step 2:** Decide who might be harmed and how.

- **Step 3:** Evaluate the risks and decide whether the existing precautions are adequate or whether more should be done.

- **Step 4:** Record your findings.

- **Step 5:** Review your assessment and revise it if necessary.

To help understand the concepts, when you read on try to put yourself in the position of an employer carrying out a risk assessment. We have worded it in that way.

Step 1: Look for the hazard

When you are doing a risk assessment, walk around the workplace. Look at what could reasonably be expected to cause harm, ignoring the trivial and concentrating on significant hazards that could result in serious harm or affect several people. Some typical examples are slippery, uneven surfaces, spillages, scrap or waste materials lying around, flammable materials, faulty or missing machine guards, electricity or faulty electrical connections damaged cables, materials that are ejected from machines, pressure systems (steam boilers), unshielded processes such as arc welding, chemicals, dust and fumes, manual handling and movement of loads and working from heights. Check with your employees as they may have noticed things that are not apparent. Manufacturers' instructions or datasheets will also identify hazards can put the risks in perspective. Another aid can be accident and ill health records.

Step 2: Decide who might be harmed and how

Don't forget categories such as:

- young workers, apprentices and trainees, new and expectant mothers etc., who may all be at particular risk
- cleaners, visitors, contractors, maintenance workers etc. who may not be in the workplace all the time
- members of the public, or people you share your workplace with, if there is a chance they could be hurt by your activities.

Step 3: Evaluate the risks

Evaluate the risks and decide whether existing precautions are adequate or more should be done. Consider how likely it is that each hazard could cause harm. This will determine whether or not more needs to be done to reduce the risk. Even after all precautions have been taken, some risk usually remains. Decide for each hazard whether this remaining risk is high, medium or low. First, ask yourself whether you have done all the things that the law says you have to do. For example, there are legal requirements on prevention of access to dangerous parts of machinery. Then ask yourself whether generally accepted industry standards are in place.

Your real aim is to make all risks small by adding precautions as necessary. If something needs to be done, draw up an 'action plan' and give priority to any remaining risks which are high and/or those which could affect most people. In taking action, ask yourself these questions.

- Can I get rid of the hazard altogether?
- If not, how can I control the risks so that harm is unlikely?

In controlling risks apply the principles below, where possible in the following order.

- Try a less risky option.
- Prevent access to the hazard (e.g. by guarding and using barriers and notices).
- Organize work to reduce exposure to the hazard.
- Provide welfare facilities (e.g. washing facilities for removal of contamination; first aid).

Improving health and safety need not cost a lot. For instance, placing a mirror on a dangerous blind corner to help prevent vehicle accidents, or putting some non-slip material on slippery steps, are inexpensive precautions considering the risks. Failure to take simple precautions can cost you a lot more if an accident happens.

If the work you do tends to vary a lot, or you or your employees move from one site to another, identify the hazards you reasonably expect and assess the risks from them.

Step 4: Record your findings

If you have fewer than five employees, you do not need to write anything down, though it is sensible and useful to keep a written record of what you have done. But if you employ five or more people you must record the significant findings of your assessment.

This means writing down the significant hazards and conclusions. You must also tell your employees about your findings.

Risk assessments must be suitable and sufficient. You need to be able to show the following.

- A proper check was made.
- You asked who might be affected.

- You dealt with all the obvious significant hazards, taking into account the number of people who could be involved.
- The precautions are reasonable, and the remaining risk is low.

Keep the written record for future reference or use; it can help you if an inspector asks what precautions you have taken, or if you have become involved in any action for civil liability.

It can also remind you to keep an eye on particular hazards and precautions and it helps to show that you have done what the law requires.

Step 5: Review your assessment

Review your assessment and revise it if necessary. Sooner or later you will bring in new machines, substances or procedures that could lead to new hazards. If there is any significant change, add to the risk assessment to take account of the new hazard. Don't amend your assessment for each trivial change, or still more, for each new job.

But if a new job introduces significant hazards of its own, you will want to consider them in their own right and do whatever you need to keep the risks down. In any case, it is good practice to review your assessment from time to time to make sure that the precautions are still working effectively.

Pages 204 and 205 show two examples of forms from *5 Steps to Risk Assessment*.

STEP 1

Hazard

Look only for hazards which you could reasonably expect to result in significant harm under the conditions in your workplace. Use the following examples as a guide:

- slipping/tripping hazards (poorly maintained floors or stairs)
- fire (from flammable materials)
- chemicals (battery acid)
- moving parts of machinery (blades)
- work at height (from mezzanine floors)
- ejection of material (from plastic moulding)
- pressure systems (steam boilers)
- vehicles (fork-lift trucks)
- electricity (poor wiring)
- dust (from grinding)
- fumes (welding)
- manual handling
- noise
- poor lighting
- low temperature.

STEP 2

Who might be harmed?

There is no need to list individuals by name – just think about groups of people doing similar work or who may be affected:

- office staff
- maintenance personnel
- contractors
- people sharing your workplace
- operators
- cleaners
- members of the public.

Pay particular attention to:

- staff with disabilities
- visitors
- inexperienced staff
- lone workers.

They may be more vulnerable.

STEP 3

Is more needed to control the risk?

For the hazards listed, do the precautions already taken:

- meet the standards set by a legal requirement?
- comply with a recognised industry standard?
- represent good practice?
- reduce risk as far as reasonably practicable?

Have you provided:

- adequate information, instruction or training?
- adequate systems or procedures?

If so, then the risks are adequately controlled, but you need to indicate the precautions you have in place. (You may refer to procedures, company rules, etc.)

Where the risk is not adequately controlled, indicate what more you need to do (the 'action list').

STEP 5

Review and revision

Set a date for a review of the assessment.

On review check that the precautions for each hazard still adequately control the risk. If not indicate the action needed. Note the outcome. If necessary complete a new page for your risk assessment.

Making changes in your workplace, e.g. when bringing in new
- machines
- substances
- procedures
may introduce significant new hazards. Look for them and follow the 5 steps.

(continued over)

RISK ASSESSMENT FOR

Company name

Company Address

Postcode

ASSESSMENT UNDERTAKEN

(date)

Signed

Date

ASSESSMENT REVIEW

Date

STEP 1

List significant hazards here:

STEP 2

List groups of people who are at risk from the significant hazards you have identified:

STEP 3

List existing controls or note where the information may be found. List risks which are not adequately controlled and the action needed:

Figure 8.01 Two sample forms from *5 Steps to Risk Assessment*

Common categories of risk

On completion of this topic area the candidate will be able to list the common categories of risk.

Risks at work

The risks from electrical work are many and varied. It is important to be familiar with the dangers from every possible method you might work in. This is the best way to be prepared for an accident and respond quickly and correctly to any incident.

There are several common categories of risk. These are:

- slips, trips and falls
- storage of goods and materials
- fire
- electricity
- mechanical handling.

Safety tip

Preparation can save lives!

Slips, trips and falls

Many of the dangers in this area come from working either at heights or from lifting and handling. We covered these factors in chapter 7.

Work in the construction industry is all about the movement of people and materials from one place to another. It is estimated that more than 25 per cent of all accidents in the workplace are due to manual handling or 'movement' type accidents. Therefore if we can eliminate or reduce the number of these types of accident, then we will also reduce the number of injured people!

Many accidents are caused by 'bad housekeeping'. This is basically a failure to keep the workplace tidy and free from hazards. While you are at work, here are some of the things you should look out for – we call it 'hazard spotting':

- fluids spilt on the floor
- nails sticking up from bits of timber
- polythene left lying around
- plastic strapping from around bricks
- 'traps' formed by unsupported scaffold boards
- broken glass
- trailing cables (this is one of the most common causes of trips)
- lack of safety rails
- projecting scaffold poles and other objects.

This list is not meant to be complete – there will certainly be more hazards you can think of.

Manual handling

Manual handling also includes the lifting, putting down, pushing, pulling, carrying or moving of any load by hand or bodily force. If this is not done in the correct manner then injury can result.

Before you lift a load you should always do the following.

- Assess the weight, size and shape of the load.
- How many people are needed to lift it?
- Look at the contents – are they hazardous or fluid?
- Is the load secure and is it the right way up?
- Has it got carrying handles and has it got sharp edges?
- Is the route free of obstructions?
- Will it fit through the doors and are the doors open?
- Have you got somewhere to put it down, either en route or when you get there?

Thinking about the above before you pick up the object can save you some embarrassment and could save you from serious injury.

Storage and handling of materials

We covered this in chapter 7.

Fire

How fire happens

In order to avoid the dangers from fire, you need to know how fires can start.

All matter exists in one of three states: solid, liquid or gas (vapour). The atoms or molecules of a solid are packed closely together, and those of a liquid are packed loosely. The molecules of a vapour are not really packed together at all and are free to move about.

In order for a substance to oxidise, its molecules must be well surrounded by oxygen molecules. The molecules of solids and liquids are packed too tightly for this to happen and therefore only vapours can burn.

When a solid or liquid is heated, its molecules move about rapidly. If enough heat is applied, some molecules break away from the surface to form a vapour just above the surface. This vapour can now mix with oxygen. If there is enough heat to raise the vapour to its ignition temperature and, if there is enough oxygen present, the vapour will oxidise rapidly and it will start to burn.

What we call burning is the rapid oxidation of millions of vapour molecules. The molecules oxidise by breaking apart into individual atoms and recombining with

Remember

A Safety Data Sheet must accompany any material supplied to you that has potentially hazardous properties

oxygen into new molecules. It is during the breaking-recombining process that energy is released as heat and light. The heat that is released is radiant heat, which is pure energy. It is the same sort of energy that the sun radiates and that we feel as heat. It radiates (travels) in all directions. Therefore, part of it moves back to the seat of the fire, to the 'burning' solid or liquid (the fuel). The heat that radiates back to the fuel is called radiation feedback.

Part of this heat releases more vapour, and part of it raises the vapour to the ignition temperature. At the same time, air is drawn into the area where the flames and vapour meet. The result is that there is an increase in flames as the newly formed vapour begins to burn.

Understanding how fire happens can lead to prevention

The fire triangle

The three things that are needed for combustion to take place are:

- fuel (to vaporise and burn)
- oxygen (to combine with fuel vapour)
- heat (to raise the temperature of the fuel vapour to its ignition temperature).

The fire triangle shows us that fire cannot exist without all three together.

- If any side of the fire triangle is missing, a fire cannot start.
- If any side of the fire triangle is removed, the fire will go out.

Classes of fire

Combustible and flammable fuels have been broken down into five categories.

- Class A fires are those involving organic solids such as paper or wood.
- Class B fires are those involving flammable liquids.
- Class C fires are those involving flammable gases.
- Class D fires are those involving metals.
- Class F fires are those involving cooking oils.

Figure 8.01 The fire triangle

General electric hazards

Before using any electrical equipment you are strongly advised to carry out a number of visual check.

- Check that the cable is not damaged or frayed.
- Check that the cable is properly secured at both ends and that none of the conductors are visible.

Did you know?

Electrical fires have no classification, as electricity is a source of ignition that will feed a fire until switched off or isolated

- Check that the plug is in good condition.
- If using an extension lead of the coiled type, uncoil all the cable before use, check it for serviceability and leave uncoiled during use. Recoil carefully after use.
- Check that any portable power tools are suitable for working from a reduced voltage (110 V or 55 V) and that a suitable transformer is supplied.
- Never carry power tools by their cables.
- Keep plugs and sockets clean and in good order.

When working with electricity, always make certain that the circuit is isolated from the supply by switching it off and removing the fuses before touching any of the conductors. Make sure that the circuit cannot be turned on while you are working on it. Use the company's Permit to Work or Safe Isolation Procedures.

Portable appliance testing (PAT)

As we have previously discussed, the Health and Safety at Work Act puts a duty of care on both employer and employee to ensure the safety of all persons using the work premises. However, the specific legal requirements regarding the use and maintenance of electrical equipment are contained in the Electricity at Work Regulations 1989 (EAWR).

The EAWR require all electrical systems to be maintained to prevent danger. This requirement covers all electrical equipment, including fixed and portable equipment. Employers must therefore maintain their electrical equipment to prevent accidents.

The majority of equipment defects can be found by a detailed visual inspection (after disconnecting the appliance). This is likely to eliminate hazards caused by damaged plugs, cuts in cable or equipment being used in the wrong circumstances. Other signs such as burn marks may show that the equipment's condition could create faults or render it a danger to the user.

However, a visual inspection alone is insufficient and will not identify all dangerous faults. Therefore a visual inspection needs to be linked to a testing programme carried out with an approved test instrument, to reveal less obvious electrical faults with earth continuity, insulation resistance and earth leakage.

There is no legal requirement to keep records regarding the frequency of testing. It would seem sensible to keep records, though, as this would allow employers to monitor PAT within their organisation. Guidance given by the Institute of Electrical Engineers (IEE) suggests the following frequencies for industry applications.

Equipment type	When	Class I		Class II	
		Visual only	Visual + test	Visual only	Visual + test
Stationary	Weekly	None	1 year	None	1 year
IT equipment	Weekly	None	1 year	None	1 year
Movable	Before use	1 month	1 year	3 months	1 year
Portable	Before use	1 month	6 months	3 months	6 months
Hand-held	Before use	1 month	6 months	3 months	6 months

Table 8.01 Recommended frequency of equipment testing of health and safety

Mechanical handling

To help overcome some of the risks at the workplace, there are a number of mechanical handling aids available to electricians to use on site. Some of these we have looked at elsewhere in this book. Mechanical handling aids can range from simple tools to more complex conveyers and lifting devices.

All of these devices are designed to 'lighten the load' for workers, and can help to alleviate some risks of injury.

However, all mechanical handling devices can have their own risks. They generally need maintenance, and it is very important to keep them in good condition. Some mechanical handling aids, such as trucks, carry a legal obligation for regular inspection and testing. Lifting machines are also tested on a regular basis.

These aids can be dangerous if used by those who do not have training – either to the user or to those nearby. It is therefore vital that, before using a mechanical handling aid, you get an introduction in how to use it.

Dealing with hazards

On completion of this topic area the candidate will be able to list common options for risk control, state the necessity for a positive personal attitude and list the safety procedures required to prevent accidents.

Risk control

It is important to remember the old saying 'prevention is better than cure'. This is as true for hazards in the workplace. It is far better to work towards preventing accidents and hazards, than only focusing on what to do after they have occurred. There are several common options for risk control.

- **Elimination** – it should be possible, after a risk assessment, to quickly identify and remove the hazard, thus completely ending the danger it presents.

- **Substitution** – replacing items or equipment with other, safer, alternatives is a simple way of controlling risks.

- **Enclosure** – encircling, or restricting access to, potentially hazardous areas automatically reduces the potential impact of the hazard by reducing the number of people who could come into contact with it.

- **Guarding** – out of working hours it is a good idea to have guards on site. This means that access to any potentially hazardous areas is controlled at all times.

- **Safe system of work** – putting together a detailed and logical system of work, focused on informing and protecting everyone working on the site, can be a major contributor to a safe working environment.

- **Supervision, training and information** – logically, the more training you receive, the more informed your decisions will be. This will lead to an increase in your knowledge and understanding of hazards and enable you to work more efficiently towards preventing accidents from happening.

- **Personal protective equipment (PPE)** – using the correct equipment to protect yourself during a job, can preserve our own health and safety.

Positive attitude

One of the best tools you can use for adopting safe systems of working is a positive personal attitude. It is always important to respond to changes and developments within the workplace with an openness and willingness to accept changes. As part of this attitude it is important to remember the need to:

- act and work responsibly and safely in order to protect both yourself, other people and the environment

- know the hazards that can occur, and be familiar with the protection available and the means of preventing accidents.

Safety procedures

There are a number of efficient safety procedures used to prevent accidents. Many of these were covered earlier in the book. We will briefly recap them here.

- Personal protection – this is through the use of PPE and safety guards.

- Electrical safety – checking and inspecting cables, leads and plugs. Care must be taken to ensure the earthing of equipment and the use of reduced voltage equipment when required.

- Emergency procedures – information must be widely communicated through the business concerning fire drills and the location and types of extinguisher (see pages 32–33). Everyone must be aware of evacuation procedures, assembly points and the routes that can safely be taken to leave buildings in emergencies.

However, on site there will be certain key substances you will work with that will require special precautions and procedures.

LPG (liquid petroleum gas)

Even small quantities of LPG when mixed with air create an explosive mixture (LPG is a gas above –42°C).

LPG is widely used in construction and building work as a fuel for burners, heaters and gas torches. The liquid, which comes in cylinders and containers, is highly flammable and needs careful handling and storage. Here are some guidelines.

- Everyone using LPG should understand the procedures to be adopted in case of an emergency.
- Appropriate fire-fighting extinguishers (dry powder) should always be available.
- Cylinders must be kept upright, whether in use or in storage.
- When not in use the valve should be closed and the protective dust cap should be in place.
- When handling cylinders do not drop them or allow them to come into violent contact with other cylinders.
- When using a cylinder with an appliance ensure it is connected properly, in accordance with the instructions you have been given, and that it is at a safe distance from the appliance or equipment that it is feeding.

It is essential to make sure the gas does not leak. LPG is heavier than air; if it leaks, it will not disperse in the air but will sink to the lowest point and form an explosive concentration that could be ignited by a spark. Leakages are especially dangerous in trenches and excavations because the gas cannot flow out of these areas. For this reason LPG cylinders should not be taken into them.

Chemical hazards

Certain chemicals can be harmful if they come into contact with your body by accident. When not contained or handled properly chemicals can be:

- inhaled as a dust or gas
- swallowed in small doses over a long time
- absorbed through skin or clothing
- touched or spilled on to unprotected skin.

Some chemicals can cause:

- damage to eyes, skin or internal organs, through direct chemical reaction
- illness that can sometimes leave you feeling fine after the exposure but cause medical problems after many months or years of exposure
- allergic skin reaction such as a rash, coughing and breathing problems
- death – some poisonous chemicals can kill outright.

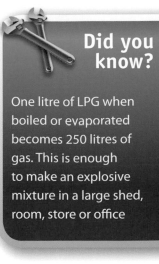

Did you know?

One litre of LPG when boiled or evaporated becomes 250 litres of gas. This is enough to make an explosive mixture in a large shed, room, store or office

Safety tip

On no account should you smoke in areas where LPG is in use

Did you know?

The HASAWA defines risk assessments for substances

Chemical accidents may be caused by:

- hurrying, overconfidence, fooling around or not following instructions
- spills and leaks that are not noticed or wiped up
- unsafe working conditions – vapours may build up in poor ventilation
- exposure of some chemicals to heat or sunlight, causing explosion, fire and poisonous reactions
- contact between a chemical and the wrong material, causing harmful reactions
- neglect or failure to dispose safely of certain old chemicals – chemical changes can happen with time, leading to a dangerous situation
- chemicals not stored in a safe manner, falling and rupturing, causing spillage.

There are four main types of chemicals. You need to know them and how to guard against their hazards:

- toxic agents
- corrosives
- flammables
- reactives.

Toxic agents

Poisons, such as hydrogen sulphide and cyanide, can cause injury, disease and death. To protect yourself:

- close containers tightly when not in use
- be sure the work area is well ventilated
- wear personal protective equipment
- wash hands often
- carry cigarettes in a protective packet
- safely dispose of contaminated clothing
- keep any proper antidotes handy
- always plan for emergencies and rescue.

Corrosives

These are irritants such as acids and alkalis, which are especially dangerous to the eyes and respiratory tract. To protect yourself:

- wear personal protective equipment, goggles, breathing devices, protective gloves
- make sure ventilation is good
- run for water if corrosives come into contact with you; use a safety shower

- if your eyes are affected, flush with water for 15 minutes and get medical aid. Most first aid kits contain eye irrigation kits

- always plan for emergencies and rescue.

Flammables

These are liquids and gases that burn steadily such as ethylene and petrol. For these to burn they require just the right amount of flammable material (fuel), oxygen and a spark or other source of ignition.

To protect yourself:

- make sure no flames, sparks or cigarette lighters are near flammables

- keep only a small amount of flammables in the work area

- store and dispose of flammables safely

- in an emergency you may have to evacuate the area; if safe to do so, turn off all flames and then call the emergency services.

Reactives

These are substances that can explode, such as nitro-compounds. To protect yourself:

- know your chemicals; read about them and test them for stability

- handle reactives with great care; for example, Toilet Duck mixed with bleach can give off hydrogen peroxide

- at the first sign of trouble close the doors and evacuate the room through doors that do not lead to the danger area.

Asbestos dust safety

Asbestos is a mineral found naturally in certain rock types. When separated from rock it becomes a fluffy, fibrous material that has many uses in the construction industry. It is used in cement production, roofing, plastics, insulation, floor and ceiling tiling and fire-resistant board for doors and partitions.

Asbestos becomes a health hazard if the dust is inhaled; some of the fine rod-like fibres may work their way into the lung tissue and remain embedded for life. This becomes a constant source of irritation.

To protect yourself:

- know the hazards, avoid exposure and always follow recommended controls

- wear and maintain any personal protective equipment provided

- practise good housekeeping; use special vacuums and dust-collecting equipment

- report any hazardous conditions, e.g. unusually high dust levels, to your supervisor.

Safety tip

When using chemicals of any sort, you must understand the dangers involved, you must follow all safety rules and procedures recommended and you must know what to do in an emergency

Remember

The removal of asbestos must only be carried out by specialist contractors. Consider your health at all times. Do not work with asbestos under any circumstances, unless trained and equipped to do so. After all it is your health and safety, and nothing is more important

Find out

How often does equipment in a car repair garage need inspecting

Carry out preventive maintenance

All electrical equipment and installations should be maintained to prevent danger. It is strongly recommended that this includes an appropriate system of visual inspection and, where necessary, testing.

By concentrating on a simple, inexpensive system of looking for visible signs of damage and faults, most electrical risks can be controlled. This will need to be backed up by testing as necessary. It is always recommended that fixed installations are inspected and tested periodically by a competent person.

The frequency of inspections and any necessary testing will depend on the type of installation, who uses it and the environment in which it is used (see BS 7671). Records of the results of inspection and testing must be maintained and used to assess the effectiveness of the maintenance system.

Work safely

Working on or even near electricity can lead to danger –people must know what they are doing before they start. Therefore you must check that:

- suspect or faulty equipment is taken out of use, labelled 'DO NOT USE' and kept secure until examined by a competent person

- where possible, tools and power socket outlets are switched off before plugging or unplugging

- equipment is switched off and/or unplugged before cleaning or making adjustments

- more complicated tasks, such as equipment repairs or alterations to an electrical installation, should only be tackled by people with a knowledge of the risks and the precautions needed

- working on or near exposed live parts of equipment is not carried out unless it is absolutely unavoidable and suitable precautions have been taken to prevent injury, both to yourself and to anyone else who may be in the area.

Other situations

Underground power cables

Although this will be rare occurrence for an electrician, always assume cables will be present when digging in the street, pavement or near buildings. Use up-to-date service plans, cable-avoidance tools and safe digging practice to avoid danger. Service plans should be available from regional electricity companies, local authorities, highway authorities etc.

Overhead power lines

When working near overhead lines, it may be possible to have them switched off if the owners are given enough notice. If this cannot be done, consult the owners about the safe working distance from the cables. Over half of the fatal electrical accidents each year are caused by contact with overhead lines.

Electrified railways and tramways

If working near electrified railways and tramways, consult the line or track operating company. Not all trains are diesel powered, and some railways and tramways use electrified rails rather than overhead cables.

Remember

Electricity can flash across from overhead lines, even though the plant and equipment do not touch them

Access and exit

On completion of this topic area the candidate will be able to state the formalised, controlled methods of establishing safe access and exit from sites

Safe access and exit

During normal on site work activities it is inevitable that the site will receive visitors, from delivery drivers to architects and clients. This should not be a problem in areas such as factories, office blocks or hospitals if a construction site is located within the boundary of these organisations, as they probably already have a visitor facility.

However, at sites without designated facilities, the motivation and co-operation of site personnel is relied on to display the correct actions to any expected visitors. Irrespective of the quality of the facilities available for receiving visitors, the reasons for putting into practice a 'visitor's procedure' remain the same:

- to meet with health and safety requirements
- to maintain site security
- to project a professional approach for the company
- to establish and maintain good client relationships.

Generally speaking, the following good practice is recommended:

- Receive visitors in a manner that fosters goodwill.
- Check the reason for their visit.
- Log their arrival time (and departure time when they leave).
- Check the validity of the visitor.
- Establish whom they wish to see.
- Brief them on site safety.

- Issue an identification badge if necessary.
- Issue them with a hard hat and high visibility jacket if necessary.
- If a request is made that is beyond your authority, contact an authorised person.

Date	Visitor's name	Company	To see	Time in	Time out	ID checked	Badge number	H&S briefed	Visitor's signature
Enter date	PRINT visitor's name	Enter name of company or organisation visitor is representing	Name of person to be visited	Time visitor is booked into the workshop	Time visitor is booked out of workshop	Type of ID used – e.g. student card, letterhead etc.	Number of visitor pass or badge issued	Enter Yes when briefed and PPE issued, if required	Ask visitor to sign here

Table 8.02 Site visitors' book

Access equipment

On completion of this topic area the candidate will be able to establish that access equipment is in safe working order for working at height and state the rules for lifting and handling.

We covered both these topics in chapter 7. Please refer back to pages 179–188 for more information and a full discussion of these health and safety topics.

Securing tools and equipment

On completion of this topic area the candidate will be able to state the need to secure tools and equipment to avoid losses and maintain insurance cover.

Insurance

Employers are responsible for the health and safety of their employees while they are at work. An employee may be injured at work, or a former employee may become ill as a result of their work while in employment. The employee might try to claim compensation from the employer, if they believe the employer is responsible.

The Employers' Liability (Compulsory Insurance) Act 1969 ensures that employers have at least a minimum level of insurance cover against any such claims. Employers' liability insurance will enable the employer to meet the cost of compensation for an employee's injuries or illness whether they are caused on or off site; any injuries or illness relating to motor accidents that occur while employees are working for an employer may be covered separately by the employer's motor insurance.

Public liability insurance is different. It covers an employer for claims made against it by members of the public or other businesses, but not for claims by employees. While public liability insurance is generally voluntary, employers' liability insurance is compulsory. Employers can be fined if they do not hold a current employers' liability insurance policy that complies with the law.

When an employer takes out employers' liability insurance, it will have an agreement with the insurer about the circumstances in which the latter will pay compensation. For example, the policy will cover the specific activities that relate to the business. There are certain conditions that could restrict the amount of money an insurer might have to pay. Employers must make sure that their contract with the insurer does not contain any of these conditions. The insurer cannot refuse to pay compensation purely because the employer:

- has not provided reasonable protection for employees against injury or disease

- cannot provide certain information to the insurer

- did something the insurer told it not to do (for example, said it was at fault)

- has not done something the insurer told it to do (for example, report the incident)

- has not met any legal requirement connected with the protection of the employees.

However, this does not mean employers can forget about their legal responsibilities to protect the health and safety of their employees. If an insurer believes that an employer has failed to meet its legal responsibilities for the health and safety of its employees and that this failure has led to the claim, the policy may enable the insurer to sue the employer to reclaim the cost of the compensation.

Employers must be insured for at least £5 million. However, they should look carefully at their risks and liabilities, and consider whether insurance cover of more than £5 million is needed. In practice, most insurers offer cover of at least £10 million. If the business is part of a group, a policy for employers' liability insurance can be taken out for the group as a whole. In this case, the group as a whole, including subsidiary companies, must have cover of at least £5 million. Employers can have more than one policy for employers' liability insurance, but the total value of the cover provided by the policies must be at least £5 million.

Tools and equipment

You will use a variety of tools and pieces of equipment in your work as an electrician. All of them are potentially dangerous if misused or neglected. Instruction in the proper use of tools and equipment will form part of your training, and you should continue to follow safe working practices.

Hand tools and manually operated equipment are often misused. You should always use the right tool for the right job, never just make do with whatever tool you may have to hand. For example, never use a hammer on a tool with a wooden handle as you may damage the wooden handle and create flying splinters. Some general guidelines follow.

- Keep cutting tools, saws, chisels, drills etc. sharp and in good condition.

Did you know?

Employers must ensure that they use an authorised insurer. If they do not, they may be breaking the law. They should check that their insurer is authorised before taking out employers' liability insurance

Remember

Employers must carry out a risk assessment, take practical measures to protect the employees and report incidents

- Ensure handles are properly fitted and secure, and free from splinters.
- Check that the plugs and cables of hand-held electrically powered tools are in good condition. Replace frayed cables and broken plugs.
- Electrically powered tools must be PAT tested in accordance with your employer's procedures.

Other common items of equipment – e.g. barrows, trucks, buckets, ropes and tackle etc. – are all likely to deteriorate with use. If they are damaged or broken they will eventually fail and may cause an accident.

Cartridge tools and high-pressure airlines

If you have to use cartridge operated tools you will be given the necessary instruction on the safe and correct methods of use. They can be dangerous, especially if you have a negligent discharge. This can cause ricochets, which can lead to a serious injury.

You may also come into contact with high-pressure airlines. Used carelessly, compressed air can be dangerous. It can cause explosions or blow tools, equipment or debris about, which may injure others. Never use an airline to blow dust away, never aim it at any part of your body and never point it at somebody else. If high-pressure air enters the body through a cut or abrasion, or through one of the body's orifices, it can cause an air embolism, which is very painful and can be fatal.

Untidy working

Tools, equipment and materials left lying about, trailing cables and air hoses, spilt oil and so on can cause people to trip, slip and fall. Clutter debris, oily rags, paper etc. should be cleared away to prevent fire hazards.

Tools and equipment left in this state become obvious targets for thieves and make it difficult for any employers to maintain effective levels of insurance cover. Upon completion of any work activity, all tools etc. should be cleared away and the workplace left in a safe condition.

On the Job: site safety

Kelly, a first-year electrical apprentice, has been left in charge of a scissor boom while the other electricians take their lunch break in a local pub. Two carpenters climb on to the scissor boom and use it for working; while doing so, they crash into a low ceiling, damaging the boom. They tell Kelly not to worry because they all work for the same company. They finish the work and leave the site. A little later the electricians return.

1. What should Kelly have done in the situation?

2. Who was ultimately responsible for the safety of the scissor boom?

3. Who would have been responsible if the carpenters had hurt themselves?

Emergency equipment

On completion of this topic area the candidate will be able to identify emergency switches, isolators, alarms and emergency equipment in the workplace and describe methods of verifying and securing (lock-off) isolation.

Dangers of electricity

Electricity can kill. Even non-fatal shocks can cause severe and permanent injury. Shocks from faulty equipment may lead to falls from ladders, scaffolds or other work platforms. Those using electricity may not be the only ones at risk, as poor electrical installations and faulty electrical appliances can lead to fires. Yet most of these accidents can be avoided with careful planning and straightforward precautions.

Hazards in harsh conditions

The risk of injury from electricity is also strongly linked to where and how it is used. The risks are greatest in harsh conditions.

- In damp/wet surroundings, such as construction sites, unsuitable equipment can easily become live and make its surroundings live.

- Out of doors, equipment may not only become wet but may be at greater risk of damage.

- In cramped spaces with a lot of earthed metalwork, such as inside a tank, if an electrical fault were to develop it could be very difficult to avoid a shock.

- Some items of equipment can also involve greater risk than others. Extension leads are particularly liable to damage: to their plugs and sockets, to their electrical connections and to the cable itself. Other flexible leads, particularly those connected to equipment that is moved a great deal, can suffer from similar problems.

Remember

Electricity is dangerous. Always take precautions

Safety precautions

There are precautions that can be taken to ensure that the electrical installation is safe.

- Install new electrical systems to a suitable standard, e.g. BS 7671 Requirements for Electrical Installations, and then maintain them in a safe condition.

- Existing installations should also be properly maintained.

- Provide enough socket outlets – overloading socket outlets by using adaptors can cause fires.

- Choose and use equipment that is suitable for its working environment.

- Risks can sometimes be reduced by using air, hydraulic, hand or battery-powered tools. These are especially useful in harsh conditions.

- Ensure that equipment is safe when supplied and maintain it in a safe condition.

- Regularly inspect and test portable electrical equipment (PAT).

- Provide an accessible and clearly identified switch near each fixed machine to cut off power in an emergency.

- For portable equipment, use socket outlets that are close by so that equipment can easily be disconnected in an emergency.

- The ends of flexible cables should always have the outer sheath of the cable firmly clamped to stop the wires (particularly the cpc) pulling out of the terminals.

- Replace damaged sections of cable completely.

- Use proper connectors or cable couplers to join lengths of cable. Do not use strip connector blocks covered in insulating tape.

- If the plug is not the moulded-on type, ensure that all the wires are properly connected to the correct terminals in the plug top.

- Protect lamps and other equipment which could easily be damaged in use. There is a risk of electric shock if they are broken.

- Check electrical equipment used in flammable/explosive atmospheres. It should be designed to stop it from causing ignition. You may need specialist advice.

Reduce the voltage

One of the best ways of reducing the risk of injury when using electrical equipment is to limit the supply voltage to the lowest needed to get the job done. For example:

- temporary lighting can be run at lower voltages, e.g. 12, 25, 50 or 110 volt

- where electrically powered tools are used, battery-operated ones are safest

- portable tools are readily available that are designed to be run from a 110 volt centre-tapped-to-earth supply.

Provide a safety device

Whether equipment operates at 110 volts, 230 volts or higher an **RCD (residual current device)** can provide additional safety. The best place for an RCD is built into the main switchboard or the socket outlet. This means that the supply cables are permanently protected. If this is not possible, a plug incorporating an RCD, or a plug-in RCD adaptor, can also provide additional safety.

RCDs for protecting people have a rated tripping current (sensitivity) of usually not more than 30 milliamps (mA). However, please remember:

- an RCD is a valuable safety device; never bypass it

- if an RCD trips, this is a sign that there is a fault

- check the system before using it again

- if the RCD trips frequently and no fault can be found in the system, consult the manufacturer of the RCD

- the RCD has a test button to check that its mechanism is free and functioning. Use this regularly.

We will look at RCDs in greater depth throughout chapter 12.

Lock-off procedures and isolation

We covered these areas earlier in the book on pages 16 – 23. Please refer back to this section for information on these two areas.

Waste procedures

On the completion of this topic area the candidate will be able to state the correct procedures for removing unused materials and disposing of waste.

Waste

Any substance or object that you discard, intend to discard or are required to discard is waste and as such is subject to a number of regulatory requirements. The term 'discard' has a special meaning. Even if material is sent for recycling or undergoes treatment in-house, it can still be waste.

Under the Controlled Waste Regulations 1998, the Duty of Care requires that you ensure all waste is handled, recovered or disposed of responsibly, that it is only handled, recovered or disposed of by individuals or businesses that are authorised to do so, and that a record is kept of all wastes received or transferred through a system of signed **Waste Transfer Notes**.

If you work as a subcontractor and the main contractor arranges for the recovery or disposal of waste that you produce, you are still responsible for those wastes under the Duty of Care.

Be aware that materials requiring recycling (such as copper cable), either on your premises or elsewhere, are likely to be waste and will be subject to the waste management regime and the Duty of Care. If in any doubt, seek advice from the Environmental Regulator.

Did you know?

If you transport your own waste, which is either building or demolition waste, you will need to be registered as a waste carrier with your Environmental Regulator

Special waste

If the material you are handling has hazardous properties, it may need to be dealt with as special waste.

Covered by the Special Waste Regulations 1996, 'special waste' is waste that is potentially hazardous or dangerous and which may therefore require extra precautions during handling, storage, treatment or disposal. In reality this means that most businesses are likely to produce some 'special wastes'. The following are examples:

- asbestos
- lead-acid batteries

- used engine oils and oil filters
- oily sludges
- solvent-based paint and ink
- solvents
- chemical wastes
- pesticides.

Disposal of these requires technical competence.

Waste disposal

If you intend to discard containers, an assessment must be made as to whether they are special waste. Containers may be special waste if they contain residues of hazardous or dangerous substances/materials. If the residue is 'special', then the whole container is special waste.

Waste disposal is commonly segregated into different types

Here are some guidelines for good waste disposal practice.

- Do not burn cable or cable drums on site. Find another method of disposal.

- Before allowing any waste haulier or contractor to remove a waste material from your site, ask where the material will be taken and ask for a copy of the waste management licence or evidence of exemption for that facility.

- Segregate the different types of waste that arise from your works. This will make it easier to supply an accurate description of the waste for waste transfer purposes.

- Minimise the quantity of waste you produce to save you money on raw materials and disposal costs.

- Label all waste skips – make it clear to everyone which waste types should be disposed of in that skip.

Hired plant or equipment

Where hired plant and equipment cause an environmental incident while on hire, the responsibility for that incident rests with the person or business that has control of that plant or equipment.

Check that the hire company supplying you with plant and equipment operates a system of Preventive Plant Maintenance – well-maintained plant and equipment is less likely to break down, will emit fewer pollutants to the air and can help prevent spillage of oil and fuel to the environment.

Activity

1. Obtain a copy of your company's risk assessment forms and complete one for the current job you are working on.

2. Have a look at the electrical tools you are using on site. Have they been PAT tested? Are they overdue for testing?

FAQ

Q Why do I have to keep doing risk assessments? The jobs I do are always the same.

A Well, the tasks you do from one day to the next may be very similar but even slight changes can increase the risk to yourselves and others. The construction site environment is an extremely variable one; changes in the weather and in the people on site are just two factors which need constant monitoring.

Q My boss has this real thing about 'tidying up as I go along'. It's really getting on my nerves. Do I need to do it?

A One of the most common types of accidents are 'slips, trips and falls' and these are usually caused by poor housekeeping. Your boss is only trying to maintain a safe working environment for you and your colleagues – you'll thank him for it one day!

Q My company insists that my electrical tools are brought into the office and PAT tested. Can they do this? After all, they are my tools.

A Yes. All equipment used in the workplace must be 'maintained in a safe condition', regardless of who owns it.

Knowledge check

1. Explain why it is important to keep the workplace clean and tidy.

2. Describe the meaning of the terms 'hazard' and 'risk'.

3. State the five steps of assessing risk in the workplace.

4. What are the three components in the fire triangle?

5. List the five different categories of fire.

6. Why do we use reduced voltage equipment on building sites?

7. What protection does an RCD provide?

chapter 9

Using technical information, specifications and data

Unit 3 Outcome 2

As an electrician, technical information, specifications and data will be very important to your everyday work. All installation and maintenance work will rely on the use of written materials to correctly complete work. In order to do this, electricians use drawings and diagrams in conjunction with related specifications to complete their work.

Technical information can be distributed in a wide variety of ways, through: written reports, BS or BS EN symbols on a diagram, email, fax machine etc. Whatever the method used, it is important to remember that the recipient of any technical information can understand and use any information that you put together. Remember, not everyone has received full training!

On completion of this chapter the candidate will be able to:

- describe the types of technical information
- prepare materials lists and requisites from specifications
- state how to take in-situ measurements
- site decontamination including receipt and checking of material
- state the relevant people involved in the use of technical information
- describe how dimensions and measurements may be transferred from a scale drawing to a site
- interpret drawings and diagrams to complete installations
- describe the need to present the right image to recipients of technical information
- state the importance of ensuring that the recipient can understand information.

Technical information

On completion of this topic area the candidate will be able to describe the types and sources of technical information.

Sources of technical information

As we saw in chapter 2 lots of different people may be involved in an electrical installation project. For them to work together they must communicate with one another, using or transmitting technical information.

There are many sources of this information, including drawings, diagrams, charts and data, British Standards, Codes of Practice, specifications, manufacturers' instructions and manuals, catalogues and reports. We covered some of these in chapter 3.

Equipment and system specifications

We covered specifications when we looked at the role of the estimator in Chapter 2 (pages 46–47). However we will briefly recap the function and features of these documents now.

Specifications are lists of certain requirements, as detailed by the end user (the client). They may also contain certain test requirements that the final product needs to reach. Specifications are generally broken down into two sections.

Section 1: General specification

This section tends to give general information about installation circumstances, such as the wiring systems, enclosures and equipment. The section gives generic requirements applicable to any project issued by the consulting engineer.

Section 2: Particular specification

This section gives the specific details on the project being tendered for. For example, Section 2 may say that for this job the installation will be carried out in galvanised steel conduit. Unless there are specific conditions applicable to the project, the contractor then needs to see what the consulting engineer's general requirements are for installing galvanised steel conduit, and this information will be contained in Section 1.

Manufacturers' data and service manuals

Almost all equipment will have the manufacturer's fitting instructions and other technical data or information sheets. These should be read and understood before fitting the item concerned.

Once installation of the item is complete, they should be kept safely in a central file so that they can be given to the customer in a handover manual when the project is completed. Sometimes you may need more details about an item. You can get this from the manufacturer's catalogue, datasheets, website or by speaking to the manufacturer directly. It may be useful to talk to the contracts engineer to make sure that he or she knows what is happening.

Working drawings

We have seen how technical information is recorded in diagrams in chapter 3 pages 60–63. Both the contractor and subcontractors will work from these documents.

User instructions

Electrical equipment and material will also come with user instructions, containing information about the operation of the device. This is in an invaluable quick reference point for using new materials and will give you a clear understanding of the technical aspects of new equipment.

Reports and schedules

Report writing

There may be occasions when it is necessary to communicate within the organisation, i.e. when there has been an incident or if an issue regarding health and safety needs to be communicated with other members of the team.

Reports usually describe a problem or an investigation. Their purpose is to help someone, or a team, to make a decision by 'reporting' all the relevant facts and perhaps making some recommendations for action. Some reports will present a solution to the problem while others may simply list historical and factual data.

In a construction project a report may be needed for many reasons, including:

- progress made since the last report
- description of an equipment installation problem
- an investigation into the different ways of solving a problem (with a recommendation for action)
- explanation of why there is an attendance problem on site.

Reports need to be factual, to the point and not long-winded. Often a report is requested by someone needing information about an issue. At other times, a report will be written by someone who has information to share.

The memo shown in Figure 9.10 (page 238) is really a short report. It may be that the contracts engineer will ask for more details to understand how the situation has arisen. When preparing a report, an electrician may have to refer to other documents, such as the site diary (see page 239).

Many companies will have a standard structure or format for reports, and you should use this whenever you can. However, a lot depends on the subject and type of the report, and there is no merit in slavishly following a format if it makes it difficult to understand what the report is about. As with any writing, it is important to know clearly what you are trying to say and then to put that into a style and format that will be acceptable to the reader.

Style, although hard to define, can make a big difference to the success of a report. A good style can help to convince the reader of the merits of the report and its recommendations. A bad style may put the reader off, even if the content of the report is appropriate.

Report structure

Companies may have a required format for a report's structure, but they will probably include most or all of the elements described below.

For a business report or memo to someone such as a **Supervisor**:

- basic identification data (who it is to, who is writing it, the date, subject, reference number)
- summary – the project or problem and the purpose of the report
- background – the history of the issue being reported
- relevant data – the evidence you have gathered
- conclusions and recommendations.

For a more formal technical report where you would wish to communicate with a **Safety Officer** or **Safety Representative**:

- title page
- acknowledgement
- change history
- contents list
- report summary
- introduction
- technical chapters
- conclusions and recommendations
- references
- appendices – these would contain additional specialised information such as copies of risk assessment forms that may have been completed.

Whatever style and format is used, the principles of good writing still apply (see Table 9.01)

Be clear	• The reader must be able to understand the report easily. • Explain symbols, tables or diagrams. • Never assume that your reader has prior knowledge of the issues. • Do not use jargon or abbreviations unless you know that the reader will be familiar with them.
Be concise	• Do not waffle. • Be succinct without making the report hard to understand. • Think about what the reader wants to know, i.e. evidence and conclusions, not a vivid description of the valiant effort you made to compile them.
Be logical	• Ensure you have a beginning, middle and end. • There should be a sensible flow between sections, chapters, paragraphs and sentences. • Avoid jumping randomly from idea to idea. • Take out unnecessary distractions such as pretty graphics that do not add any extra information.
Be accurate and objective	• Base your report on honest facts – an inaccurate report is at best pointless and at worst harmful. • Remember that someone may make an important decision based on your recommendations. • If you have to make assumptions, make it clear that you have done so. • Be tactful. What you say may contradict others or what the reader thinks. • If you feel very strongly about something, do not send the report until you have had some time to think about it, and make sure it contains only facts. • Do not use reports as a sounding-off platform. It is very easy to destroy a relationship by sending an angry, ill-conceived letter or report.

Table 9.01 Good writing practice

Critical path analysis

In larger projects, many tasks must be completed before the project is finished. Not all of these activities can be done at the same time, and some cannot begin until others are completed. We need a way of working out how best to organise the project efficiently, and critical path analysis (CPA) is one solution.

Critical path networks (CPNs) are diagrams that represent each task and how they relate to one another. If we know the time needed for each activity, the overall project completion time can be calculated. The chart also helps us to see what happens if a task is delayed unexpectedly.

The critical path is the sequence of activities that fix the duration of a project. In many projects there will be some activities that are not on the critical path. The first step in constructing a critical path network is to list all the activities, what must be done before they can start (the sequence) and the expected time required for each.

The circles are 'events'; they have no duration but represent the time at which the activity starts or finishes. The arrow represents the activity and always goes from left to right. It starts and ends with an event. Figure 9.01 represents an activity (A) which lasts for two days.

Figure 9.01 Critical path diagram

We build the network by linking the activities from left to right at their start and end events. One rule of CPN is that no two activities can begin and end on the same two events. To explain this, we will use an example.

We have four activities (A, B, C and D) to complete. Activities A and B can run at the same time, but C cannot begin until Activities A and B have been completed, and Activity D cannot begin until Activity B has been completed.

Activity	Activities that need to be done before this activity
A	None
B	None
C	A and B
D	B

Table 9.02 Sequence of activities for a critical path analysis

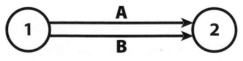

Figure 9.02 Timing of activities A and B

As the first two activities (A and B) can begin at the same time, the temptation could be to show them as in Figure 9.02.

However, to comply with the rule that no two activities can start and end on the same events, we introduce a 'dummy activity'. Shown as a dotted line (see Figure 9.03), a dummy activity doesn't take any time.

This shows that Activity C can only start at Event 3, when both A and B are complete. Now let's look at a more detailed example.

Here is information we have about the activities.

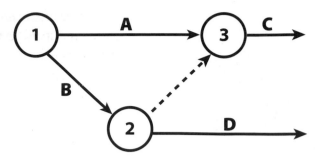

Figure 9.03 Introducing a dummy activity

- **Activity A** starts on Day 1 and lasts two days.

- **Activity B** starts on Day 1 and lasts three days.

- **Activity C** can only begin once Activity B is complete and lasts three days.

- **Activity D** can only begin once Activity C is complete and lasts three days.

- **Activity E** cannot start until Activity A is complete; it will take six days.

- **Activity F** can start at any time and lasts two days.

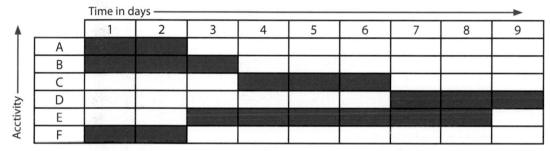

Figure 9.04 Activities shown as a bar chart

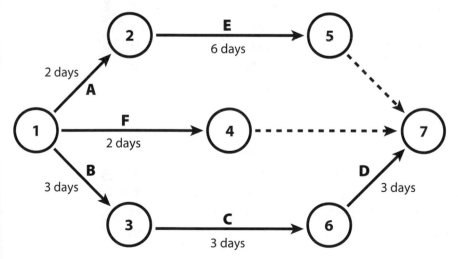

Figure 9.05 Critical path network

This may need a little careful study to see how it all fits together.

The minimum time for project completion is taken as the longest time path through the network. In this example it is through points 1, 3, 6 and 7, giving us a project time of nine days.

Remember

Both the bar chart and the critical path network are management planning tools; their purpose is to help a contract run smoothly

Remember

The main contractor acts as co-ordinator for all the individual subcontractors, checks that they carry out their work properly and on time, and pays them for the work done

Remember

Prompt delivery and competitive pricing will be important factors in the choice of non-nominated suppliers

Find out

When would you need to use a variation order?

The bar chart may seem easier to understand but it can't readily show when several activities have to be completed before another can begin. With practice, using the critical path method will give you greater control over establishing when activities may start.

Bar charts and critical path networks will help you to see which activities affect your work and how they fit in with everyone else's working schedule. In this way you can:

- plan which areas to work in
- see when you can start each activity
- make sure you have the correct materials and equipment ready at the right time
- avoid causing delays to others and the overall contract.

Site documentation

Job sheets

Job sheets give detailed and accurate information about a job to be done. Some electrical contracting companies issue them to their electricians. They will include:

- the customer's name and address
- a clear description of the work to be carried out
- any special instructions or special conditions (e.g. pick up special tools or materials).

Sometimes extra work is done which is not included in the job sheet. In this case it is recorded on a day worksheet so that the customer can be charged for it. Job sheets frequently require the customer to sign them to indicate that the work has been done. Returning the job sheet to the office triggers the invoice being sent to the client for payment.

Job Sheet — **Evan Dimmer** Electrical contractors

Customer: Dave Wilkins

Address: 2 The Avenue
Townsville
Droopshire

Work to be carried out:
Install 1 x additional 1200mm
fitting to rear of garage

Special conditions/instructions:
Exact location to be specified
by client

Figure 9.06 A typical job sheet

Variation order

A **variation order** (VO) is issued ('raised') when the work done varies from the original work agreed in the contact and listed in the job sheet. If this situation arises, it is important for the site electrician to tell his supervisor immediately. A variation order can then be made out to enable the new work to be done without breaking any of the terms of the contract. The purpose of the VO is to record the agreement of the client (or the consulting engineer representing the client) for the extra work to be done, as well as any change that this will make to the cost and completion date of the project.

Day worksheets

Work done outside the original scope of the contract, perhaps as a result of a VO initiated by the architect, engineer or main contractor, is known as day work. When the work is completed the electrician or supervisor fills out a day worksheet and gets a signature of approval from the appropriate client representative. Day work is normally charged at higher rates than the work covered by the main contract, and these charges are usually quoted on the initial tender. Typical day work charges are:

- labour: normal rates plus 130 per cent

- materials: normal costs plus 25 per cent

- plant: normal rate plus 10 per cent.

Disputes over day work can easily arise, so it is important that the installation team on site records any extra time, plant and materials used when doing day work. Completed day worksheets, when signed by the client, must be returned to the office. Receipt of the day worksheet triggers the issuing of an invoice for the additional work.

Day Worksheet

Evan Dimmer
Electrical contractors

Customer			Job no.	
Date	Labour	Start time	Finish time	Notes

Quantity	Description		Office use

Supervisor's signature	Customer's signature
Date	

Figure 9.07 A typical day worksheet

Timesheets

Timesheets are very important to you and your company. They are a permanent record of the labour on a site and include details of:

- each job

- travelling time

- overtime

- expenses.

This information allows the company to track its costs and make up your wages. If you work on several sites over the course of a week, you may need to fill in a separate timesheet for each job.

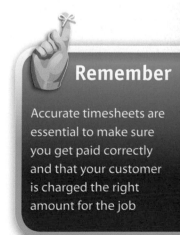

Remember

Accurate timesheets are essential to make sure you get paid correctly and that your customer is charged the right amount for the job

Timesheet					**Evan Dimmer** Electrical contractors		
Employee				Project/site			
Date	**Job no.**	**Start time**	**Finish time**	**Total time**	**Travel time**	**Expenses**	
Mon							
Tue							
Wed							
Thu							
Fri							
Sat							
Sun							
Totals							

Employee's signature

Supervisor's signature

Date

Figure 9.08 Typical time sheet

Purchase orders

Before a supplier dispatches any materials or equipment, they will require a written purchase order. This will include details of the material, quantity required and (sometimes) the manufacturer; it may also specify a delivery date and place. In many cases the initial order is made via telephone, email or the Internet, and a written confirmation is sent immediately afterwards. The company keeps a copy of the original order in case there are any problems.

Usually the purchasing department sends out these orders but sometimes an order is raised directly from site when there is a need for immediate action.

Delivery notes

Delivery notes are usually forms with several copies that record the delivery of materials and equipment to the site. Materials delivered directly to the site will arrive with a delivery note. As the company representative on site this is the form you are most likely to deal with.

The delivery note should give the following information:

- the name of the supplier
- to whom the materials are being sent
- a list of the type, quantity and description of materials that are being delivered to the site in this particular load
- the time period allowed for claims for damage.

When materials arrive on site you should do the following.

- Ensure that they are unloaded and stored correctly.
- Check each item against the delivery note.
- Check for obvious signs of damage.
- If everything is OK, sign the note. If not, note any missing/rejected items on the delivery note, then you and the delivery driver should both sign it.
- Store your copy of the note safely until you return it to the office. When the office receives the delivery note from site this is the confirmation they need that the goods have been received and payment of the bill is authorised.
- The project manager is also able to see from the delivery notes that the materials are being delivered to the site. He or she is able then to monitor the progress being made.
- Check the materials thoroughly for damage within the time stated in the delivery note (usually three days) and inform the supplier immediately if there are any problems.

A delivery will not always contain all the materials listed in the purchase order. Sometimes not all of the material is required on site at once and is delivered in several loads. This helps to reduce the need for on-site storage and minimises the risk of damage or loss.

Incomplete deliveries may also occur if the supplier is out of stock, or if some of the order is coming directly from the manufacturer.

Completion orders are a record that all the material on an original purchase order has been delivered.

Delivery note	**A. POWERS** *Electrical wholesalers*	
Order no.	Date	
Delivery address	Invoice address	
2 The Avenue Townsville Droopshire	Evan Dimmer Electrical Contractors	
Description	Quantity	Catalogue No.
Thorn PP 1200 mm fit fitting	1	
1.5 mm T/E cable	50 m	
Comments		
Date and time of receiving goods		
Name of recipient	Signed	

Figure 9.09 Typical delivery note

On the Job: Receiving deliveries on site

Alex works for Gurnards Electrical Services. He is left to man the site while his colleagues go to lunch. A delivery arrives while he is alone. He accepts the delivery and signs the delivery note without checking the contents. When his boss returns he checks the delivery and finds that some items are missing; he calls the company, but he is told that all items were delivered as specified.

1. How could this situation have been avoided?

2. Outline the steps Alex should have taken to receive the delivery.

3. What would have happened if Alex had lost the delivery note?

Site reports and memos

Most companies require regular reports about the progress on site. The site foreman, supervisor or engineer in charge usually compiles them. Site reports contain details of work progress, defects, problems and delays. Sometimes other reports into specific problems or incidents will be made. The reports enable the company to:

- provide evidence when they make claims for progress payments
- take prompt action to avoid any difficulties that may be building up
- spot problems that keep being repeated and take action to eliminate them.

A memo is usually a short document about a single issue, for example a problem installing a piece of equipment or materials not being delivered on time.

INTERNAL MEMO

To: D. Boss, Contracts Engineer

From: A. Foreman, Site Supervisor

Project: The New Hospital

Following the arrival of the new essential services generator on site, we have found that it is too big for the entrance to the existing generator house. I have spoken to the Main Contractor, and we believe that a section of roof could be removed easily and the generator craned into position.

Please advise.

A. Foreman

Figure 9.10 A short memo from a site supervisor to a contracts engineer

Site diaries

As a project develops, the **labour force** increases, the work area grows and the number of ongoing tasks increases. With more meetings and discussions taking place, it becomes more difficult for the person in charge – whether the foreman, chargehand or electrician – to keep control of what is going on. This is where the site diary is used. All information relating to the running of the site should be kept as a daily record, which can be referred to in the event of a problem.

A site diary should be used to record:

- dates and times of meetings, both official and unofficial
- notes on the outcomes of those meetings
- details and dates of any actions that needed to be undertaken
- areas worked in and completed with details of the workforce used in those areas
- notes of any problems encountered
- staffing issues, lateness, absence, discipline, holidays
- dates on which materials were ordered
- dates on which materials were received
- detail of any discrepancies in deliveries and action taken
- dates of proposed deliveries
- agreed variations
- contact details for staff, consultant engineers, health and safety advisers etc.

Relevant people

Technical information will be needed by all people working on an installation or work site. As we have seen, the correct interpretation and use of technical information is vital for the successful completion of a job. There are several key individuals who need to be involved and consulted in the use of technical information, as below.

- **Operative** – the craftsmen and women on site, who carry out the installation and construction work.
- **Supervisor** – the contractor's representative on site, with responsibility for overseeing day-to-day operations. He or she is responsible for the supervision of the electricians.
- **Contractor** – those taking on the bulk of the work to be carried out. They employ subcontractors to carry out different parts of the project.
- **Site agent** – the person in charge of the building contract, with responsibility for all work carried out on the site. They will ensure that the original aims of the contract are being fulfilled and that the work is running to schedule and budget.
- **Client** – the person commissioning the work who specified the purpose of the work.

Interpreting diagrams

On completion of this topic area the candidate will be able to interpret diagrams to produce, locate or install systems, describe how measurements can be transferred from a scaled drawing to a workplace and state how to take in-situ measurements.

Using diagrams

We looked in depth at the interpretation of diagrams and drawings in chapter 3 pages 60–63. Please refer back to this chapter for more information. A diagram communicates information in a clearer and easier way than can be expressed in words.

Transferring measurements

Scale drawings are used for planning and preparing an installation. You will remember that scaled drawings used fixed ratios, for example 1:100. This would mean that every 10 mm on the drawing is 100 times bigger in reality. We can transfer measurements from scaled drawings to a workplace by continuing this multiplication. Scaled drawings can therefore be used to plan detailed installations in advance.

Taking in-situ measurements

When carrying out measurement work you will need to have a thorough understanding of the intended installation and any special requirements. On some larger installations the measurements will already have been calculated in advance, and your role will be fitting your installation to them.

When arriving to take measurements you must carefully record them in a notebook, and keep this safe. If you to mislay this, or incorrectly remember the measurements you have taken, it will have a negative effect on the material you require for the installation – you could either order too much material or too little.

When carrying out in-situ measurements there are a number of tools that you can use to assist you.

Folding rules

Folding rules are 1 metre long when unfolded and made of wood or plastic. They can show both metric and imperial units.

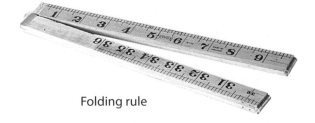

Folding rule

Retractable steel tape measures

Retractable steel tape measures, often referred to as spring tapes, are available in a variety of lengths. They are useful for marking out lengths and other materials. They have a hook at right angles at the start of the tape to hold over the edge of the material. On some tapes this will slide, so that it is not in the way when measuring from an edge.

Retractable steel tape measure

Metal steel rules

Metal steel rules, often referred to as bar rules, are used for fine, accurate measurement work. They are generally 300 mm or 600 mm long and can also serve as a short straight edge for marking. The rule can also be used on its edge for greater accuracy.

They may become discoloured over time. If so, give them a gentle rub with very fine emery paper and a light oil. If they become too rusty, replace them.

600 mm steel rule

Materials list

On completion of this topic area the candidate will be able to prepare materials lists and requisites from drawings and specifications.

Preparing a materials list

At some stage, you are going to have to put together a materials list for a project or part of a project. The estimator and contracts engineer will usually have ordered most of the systems and equipment, but there will always be smaller projects, or parts of a big project, where you will be asked to do it for yourself. We will now work through an example, to see how these lists are compiled.

Figure 9.11 is an extract from a specification for a warehouse that is part of a large hospital project. It is going to be used for storing information and leaflets for medical staff. You will probably be able to order the materials you need from a central store or, if not, directly from a local wholesaler. Either way, the process of working out what you need is just the same. The first thing to do is to get a copy of the layout drawing. It must be a scaled drawing (Figure 9.12).

Remember

As well as compiling a materials list, you will need to work out what wiring is required to control the installation and feed the lighting distribution board

> The warehouse lighting and sub-main installation will be completed in PVC single core cable contained within galvanised steel conduit and galvanised steel trunking where required. All fluorescent lighting fittings will be 1700mm x 58 W of type Tamlite TM58 and suspended on chain from back boxes. All light switches will be of MK type Metalclad Plus with an aluminium front plate. The installation will be fed from the new lighting distribution board (MEM), which shall be fed from the existing MCB DB located in the warehouse.

Figure 9.11 Extract from specification

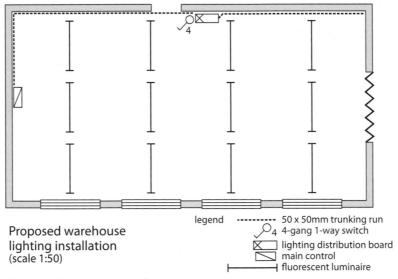

Proposed warehouse lighting installation (scale 1:50)

legend
----------- 50 x 50mm trunking run
4-gang 1-way switch
lighting distribution board
main control
fluorescent luminaire

Figure 9.12 Scale layout drawing

Take a good look at Figure 9.12. It shows the layout of the building, the location of specific services and how some items are to be installed and connected. As it is to scale (1:50), you can measure it to find the actual dimensions of the building and prepare a materials list for the job. You can also scale up positions shown on the drawing and mark them for real inside the building itself. However, what the drawing does not tell us is anything about the fitments or how they should be installed. This information will be in the specification, so you will need a copy of that too. The extract from the specification at Figure 9.11 tells us what we need to know.

Step 1: Count up all the major pieces of equipment needed

Looking at the plan and referring to the specification, we can see that we need:

- 12 Tamlite TM58 fittings, plus tubes

- 1 MK Metalclad Plus four-Gang one-way surface switch with aluminium front plate

- 1 MEM three-way SPN surface mounted DB (distribution board).

Step 2: Decide the best runs of conduit and trunking

A logical way would be to install the 50 × 50mm trunking as shown on the drawing, then install separate conduits from the trunking up to each row of lights and then along each row of lights, fixing the conduit to the roof structure.

Step 3: Calculate lengths of trunking, conduit and cable required

We measure these from the drawing and calculate the actual distance using the scale provided (1:50). You will also need to know the height of the building, which is not shown on the layout diagram.

Step 4: Include accessories, fixings etc.

We will need to allow back boxes and hook and chain arrangements at each luminaire and order sufficient fixings (screws, bolts, saddles for conduit etc.) of various types for the installation.

Step 5: Consider special access equipment

Do you need special ladders, scaffolding etc? It's no use having the materials if you cannot get to the right places!

As well as compiling a materials list, you will need to work out what wiring is required to control the installation and feed the lighting distribution board.

Presenting information

Did you know?

Even something as straightforward and simple as a smile can make a big difference when communicating with other people

On completion of this topic area the candidate will be able to describe the need to present the right image to recipients of technical information and ensure that the information they have distributed is understood.

Presentation

People will often ask you for information. The principles of giving information to people are very similar to those you use when report writing. You can communicate a great deal by how you look, the gestures and facial expressions you use, and the way you behave.

Good communication will have a hugely positive effect on your relationships with both other workers and clients. Clearly conveyed thoughts and ideas, backed with good working practices and procedures, will improve working and lead to satisfied customers.

When presenting information it is important to remember that some recipients do not have technical knowledge. Some may even find detailed explanations of electrical matters rather intimidating.

As an electrician you will need to remember four key tips for presenting information.

1. **Appearance** – make sure you look smart and businesslike when meeting clients. Remember the importance of first impressions. Many people will form their opinions about you based on your initial appearance. They might not mean to, but this will affect their reactions to you.

2. **Attitude** – be friendly and helpful when explaining installations. The attitude and manner you show when talking to clients can go a long way towards reassuring them and making then trust and listen to you. For example, would you rather work with someone who listens to what you say and offers their advice when needed or with someone who ignores what you say and avoids talking to you?

3. **Confidence** – if you do not sound too sure yourself, then you cannot expect other people to be reassured (or impressed!) either. Your confidence should come from the fact that you know and understand the information you are giving and that it is complete and correct at every point.

4. **Knowledge** – this is perhaps the most important part of communication. Having a full and detailed knowledge of electrical installation will allow you to answer any unexpected questions that people might have, as well as give a firm grounding to the information you are presenting. Without a full knowledge of electrical installations, you are no more qualified to give the information than the person who needs it.

Having confidence, knowledge and the right attitude are essential when working with others

Establishing understanding

After communicating information to people it is vital to make sure that the details of the installation have been fully understood by them. The information you have given may be used during potentially hazardous installation and electrical work, or may be used to place orders for materials. In either circumstance, incorrect information – and the results – will be your responsibility. At all times, information must be full, detailed, correct and understood.

Activity

The next time a delivery comes in, see if you can check the items against the delivery note – there should be the right number of each item ordered and each item should be undamaged.

FAQ

Q Why is it important to tell my company about any extra work needed to finish a job?

A Often very minor extra work or materials will be included in the overall costing of the job. However any significant variations should always be reported as they may have a significant effect on the cost and time of the job. Alterations to the contract may need the client's approval before you start work.

Q Why can't I just trust a delivery note? It takes so long to check all the materials that have been delivered.

A In all walks of life mistakes are made and things get damaged or go missing. Checking the items now will mean that items can be chased, re-ordered or replaced, hopefully in time to complete the job – this will save time in the long run.

Knowledge check

1. State the main features of a technical report.

2. Describe the possible benefits of conducting a critical path analysis for a multiple activity job.

3. When is a memo more likely to be used than a letter or report?

4. What are the five main steps to producing a materials list?

chapter 10

Electrical machines and alternating current theory

Unit 3 Outcome 3

As we saw earlier in the book knowledge of the basic circuitry is vital in order to successfully begin an installation. We will now look at the electrical machines and current theory through which electronic devices arc powered.

Alternating current theory was invented by a man named Nikola Tesla in the late nineteenth century, and differs from direct current because the magnitude and direction of the current varies rather than staying constant. Alternating current (or a.c.) is the principle method used in modern electrical devices.

This chapter will build upon much of the knowledge you acquired in chapters 5 and 6, as well as introducing new measurements and theories for the first time.

On completion of this chapter the candidate will be able to:

- state the basic principles of operation of a.c. motors, d.c. machines, relays and fluorescent luminaries

- state the operating principles of basic transformers

- recognise the effects of resistance, inductance, capacitance, reactance and impedance in a.c. circuits

- define and measure power factor

- state why power factor correction is required.

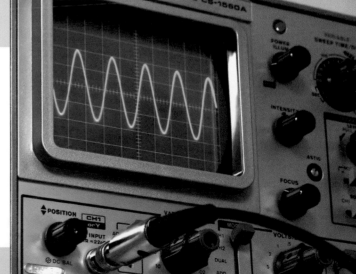

Machines and relays

In completion of this topic area the candidate will be able to state the basic principles of operation of a.c. machines, d.c. machines, fluorescent luminaries and relays.

The effects of magnetism

As most alternating current is produced by magnetism let us start by exploring the movement of a conductor through a magnetic field as the single-phase a.c. generator (or alternator).

If we were to form our conductor into a loop and then mount it so that it could rotate between two permanent magnets, it would look like Figure 10.01.

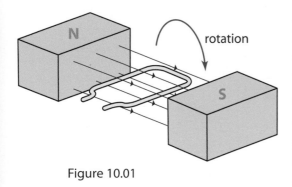

Figure 10.01

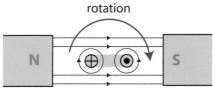

Figure 10.02

In this position, it could be said that the loop is lying in between the lines of magnetism (magnetic flux). If we were to look at a side view of this arrangement towards the ends of our loop, Figure 10.02 shows what it would look like.

Therefore, as the loop is not interfering with any lines of flux, we say that it is not 'cutting' any lines of flux. However, as we slowly start to rotate the loop in the direction indicated, it will start to pass through the lines of flux. When this happens, we say that we are cutting through the lines of flux and, as we do so, we start to induce an e.m.f.

The maximum number of lines that are cut through will occur when the loop has moved through 90 degrees and the maximum induced e.m.f. in this direction will therefore occur at this point. Keep rotating the loop and the number will once again reduce to zero as we are again lying between the lines of flux. The loop has now completed what is known as the positive half cycle.

Repeat the process and an e.m.f. will be induced in the opposite direction (the negative half cycle) until the loop returns to its original starting position.

If we were to plot this full 360 degree revolution (cycle) of the loop as a graph, we would see the e.m.f. induced in the loop as in Figure 10.03. This shape is known as a **sine wave**.

As we can see, the sine wave shows the e.m.f. rising from zero, as we start to cut through more lines of flux, to its maximum after 90 degrees of rotation. This is known as the **peak value**.

However, after completing 180 degrees (half a rotation or cycle), the e.m.f. passes through zero and then changes direction. We sometimes refer to these as being the positive and negative half cycles.

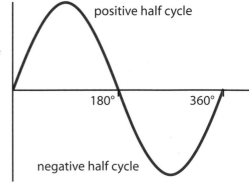

Figure 10.03 Sine wave

The opposite directions of the induced e.m.f. will still drive a current through a conductor, but that current will alternate as the loop rotates through the magnetic field. It will be flowing in the same direction as the induced e.m.f. Consequently, the current will rise and fall in the same way as the induced e.m.f. and, when this happens, we say that they are in phase with each other. To access this a.c. output, the ends of the loop are connected via slip rings as shown in Figure 10.04.

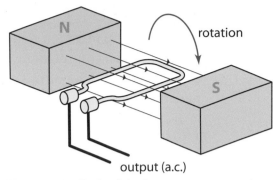

Figure 10.04 Ends of the loop are connected via slip rings

Such a device is called an a.c. generator or alternator and is producing an alternating current (**a.c.**). The number of complete revolutions (cycles) that occur each second is known as the frequency, measured in hertz (Hz) and given the symbol f. The frequency of the supply in this country is 50 Hz.

Alternating current theory

What is alternating current?

Alternating current (a.c.) is a flow of electrons, which rises to a maximum value in one direction and then falls back to zero before repeating the process in the opposite direction. In other words, the electrons within the conductor do not drift (flow) in one direction, but actually move backwards and forwards.

The journey taken, i.e. starting at zero, flowing in both directions and then returning to zero is called a cycle. The number of cycles that occur every second is said to be the frequency and this is measured in hertz (Hz).

Values of an alternating waveform

If we look at the graph of a sine wave (Figure 10.03), there are several values that can be measured from such an alternating waveform.

Instantaneous value

If we take a reading of induced electromotive force (e.m.f.) from the sine wave at any point in time during its cycle, this would be classed as an instantaneous value.

Average value

Using equally spaced intervals in our cycle (say every 30 degrees) we could take a measurement of current as an instantaneous value. To find the average we would add together all the instantaneous values and then divide by the number of values used. As with the average of anything, the more values used the greater the accuracy. For a sine wave only, we say that the average value is equal to the maximum value multiplied by 0.637. As a formula:

Average current = Maximum (peak) current × 0.637

Peak value

When the loop in an a.c. generator has rotated for 90 degrees it is cutting the maximum lines of magnetic flux and, therefore, the greatest value of induced e.m.f. is experienced at this point. This is known as the peak value and both the positive and negative half cycles have a peak value.

Peak to peak value

The peak is the maximum value of induced flow in either direction. The voltage measured between the positive and negative peaks is known as the peak to peak value. The graph in Figure 10.05 shows this more clearly.

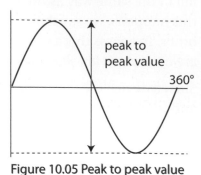

Figure 10.05 Peak to peak value

Frequency and period

The number of cycles that occur each second is referred to as the frequency of the waveform and this is measured in hertz (Hz). The frequency of the UK supply system is 50 Hz.

In the basic arrangement of the a.c. generator loop, if one cycle of e.m.f. was generated with one complete revolution of the loop over a period of one second, then we would say the frequency was 1 Hz. If we increased the speed of loop rotation so that it was producing five cycles every second, then we would have a frequency of 5 Hz.

We can therefore say that the frequency of the waveform is the same as the speed of the loop's rotation, measured in revolutions per second. We can express this using the following equation:

Frequency (f) = Number of revolutions (n) × Number of **pole pairs**

Definition

Pole pair – any system consisting of a north and south pole

If we apply this to the simple a.c. generator in Figure 10.04, and rotate the loop at 50 revolutions per second, then:

Frequency = 50 × 1 (there is 1 × pole pair) = 50 Hz

The amount of time taken for the waveform to complete one full cycle is known as the periodic time (T) or period. Therefore, if 50 cycles are produced in one second, one cycle must be produced in a fiftieth of one second. This relationship is expressed using the following equations:

$$\text{Frequency (f)} = \frac{1}{\text{Periodic time}} = \frac{1}{T} \qquad \text{Periodic time (T)} = \frac{1}{\text{Frequency}} = \frac{1}{f}$$

Root mean square (r.m.s.) or effective value of a waveform (voltage and current)

In direct current (d.c.) circuits, the power delivered to a resistor is given by the product of the voltage across the element and the current through the element.

However, this is only true of the instantaneous power to a resistor in an a.c. circuit.

In most cases the instantaneous power is of little interest, and it is the average power delivered over time that is of most use. In order to have an easy way of measuring power, the **root mean square (r.m.s.) method** of measuring voltage and current was developed.

The r.m.s. or effective value is defined as being the a.c. value of an equivalent d.c. quantity that would deliver the same average power to the same resistor.

When current flows in a resistor, heat is produced. When direct current is flowing in a resistor, the amount of electrical power converted into heat is expressed by the formulae:

$$P = I^2 \times R \quad \text{or} \quad P = V \times I$$

However, an alternating current having a maximum (peak) value of 1 A does not maintain a constant value (see Figure 10.05). The alternating current will not produce as much heat in the resistance as will a direct current of 1 A. Consider the circuits in Figure 10.06.

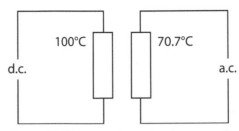

Figure 10.06 d.c. and a.c. circuits

In both the circuits in the figure, the various supplies provide a maximum (peak) value of current of 1 A to a known resistor. However, if the heat produced by 1 ampere of direct current raised the temperature of the resistor to 100°C, then 1 ampere of alternating current would only raise it to 70.7°C. We can express this by using the following formula.

$$\frac{\text{Heating effect of 1A maximum a.c.}}{\text{Heating effect of 1A maximum d.c.}} = 70.7 = 0.707$$

Therefore, **the effective or r.m.s. value of an a.c. = 0.707 × I_{max}**

where I_{max} = the peak value of the alternating current.

We can also establish the maximum (peak) value from the r.m.s. value with the following formula.

$$I_{max} = I_{r.m.s.} \times 1.414$$

The rate at which heat is produced in a resistor is a convenient way of establishing an effective value of alternating current, and is known as the 'heating effect' method.

An alternating current is said to have an effective value of one ampere when it produces heat in a given resistance at the same rate as one ampere of direct current.

To see where the r.m.s. value sits alongside the peak value, look at Figure 10.07.

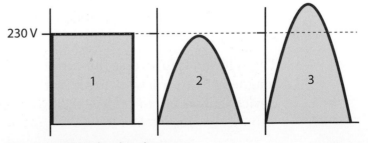

Figure 10.07 Peak value diagrams

The wave in Figure 10.07 (1) represents a 230 V d.c. supply, running for a set period of time, with the heating effect produced shown as the shaded area.

The wave in Figure 10.07 (2) represents an a.c. 50 Hz supply in which the voltage peaks at 230 V during one half-cycle, with the heating effect produced shown as the shaded area. As the wave only reaches 230 V for a small period of time, less heat is produced overall than in the d.c. wave.

The wave in Figure 10.07 (3) represents an increased peak value of voltage to give the same amount of shaded area as in the d.c. example. In this diagram 230 V has become our r.m.s. value.

Important note: Unless stated otherwise, all values of a.c. voltage and current are given as r.m.s. values.

The d.c. motor

The operating principle of a d.c. motor is fairly straightforward. Looking at Figure 10.08, if we take a conductor, form it into a loop that is pivoted at its centre, place it within a magnetic field and then apply a d.c. supply to it, a current will pass around the loop. This will cause a magnetic field to be produced around the conductor which will interact with the magnetic field between the two poles (pole pair) of the magnet.

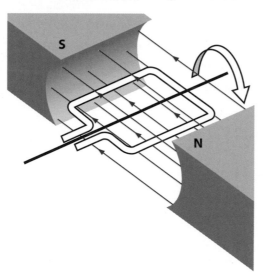

Figure 10.08 Single current-carrying loop in a magnetic field

The magnetic field between the pole pair becomes distorted and, as mentioned earlier, this interaction between the two magnetic fields causes a bending or stretching of the lines of force. The lines of force behave a bit like an elastic band and consequently are always trying to find the shortest distance between the pole pair. These stretched lines, in their attempt to return to their shortest length, therefore exert a force on the conductor that tries to push it out of the magnetic field, and thus they cause the loop to rotate.

In reality there isn't just one loop, there are many loops wound on to a central rotating part of the d.c. motor known as the armature. When the armature is positioned so that the loop sides are at right angles to the magnetic field, a turning force is exerted. But when the coil has rotated 180 degrees the magnetic field in the loop is opposite to that of the field. This will tend to push the armature back the way it came, thus stopping the rotating motion.

The solution is to reverse the current in the armature every half-rotation so that the magnetic fields will work together to maintain a continuous rotating motion. The device that we use to achieve this switching of polarity is known as the **commutator**.

The commutator

Figure 10.09 now gives a different perspective of the same arrangement, showing the device that enables the supply to be connected to the loop whereby the loop can rotate continuously. This is the commutator. This simplified version has only two segments, connecting to either side of the loop, but in actual machines there may be any number. Large machines would have in excess of 50, and the numbers of loops can be in the hundreds. The commutator is made of copper, with the segments separated from each other by insulation.

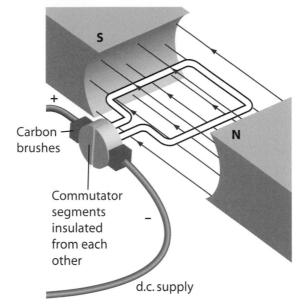

Figure 10.09 Single loop with commutator

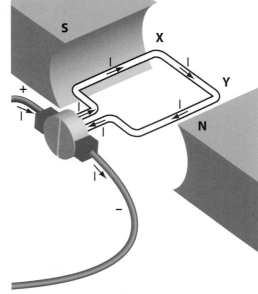

Figure 10.10 Single loop with d.c. flowing

Figure 10.10 now shows the direction of current around the loop when connected to a d.c. supply. The brushes remain in a fixed position against the copper segments of the commutator. Note the direction of current in those sections indicated by X and Y in the diagram.

Figure 10.11 now shows the wire loop having rotated 180 degrees. X and Y have changed positions but, as can be seen from the arrows, current flow remains as it was in Figure 10.10. As the poles of the armature electromagnet pass the poles of the permanent magnets, the commutator reverses the polarity of the armature electromagnet. In that instant of switching polarity, inertia keeps the motor going in the proper direction and thus the motor continues to rotate in one direction.

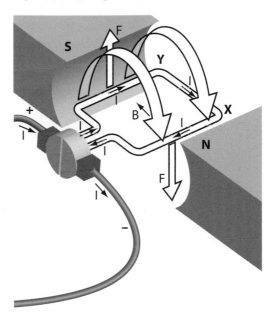

Figure 10.11 Single loop rotated 180 degrees

In the previous figures the armature is shown rotating between a pair of magnetic poles. Practical d.c. motors use electromagnets instead of permanent magnets. The electromagnet has two advantages over the permanent magnet.

1. By adjusting the amount of current flowing through the wire the strength of the electromagnet can be controlled.

2. By changing the direction of current flow the poles of the electromagnet can be reversed.

Reversing a d.c. motor

The direction of rotation of a d.c. motor may be reversed by either:

- reversing the direction of the current through the field; hence changing the field polarity

- reversing the direction of the current through the armature.

Common practice is to reverse the current through the armature, and this is normally achieved by reversing the armature connections only.

Types of d.c. motor

There are three basic forms of d.c. motor:

- series

- shunt

- compound.

They are very similar to look at, the difference being the way in which the field coil and armature coil circuits are wired.

The series motor has the field coil wired in series with the armature. It is also called a universal motor because it can be used in both d.c. and a.c. situations, has a high starting torque (rotational force) and a variable speed characteristic. The motor can therefore start heavy loads, but the speed will increase as the load is decreased.

The shunt motor has the armature and field circuits wired in parallel, and this gives constant field strength and motor speed.

The compound motor combines the characteristics of both the series and the shunt motors, and thus has high starting torque and fairly good speed torque characteristics. However, because it is complex to control, this arrangement is usually only used on large bi-directional motors.

The a.c. and d.c. generator

We now know that when a current is present in a conductor a magnetic field is set up around that conductor, always in a clockwise direction in relation to the direction of current flow.

We also know that, when a conductor passes at right angles through a magnetic field, current is induced into the conductor. The direction of the induced current will depend on the direction of movement of the conductor, and the strength of the current will be determined by the speed at which the conductor moves.

It therefore follows that if we were to take this arrangement and connect it to some device that would spin the wire loop within the permanent magnetic field, we would then induce into the wire loop an e.m.f. If we were to connect a load to the armature via the commutator and brushes, a current would flow around the circuit and the load would work. In other words, we would have created a generator.

A variety of sources can be used mechanically to turn the generator's armature, such as steam, wind, water fall or petrol/diesel driven motors. Figure 10.12 below shows just such an arrangement and the voltage output for one complete revolution.

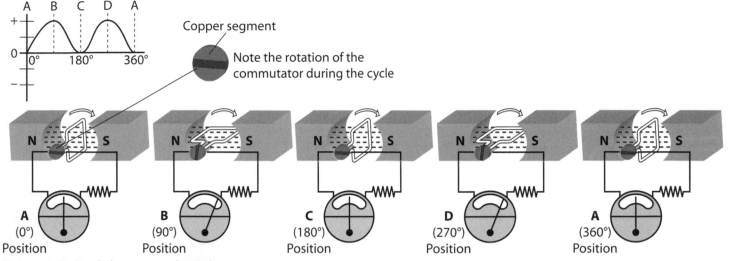

Figure 10.12 Single loop rotated 180 degrees

As you can see from Figure 10.12, the output from such a generator has no negative parts in its cycle. This type of generator produces a voltage/current that alternates in magnitude but flows in one direction only. In other words it is a direct current (d.c.).

When a d.c. generator contains only a single coil it provides a pulsating d.c. output, as shown by the wave form in Figure 10.12. Consequently, in general use a number of coils are used to produce a more stable output.

The operating principle of the a.c. generator is much the same as that of the d.c. version. However, instead of the loop ends terminating at the commutator, they are terminated at slip rings. The a.c. generator normally has a stationary armature known as the stator and a rotating magnetic field known as the rotor.

Originally d.c. was used as the main method of transmitting electricity. But as the technology developed, a.c. became the preferred choice for two main reasons.

- **Reason 1**
  Transformers make it easy to adjust a.c. to a higher or lower voltage very efficiently. This is useful, because to transmit at high voltage reduces the current and power loss and therefore allows smaller cable sizes and a reduction in costs. Transformers do not work for d.c. Adjusting d.c. voltages requires converting the d.c. to a.c., adjusting the resulting a.c. voltage with a transformer, and then converting the adjusted a.c. voltage to a corresponding d.c. voltage. Clearly, adjusting the d.c. voltage is more complicated, and not surprisingly more expensive, than adjusting a.c. voltages.

- **Reason 2**
  Good a.c. motors (and generators) are easier and cheaper to build than good d.c. motors. Although motors are available for either a.c. or d.c., the structure and

characteristics of a.c. and d.c. motors are quite different. a.c. makes it easy to produce a magnetic field whose direction rotates rapidly in space. Any electric conductor placed within the rotating magnetic field rotates with the field. Consequently, a metal armature rotates with the rotating magnetic field with little slippage and, through a shaft attached to the armature, can deliver mechanical power to a mechanical load such as a fan or a water pump.

Called an a.c. induction motor, it is a reasonably simple means of converting electric power to mechanical power. However, d.c. motors rely on a complex mechanical system of brushes and commutator switches. The mechanical complexity of d.c. motors, consequently, not only makes them more expensive to manufacture than a.c. motors, but also more expensive to maintain.

The a.c. motor

As we described earlier, there are three general types of d.c. motor. However, there are many types of a.c. motor, each one having a specific set of operating characteristics such as **torque**, speed, single-phase or three-phase, and this determines their selection for use. We can essentially group them into two categories: single-phase and three-phase. Single and three-phase motors are covered in greater depth in chapter 11.

A simple a.c. motor

In a d.c. motor, electrical power is conducted directly to the armature through brushes and a commutator. Due to the nature of an alternating current, an a.c. motor does not need a commutator to reverse the polarity of the current. Whereas a d.c. motor works by changing the polarity of the current running through the armature (the rotating part of the motor), the a.c. motor works by changing the polarity of the current running through the stator (the stationary part of the motor).

Relays

To get the basic concept across, let's look at the world famous one-way switch. In any typical lighting circuit, if we want to put the light on in a room, the switch is operated by our finger. When we do this, we are closing the internal switch contact and the contact is mechanically held in place across the terminals. Consequently when we take our finger off the switch, it remains in position and the light stays on.

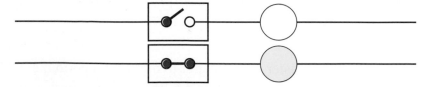

Figure 10.13 One-way switch – on position

But let's say that we don't want that sort of switch. Instead we want to control the light by using a relay. The concept is similar if you think of a relay as being an assembly that contains a one-way switch and a coil.

We'll draw the switch contact in a slightly different way this time, but the idea is exactly the same, i.e. electricity will pass from one terminal to the other when the contact is closed.

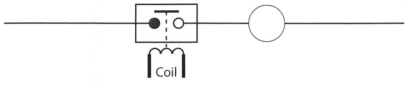

Figure 10.14 One-way switch and coil – off position

If we were now to energise the coil, the resulting magnetic field would pull the contact across the terminals, thus closing the circuit and the light would come on. Except this time, instead of the switch contact being held in place mechanically, it is being held in place by the magnetic field produced by the coil in the relay. It will only remain this way while the coil is energised.

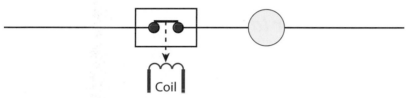

Figure 10.15 When the coil is energised the switch is on

We describe this type of relay as having 'normally open' contacts, in that when the coil is de-energised the contact opens and no electricity can pass through the relay.

It is possible to have a relay where the exact opposite function takes place, i.e. when the coil is energised the contact is pulled away from the terminals. In such a relay the supply would normally be passing through the closed contact and operating the coil will break the circuit. We say that such a relay has 'normally closed' contacts.

Now that we understand the concept, we can accurately say that a relay is an electro-mechanical switch that uses an electromagnet to create a magnetic field to open or close one or many sets of contacts.

Applications

Relays can be used to:

- control a high-voltage circuit with a low-voltage signal, as in some types of modem
- control a high-current circuit with a low-current signal, as in all the lights in the hall of a leisure centre being controlled from a 5 A switch in reception
- control a mains-powered device from a low-voltage switch.

Remember

Relays are powerful as they can be used to switch current between circuits or turn a circuit on and off

Relay selection

When choosing a relay there are several things to consider.

- **Coil voltage** – this indicates how much voltage (230 V, 24 V) and what kind (a.c. or d.c.) must be applied to energise the coil. Make sure that the coil voltage matches the supply being fed into it.

- **Contact ratings** – this indicates how heavy a load the relay can run.

- **Contact arrangement** – there are many kinds of switches, so there are many kinds of relays. The contact geometry indicates how many poles there are, and how they open and close.

For example, a changeover relay has one moving contact and two fixed contacts. One of these is normally closed when the relay is switched off, and the other is normally open. Energising the coil causes the normally open contact to close and the normally closed contact to open.

Fluorescent luminaire

This is a term that refers to illumination derived from the ionisation of gas.

Low-pressure mercury vapour lamps

The fluorescent lamp, or, more correctly, the low-pressure mercury vapour lamp, consists of a glass tube filled with a gas such as krypton or argon and a measured amount of mercury vapour. The inside of the glass tube has a phosphor coating and at each end there is a sealed set of oxide-coated electrodes, known as cathodes.

When a voltage is applied across the ends of a fluorescent tube the cathodes heat up, and this forms a cloud of electrons which ionise the gas around them. The voltage to carry out this ionisation must be much higher than the voltage required to maintain the actual discharge across the lamp. Manufacturers use several methods to achieve this high voltage, usually based on a transformer or choke. This ionisation is then extended to the whole length of the tube so that the arc strikes and is then maintained in the mercury, which evaporates and takes over the discharge. The mercury arc, being at low pressure, emits little visible light but a great deal of ultraviolet, which is absorbed by the phosphor coating and transformed into visible light.

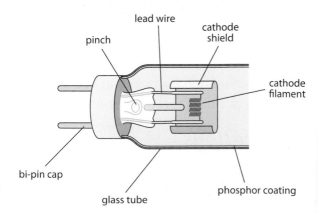

Figure 10.16 Detail of one end of a fluorescent tube

The cathodes are sealed into each end of the tube and consist of tungsten filaments coated with an electron-emitting material. Larger tubes incorporate cathode shields – iron strips bent into an oval shape to surround the cathode. The shield traps material given off by the cathodes during the tube life and thereby prevents the lamp ends blackening.

The gas in standard tubes is a mercury and argon mix, although some lamps (the smaller ones and the new slim energy-saving lamps) have krypton gas in them. The phosphor coating is a very important factor affecting the quantity and quality of light output. When choosing different lamps there are three main areas to be considered:

- lamp efficacy
- colour rendering
- colour appearance.

Lamp efficacy

This refers to the lumen output for a given wattage. For fluorescent lamps this varies from between 40 and 90 lumens per watt.

Colour rendering

This describes a lamp's ability to show colours as they truly are. This can be important, depending on the building usage. For example, it would be important in a paint shop but less so in a corridor of a building. The rendering of colour can affect people's attitude to work etc. – quite apart from the fact that, in some jobs, true colour may be essential. By restoring or providing a full colour range the light may also appear to be better or brighter than it really is.

Colour appearance

This is the actual look of the lamp. The two ends of the scale are warm and cold. These extremes are related to temperatures: the higher the temperature, the cooler the lamp. This is important for the overall effect, and generally warm lamps are used to give a relaxed atmosphere while cold lamps are used where efficiency and businesslike attitudes are a priority. More recently the subject of lamp choice has become complicated, and a programme of lamp rationalisation has begun. The intention is that the whole range currently available will be reduced. Also, new work has resulted in lamps with high lumen outputs and good colour rendering possibilities.

Did you know?

The most economic tube life is limited to around 5000 or 6000 hours. In industry, tubes are changed at set time intervals and all the tubes, whether still working or not, are replaced. This saves money on maintenance, stoppage of machinery and scaffolding erection etc.

Transformers

On the completion of this topic area the reader will be able to state the operating principles of transformers.

Transformer principles

Transformers were introduced in chapter 5 pages 114–117. In that section we looked at how transformers are constructed and the role they play in electric circuits. Here we will look a bit more deeply at the operation of transformers, their losses, some different types and, finally, do some example calculations. The **transformer** is one of the most widely used pieces of electrical equipment and can be found in situations such as electricity distribution, construction work and electronic equipment. Its purpose, as the name implies, is to transform something – the something in this case being the voltage, which can enter the transformer at one level (input) and leave at another (output).

When the output voltage is higher than the input voltage this is called a step-up transformer. When the output voltage is lower than the input, this is called a step-down transformer.

Mutual inductance

In their operation, transformers make use of an action known as mutual inductance. Two coils, primary and secondary, are placed side by side but not touching each other. The primary coil is connected to an a.c. supply and the secondary coil is connected to a load, such as a resistor.

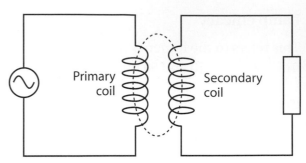

Figure 10.17 Mutual inductance

If we allow current to flow in the primary coil, this will create a magnetic field. As the current in the primary coil increases up to its maximum value, it creates a changing magnetic flux. As long as there is a changing magnetic flux, there will be an e.m.f. 'induced' into the secondary coil, which would then start flowing through the load. This effect, where an alternating e.m.f. in one coil causes an alternating e.m.f. in another coil, is known as mutual inductance.

In transformers we are only really interested in mutual inductance, and it is the rising and falling a.c. current that causes the change of magnetic flux. In other words: **an a.c. supply is necessary to allow transformers to operate correctly**.

If the two coils were now wound on an iron core, we would find that the level of magnetic flux is increased and consequently the level of mutual inductance is also increased.

Transformer types

Core-type transformers

Figure 10.18 shows a double wound, core-type transformer, which we will use to cover the principles of transformers. The supply is wound on one side of the iron core (primary winding) and the output is wound on the other (secondary winding). In other words, double wound means that there is more than one winding.

The number of turns in each winding will affect the induced e.m.f., with the number of turns in the primary being referred to as N_p, and those in the secondary referred to as (N_s). We call this the turns ratio. When voltage (V_p) is applied to the primary

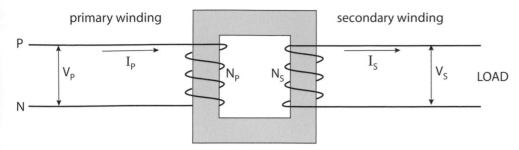

Figure 10.18 Double wound, core type transformer

winding, it will cause a changing magnetic flux to circulate in the core. This changing flux will cause an e.m.f. (V_s) to be induced in the secondary winding.

Assuming that we have no losses or leakage (i.e. 100 per cent efficiency) then power input will equal power output and the ratio between the primary and secondary sides of the transformer can be expressed as follows:

$$\frac{V_P}{V_S} = \frac{N_P}{N_S} = \frac{I_S}{I_P}$$

(where I_P represents the current in the primary winding and I_S the current in the secondary winding).

As we can see from Figure 10.18, a transformer has no moving parts. Consequently, provided that the following general statements apply, it ends up being a very efficient piece of equipment.

- Transformers use laminated (layered) steel cores, not solid metal. In a solid metal core 'eddy currents' are induced which cause heating and power losses.

- Laminated cores, where each lamination is insulated, help to reduce this effect.

- Soft iron with high magnetic properties is used for the core.

- Windings are made from insulated, low resistance conductors. This prevents short circuits occurring either within the windings or to the core.

The losses that occur in transformers are normally classed as copper and iron losses.

Copper losses

Although windings should be made from low resistance conductors, the resistance of the windings will cause the currents passing through them to create a heating effect and subsequent power loss. This power loss can be calculated using the formula:

$$P_C = I^2 \times R \text{ watts}$$

(where I is the current flowing in amps and R is the resistance of the winding in ohms).

Remember

No transformer can be 100 per cent efficient. There will always be power losses

Example

Calculate the copper loss of a secondary winding in a small step down transformer that has a resistance of 0.35 Ω when the connected external load is drawing 10 amps.

$$P_C = I^2 \times R \quad \text{therefore} \quad P_C = 10^2 \times 0.35 = 100 \times 0.35 = \textbf{35 watts}$$

What are the copper losses in the primary winding of a step up transformer, where the resistance of the winding is 0.02 Ω and the transformer is drawing 6 amps from the supply?

$$P_C = I^2 \times R \quad \text{therefore} \quad P_C = 6^2 \times 0.02 = 36 \times 0.02 = \textbf{0.72 watts}$$

Definition

Hysteresis – a generic term meaning a lag in the effect of a change of force

Iron losses

These losses take place in the magnetic core of the transformer. They are normally caused by eddy currents (small currents which circulate inside the laminated core of the transformer) and **hysteresis**. To demonstrate, let us say that you push on some material and it bends. When you stop pushing, does it return to its original shape? If it does not the material is demonstrating hysteresis. Let us look at this in context.

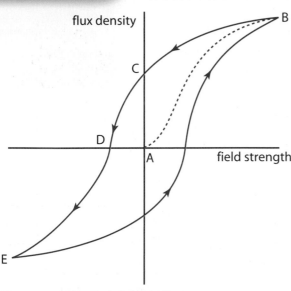

Figure 10.19 Hysteresis loop diagram

Figure 10.19 shows the effect within ferromagnetic materials of hysteresis. Starting with an unmagnetised material at point A, both field strength and flux density are zero. The field strength increases in the positive direction and the flux begins to grow along the dotted path until we reach saturation at point B. This is called the initial magnetisation curve.

If the field strength is now relaxed, instead of retracing the initial magnetisation curve the flux falls more slowly. In fact, even when the applied field has returned to zero, there will still be a degree of flux density (known as the **remanence**) at point C. To force the flux to go back to zero (point D), we have to reverse the applied field. The field strength that is necessary to drive the field back to zero is known as the **coercivity**. We can then continue reversing the field to get to point E, and so on. This is known as the hysteresis loop.

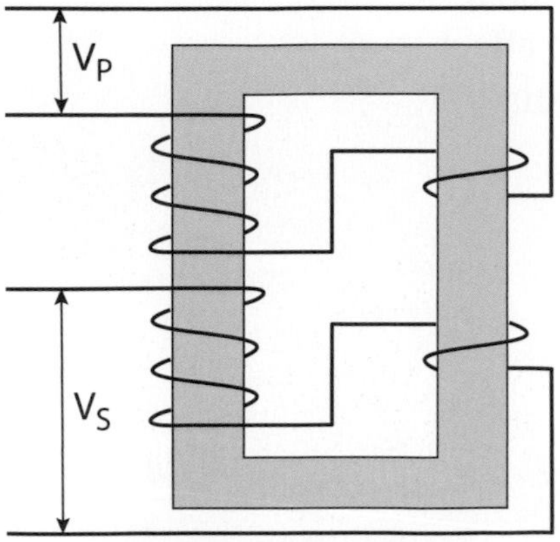

As we have already said, we can help reduce eddy currents by using a laminated core construction. We can also help to reduce hysteresis by adding silicon to the iron from which the transformer core is made.

The version of the double wound transformer that we have looked at so far makes the principle of operation easier to understand. However, this arrangement is not very efficient as some of the magnetic flux being produced by the primary winding will not react with the secondary winding and is often referred to as 'leakage'. We can help to reduce this leakage by splitting each winding across the sides of the core (see Figure 10.20).

Figure 10.20 Reducing 'leakage'

Other transformer types

Shell-type transformers

We can reduce the magnetic flux leakage a bit more by using a shell-type transformer.

In the shell-type transformer, both windings are wound on to the central leg of the transformer and the two outer legs are then used to provide parallel paths for the magnetic flux.

Figure 10.21 Shell-type transformer

The autotransformer

The autotransformer uses the principle of 'tapped' windings in its operation. Remember that the ratio of input voltage to output voltage will depend upon the number of primary winding turns and secondary winding turns (the turns ratio). But what if we want more than one output voltage?

Some devices are supplied with the capability of providing this, such as small transformers for calculators, musical instruments or doorbell systems. Tapped connections are the normal means by which this is achieved. A tapped winding means that we have made a connection to the winding and then brought this connection out to a terminal. Now, by connecting between the different terminals, we can control the number of turns that will appear in that winding and we can therefore provide a range of output voltages.

An autotransformer has only one tapped winding and the position of the tapping on that winding will dictate the output voltage.

One of the advantages of the autotransformer is that, because it only has one winding, it is more economical to manufacture. However, on the down side, we have made a physical connection to the winding. Therefore, if the winding ever became broken between the two tapping points, the transformer would not work and the input voltage would appear on the output terminals. This would then present a real hazard.

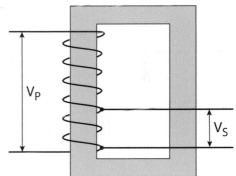

Figure 10.22 Autotransformer

Instrument transformers

Instrument transformers are used in conjunction with measuring instruments because it would be very difficult and expensive to design normal instruments to measure the high current and voltage that we find in certain power systems. We therefore have two types of instrument transformer, both double wound.

The current transformer

The current transformer (c.t.) normally has very few turns on its primary winding so that it does not affect the circuit to be measured, with the actual meter connected across the secondary winding.

Current transformer

Care must be taken when using a c.t. Never open the secondary winding while the primary is 'carrying' the main current. If this happened, a high voltage would be induced into the secondary winding. Apart from the obvious danger of electric shock, the heat build up could cause the insulation on the c.t. to break down.

The voltage transformer

This is very similar to our standard power transformer, in that it is used to reduce the system voltage. The primary winding is connected across the voltage that we want to measure and the meter is connected across the secondary winding.

Step-up and step-down transformers

Step-up transformers

A step-up transformer is used when it is desirable to step voltage up in value.

The primary coil has fewer turns than the secondary coil. We already know that the number of turns in a transformer is given as a ratio. When the primary has fewer turns than the secondary, voltage and impedance are stepped up. In Figure 10.23, voltage is stepped up from 120 V a.c. to 240 V a.c. Since impedance is also stepped up, current is stepped down from 10 amps to 5 amps.

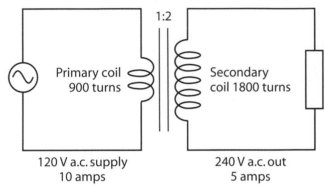

Figure 10.23 Step-up transformer

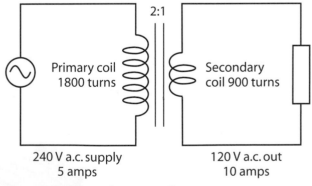

Figure 10.24 Step-down transformer

Step-down transformers

A step-down transformer is used when it is desirable to step voltage down in value.

The primary coil has more turns than the secondary coil. In Figure 10.24 the step-down ratio is 2:1. Since the voltage and impedance are stepped down, the current is stepped up, in this case to 10 A.

Safety isolating transformer

Another use of a transformer is to isolate the secondary output from the supply. In a bathroom, the shower socket should be supplied from a 1:1 safety isolating transformer. The output of the transformer has no connection to earth, thereby ensuring that output from the transformer is totally isolated from the supply.

In areas of installations where there is an increased risk of electric shock, the voltage is reduced to less than 50 V and is supplied from a safety isolating transformer; we know this as a **Separate Extra Low Voltage (SELV)** supply.

Transformer calculations

Transformers are rated in kVA (kilovolt-amps) rather than in watts. This rating has to be used as the transformer is made of copper coils. When a conductor is formed into a coil and connected to an a.c. supply, the coil takes on the property of an inductor. In this state it is not the pure resistance of the conductor that opposes the flow of current, it is the combined values of inductance and resistance – the impedance (Z).

We also know that in an a.c. circuit:

Power = IVCosθ.

We know the output voltage of the transformer from its rating, but not the current. This is because current is determined by the load connected to the transformer. The power factor Cosθ is also an unknown, as this will also be affected by the load. However, the VA rating can be used to give an indication of the transformer's performance, including the current that it can deliver for known voltage. Similarly, if we know the output voltage and current we can find the rating.

Example 1

What is the current that can be drawn from a 100 VA transformer if it has an output voltage of 20 V?

We can find the current by dividing the VA rating by the current.

$$I = \frac{VA}{V} = \frac{100}{20} = \mathbf{5\ A}$$

If a transformer has a maximum output voltage of 200 V and supplies a current of 12 amp what is the kVA rating of the transformer?

$$kVA = \frac{VA}{1000} = \frac{200 \times 12}{1000} = 2.4\ kVA$$

Calculate the line current delivered by an 11 kVA three-phase transformer and a line voltage of 400 V.

$$kVA = \frac{\sqrt{3}V_L I_L}{1000} \quad \text{therefore } 11 = \frac{\sqrt{3} \times 400 \times I_L}{1000}$$

$$\text{Transpose to find } I_L = \frac{11 \times 1000}{\sqrt{3} \times 400} = \mathbf{15.88\ A}$$

Example 2

A transformer having a turns ratio of 2:7 is connected to a 230 V supply. Calculate the output voltage.

When giving transformer ratios they appear in the order primary then secondary. Therefore in this example we are saying that for every two windings on the primary winding there are seven on the secondary. Therefore using our formula:

$$\frac{V_P}{V_S} = \frac{N_P}{N_S}$$

We should now transpose this to get:

$$V_S = \frac{V_P \times N_S}{N_P}$$

However, we do not know the exact number of turns involved. But do we need to, if we know the ratio? Let us find out.

The ratio is 2:7, meaning for every two turns on the primary there will be seven turns on the secondary. Therefore, if we had six turns on the primary, this would give us 21 turns on the secondary, but the ratio of the two has not changed. It remains 2:7.

This means we can just insert the ratio rather than the individual number of turns into our formula:

$$V_S = \frac{V_P \times N_S}{N_P} = \frac{230 \times 7}{2} = \textbf{805 V}$$

Now, to prove our point about ratios, let us say that we know the number of turns in the windings to be 6 in the primary and 21 in the secondary (which is still giving us a 2:7 ratio). If we now apply this to our formula we get:

$$V_S = \frac{V_P \times N_S}{N_P} = \frac{230 \times 21}{6} = \textbf{805 V}$$

The same answer.

Example 3

A single-phase transformer, with 2000 primary turns and 500 secondary turns, is fed from a 230 V a.c. supply. Find:

(a) the secondary voltage

(b) the volts per turn.

Secondary voltage

$$\frac{V_P}{V_S} = \frac{N_P}{N_S}$$

Using transposition, re-arrange the formula to give:

$$V_s = \frac{V_p \times N_s}{N_p}$$

$$V_s = \frac{230 \times 500}{2000} = \frac{115,000}{2000} = \textbf{57.5 V}$$

Volts per turn

This is the relationship between the volts in a winding and the number of turns in that winding. To find volts per turn we simply divide the voltage by the number of turns.

Therefore, in the primary:

$$\frac{V_p}{N_p} = \frac{230}{2000} = \textbf{0.115 volts per turn}$$

In the secondary :

$$\frac{V_s}{N_s} = \frac{57.5}{500} = \textbf{0.115 volts per turn}$$

Example 4

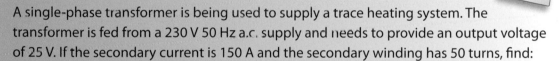

A single-phase transformer is being used to supply a trace heating system. The transformer is fed from a 230 V 50 Hz a.c. supply and needs to provide an output voltage of 25 V. If the secondary current is 150 A and the secondary winding has 50 turns, find:

(a) the output kVA of the transformer

(b) the number of primary turns

(c) the primary current

(d) the volts per turn.

The output kVA:

$$kVA = \frac{volts \times amperes}{1000} = \frac{V_s \times I_s}{1000} = \frac{25 \times 150}{1000} = \textbf{3.75 kVA}$$

The number of primary turns

If: $\dfrac{V_p}{V_s} = \dfrac{N_p}{N_s}$ then by transposition:

$$N_p = \frac{V_p \times N_s}{V_s} = \frac{230 \times 50}{25} = \textbf{460 turns}$$

The primary current:

If: $\dfrac{V_P}{V_S} = \dfrac{I_S}{I_P}$ then by transposition:

$$I_P = \dfrac{V_S \times I_S}{V_P} = \dfrac{25 \times 150}{230} = \textbf{16 A}$$

The volts per turn:

In the primary: $\dfrac{V_P}{N_P} = \dfrac{230}{460} = \textbf{0.5 volts per turn}$

In the secondary: $\dfrac{V_S}{N_S} = \dfrac{25}{50} = \textbf{0.5 volts per turn}$

Example 5

A step-down transformer, having a ratio of 2:1, has an 800-turn primary winding and is fed from a 400 V a.c. supply. The output from the secondary is 200 V and this feeds a load of 20 Ω resistance. Calculate:

(a) the power in the primary winding

(b) the power in the secondary winding.

We know that the formula for power is: $P = V \times I$

We also know that we can use Ohm's law to find current: $I = \dfrac{V}{R}$

Therefore, if we insert the values that we have, we can establish the current in the secondary winding:

$$I_S = \dfrac{V_S}{R_S} = \dfrac{200}{20} = \textbf{10 A}$$

Now that we know the current in the secondary winding, we can use the power formula to find the power generated in the secondary winding:

$$P = V \times I = 200 \times 10 = 2000 \text{ W} = \textbf{2 kW}$$

We now need to find the current in the primary winding. To do this we can use the following formula:

$$\dfrac{V_P}{V_S} = \dfrac{I_S}{I_P}$$

However, we need to transpose the formula to find I_P. This would give us:

$$I_P = \dfrac{I_S \times V_S}{V_P}$$

Which, if we now insert the known values, gives us:

$$I_p = \frac{I_s \times V_s}{V_p} = \frac{10 \times 200}{400} = \frac{2000}{400} = \textbf{5 A}$$

Now that we know the current in the primary winding, we can again use the power formula to find the power generated in the primary winding:

$$P = V_p \times I_p = 400 \times 5 = 2000 \text{ W} = \textbf{2 kW}$$

Resistance, inductance, capacitance, reactance and impedance

On completion of this topic area the candidate will be able to recognise the effects of resistance, inductance, capacitance, reactance and impedance in a.c. circuits.

Resistance (R) and phasor representation

The a.c. current is commonly represented by the sine wave diagram. However, this can be difficult and time consuming to draw. We can therefore also represent a.c. by the use of phasors. A **phasor** is a straight line whose length is a scaled representation of the size of the a.c. quantity and whose direction represents the relationship between the voltage and current, this relationship being known as the phase angle.

To see briefly how we use phasors, let us look at the circuit diagram in Figure 10.25, where a tungsten filament lamp has been included as the load.

Circuits like this are said to be **resistive**, and in this type of circuit the values of e.m.f. (voltage) and current actually pass through the same instants in time together. In other words, as voltage reaches its maximum value so does the current (see Figure 10.25, sine wave diagram).

This happens with all resistive components connected to an a.c. supply and, as such, the voltage and current are said to be '**in phase**' with each other, or possess a zero phase angle.

The graph in Figure 10.25 shows this when represented by a sine wave. However, we could also show this by using a phasor diagram as shown in Figure 10.26.

We can, therefore, say that a resistive component will consume power and we would carry out calculations as for a d.c. circuit (i.e. using $P = V \times I$).

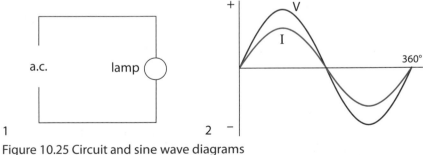

Figure 10.25 Circuit and sine wave diagrams

Figure 10.26 Phasor diagram for zero phrase angle

We can also say that resistive equipment (e.g. filament lamps, fires, water heaters) uses this power to create heat, but such a feature in long cable runs, windings etc. would be seen as unsuitable power loss in the circuit (i.e. using $P = I^2R$).

Inductance (L)

If the load in our circuit were not a filament lamp but a motor or transformer (something possessing windings), then we say that the load is **inductive**.

With an inductive load the voltage and current become '**out of phase**' with each other. This is because the windings of the equipment set up their own induced e.m.f., which opposes the direction of the applied voltage, thus forcing the flow of electrons (current) to fall behind the force pushing them (voltage). When this happens it is known as possessing a lagging phase angle or power factor. As we can see from Figure 10.27, the current is lagging the applied voltage by 90 degrees.

To make things easier in this exercise, we assumed that the above circuit is purely inductive and, in this case, over one full cycle we would see that no power is consumed. However, in reality this is not possible as every coil is made of wire and that wire will have a resistance. The opposition to current flow in a resistive circuit is resistance.

The sine wave and phasors used to represent this inductive circuit would look like this

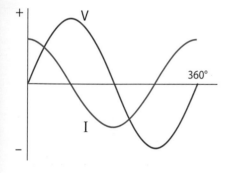

If we represented this as a phasor diagram we end up with

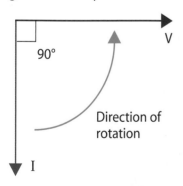

Figure 10.27 Sine wave and phasor diagrams for inductive circuit

The limiting effect to the current flow in a pure inductor is called the inductive reactance, which we are able to calculate with the following formula:

$X_L = 2\pi fL$ (Ω)

where:

X_L = inductive reactance (ohms – Ω)
f = supply frequency (hertz – Hz)
L = circuit inductance (henrys – H).

Let us now look at inductive and resistive circuits and see if the current is affected by a lagging phase angle.

To recap, we said that power factor is the relationship between voltage and current and that the ideal situation would seem to be the resistive circuit, where both these quantities are perfectly linked.

In the resistive circuit we know that the power in the circuit could only be the result of the voltage and the current ($P = V \times I$). This is known as the apparent power and possesses what we call unity power factor, to which we give the value one (1.0).

However, we now know that depending upon the equipment, the true power (actual) in the circuit must take into account the phase angle and will often be less than the apparent power but never greater.

True power (in watts) is calculated using the cosine of the phase angle (cos $\emptyset$). The formula is:

P = VI cos $\emptyset$

When there is no phase lag, $\varnothing = 0$ and $\cos \varnothing = 1$, a purely resistive circuit. To prove our previous points, let us consider the following.

Example

If we have an inductive load, consuming 3 kW of power from a 230 V supply with a power factor of 0.7 lagging, then the current (amount of electrons flowing) required to supply the load is:

$$P = V \times I \times \cos \varnothing$$

$\cos \varnothing$ = Power factor. Therefore by transposition:

$$I = \frac{P}{V \times \cos \varnothing} \text{ or in other words: } I = \frac{P}{V \times PF} = \frac{3000}{230 \times 0.7} = 18.6 \text{ A}$$

However, if the same size of load was purely resistive then $\varnothing = 0$, thus the power factor would be 1.0, and thus:

$$I = \frac{P}{V \times PF} = \frac{3000}{230 \times 1} = 13 \text{ A}$$

In other words, the lower the power factor of a circuit, the higher the current will need to be to supply the load's power requirement.

It, therefore, follows that, if power factor is low, then it will be necessary to install larger cables and switchgear etc. to be capable of handling the larger currents. There will also be the possibility of higher voltage drop due to the increased current in the supply cables.

Consequently, local electricity suppliers will often impose a financial fine on premises operating with a low power factor. Fortunately, we have a component that can help. It is called the capacitor.

Capacitance (C)

Simply put, a **capacitor** is a component that stores an electric charge if a potential difference (p.d.) is applied across it. The capacitor's use is then normally based on its ability to return that energy back to the circuit. When a capacitor is connected to an a.c. supply, it is continuously storing then discharging the charge as the supply moves through its positive and negative cycles. But, as with the pure inductor, no power is consumed.

The sine wave and phasors used to represent this would look as in Figure 10.28. From this you can see that the current is leading the voltage. This means that in a capacitive circuit there is a leading phase angle or power factor.

The current leads the voltage by 90 degrees. Consequently, the capacitor is able to help because, if we connect it in parallel across the load, it can help neutralise the effect of a lagging power factor.

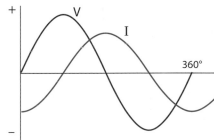

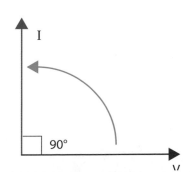

Figure 10.28 Sine wave and phasor diagrams for capacitive circuits

The opposition to the flow of a.c. to a capacitor is termed capacitive reactance, which like inductive reactance is measured in ohms and calculated using the following formula:

$$X_C = \frac{1}{2\pi fC} \ (\Omega)$$

where:

X_C = capacitive reactance (ohms – Ω)
f = supply frequency (hertz – Hz)
C = circuit **capacitance** (farads – F).

Since, in this type of circuit, we have voltage and current but no real power (in watts), the formula of $P = V \times I$ is no longer accurate. Instead, we say that the result of the voltage and current is reactive power, which is measured in reactive volt amperes (VAr).

The current to the capacitor, which does not contain resistance or consume power, is called reactive current.

Phasors

Some circuits contain combinations of all of the above components. To work out these calculations you will need to know about the addition of phasors.

When sine waves for voltage and current are drawn, the nature of the wave diagram can be based upon any chosen alternating quantity within the circuit. In other words, we can start from zero on the wave diagram with either the voltage or the current.

In electrical science we often need to add together alternating values. If they were 'in phase' with each other, then we would simply add the values together. However, when they are not in phase we cannot do this, hence the need for phasor diagrams.

When we use phasor diagrams the chosen alternating quantity is drawn horizontally and is known as the reference.

When choosing the reference phasor, it makes sense to use a quantity that has the same value at all parts of the circuit. For example in a series circuit the same current flows in each part of the circuit, therefore use current as the reference phasor. In a parallel circuit the voltage is the same through each branch of the circuit and therefore we use voltage as the reference phasor.

Using the knowledge gained in the previous section, we can now measure all phase angles from this reference phasor.

Our answer, or resultant, is then found by completing a parallelogram. Using Figure 10.29 as an example, we have been given the values of phasor A and phasor B and their power factors, which gives us the angles by which they lag or lead the reference phasor. Therefore, the result of adding A and B together will be phasor C.

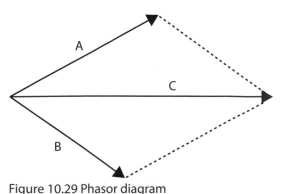

Figure 10.29 Phasor diagram

Impedance

Earlier, we considered individual components within an a.c. circuit. What those components actually are is opposition to the flow of current. Here's a summary.

- The opposition to current in a resistive circuit is called resistance (R). It is measured in ohms and the voltage and current are in phase with each other.

- The opposition to current in an inductive circuit is called inductive reactance (X_L). It is measured in ohms and the current lags the voltage by 90 degrees.

- The opposition to current in a capacitive circuit is called capacitive reactance (X_C). It is measured in ohms and the current leads the voltage by 90 degrees.

However, we know that circuits will contain a combination of these components. When this happens we say that the total opposition to current is called the **impedance** (Z) of that circuit.

Here's a summary.

- The power consumed by a resistor is dissipated in heat and not returned to the source. This is called the **true power**.

- The energy stored in the magnetic field of an **inductor** or the plates of a capacitor is returned to the source when the current changes direction.

- The power in an a.c. circuit is the sum of true power and reactive power. This is called the **apparent power**.

- **True power is equal to apparent power in a purely resistive circuit** because the voltage and current are in phase. Voltage and current are also in phase in a circuit containing equal values of inductive reactance and capacitive reactance. If the voltage and current are 90 degrees out of phase, as would be the case in a purely capacitive or purely inductive circuit, the average value of true power is equal to zero. There are high positive and negative peak values of power, but when added together the result is zero.

- Apparent power is measured in volt-amps (VA) and has the formula **P = VI**.

- True power is measured in watts and has the formula **P = VI cos $\varnothing$** where Cos $\varnothing$ is the power factor

- **In a purely resistive circuit** where current and voltage are in phase, there is no angle of displacement between current and voltage. The cosine of a zero degree angle is one, so the power factor is one. This means that all the energy that is delivered by the source is consumed by the circuit and dissipated in the form of heat.

- **In a purely reactive circuit**, voltage and current are 90 degrees apart. The cosine of a 90-degree angle is zero, so the power factor is zero. This means that the circuit returns all the energy it receives from the source back to the source.

- **In a circuit where reactance and resistance are equal**, voltage and current are displaced by 45 degrees. The cosine of a 45-degree angle is 0.7071, and so the

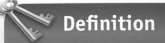

Definition

Impedance – total opposition to current in an a.c. circuit

True or active power – the rate at which energy is used

Apparent power – in an a.c. circuit the sum of the true or active power and the reactive power

power factor is 0.7071. This means that such a circuit uses approximately 70 per cent of the energy supplied by the source and returns approximately 30 per cent back to the source.

Power

On completion of this topic area the candidate will be able to define the terms 'power' and 'power factor', and state why correction is required and how this may be achieved.

Energy, work and power

We know that electrons are pushed along a conductor by a force called the e.m.f. Now consider the electrical units of work and power. Energy and work are interchangeable, in that we use up energy to complete work. Both are measured in terms of force and distance.

If a force is required to move an object some distance, then work has been done and some energy has been used to do it. The greater the distance and the heavier the object, then the greater the amount of work done.

We already know that:

Energy (or Work done) = Distance moved × Force required

Power may then be stated as being, 'the rate at which we do work' and it is measured in watts. So:

$$\text{Power} = \frac{\text{Energy (or Work done)}}{\text{Time taken}} = \frac{\text{Distance moved} \times \text{Force required}}{\text{Time taken}}$$

For example, we could drill two holes in a wall – one using a hand drill, and the other with an electric drill. When we have finished, the work done will be the same: there will be two identical holes in the wall. However, the electric drill will do it more quickly because its power is greater.

If power is therefore considered to be the ratio of work done against the time taken to do the work, we may express this as follows.

$$\text{Power (P)} = \frac{\text{Work done (W)}}{\text{Time taken (t)}} = \frac{\text{Energy used}}{\text{Time taken}}$$

The units are:

$$\text{Watts} = \frac{\text{joules}}{\text{seconds}}$$

The e.m.f. is the amount of joules of work necessary to move one coulomb of electricity around the circuit. It is measured in joules per coulomb, also known as the volt. **Noting that 1 volt = 1 joule/coulomb** and arranging the formula, this could be expressed as:

Joules = volts × coulombs

and since

coulombs = amperes × seconds

we can substitute this into:

joules = volts × amperes × seconds.

Since joules are the units of work:

Work = V × I × t (joules).

Taking this one step further, we can show how we arrive at some of our electrical formulae. It goes as follows.

If:

$$Power = \frac{joules}{seconds}$$

this means:

$$P = \frac{V \times I \times t}{t}$$

So cancelling the tees:

$$P = V \times I = I \times V$$

and in Ohm's law:

$$V = I \times R$$

then:

$$P = I \times (I \times R) = I^2 \times R$$

or:

$$P = I^2 \times R$$

Finally, if:

$$P = I \times V$$

And in Ohm's law:

$$I = \frac{V}{R}$$

Then:

$$P = \frac{V \times V}{R} = \frac{V^2}{R}$$

or:

$$P = \frac{V^2}{R}$$

Simple when you know how!

Example 1

A 100 Ω resistor is connected to a 10 V d.c. supply. What will be the power dissipated in it?

$$P = \frac{V^2}{R} \text{ therefore } \frac{10 \times 10}{100} = \textbf{1 W}$$

Example 2

How much energy is supplied to a 100 Ω resistor that is connected to a 150 V supply for one hour?

$$P = \frac{V^2}{R} \text{ therefore } \frac{150 \times 150}{100} = \textbf{225 W}$$

Now if E is the energy supplied by the power over a period of time, t:

$E = P \times t$

As all time measurements are given in seconds, we have to change hours or minutes into seconds. So:

t = one hour = 60×60 seconds = 3600 s

Therefore:

E = energy supplied = 225×3600 joules

so:

E = 810,000 joules

Kilowatt hour

It should be noted that the joule is far too small a unit for sensible energy measurement. For most applications, we use something called the kilowatt hour.

The kilowatt hour could be defined as the amount of energy used when one kilowatt (1,000 watts) of power has been used for a time of one hour (3,600 seconds).

From this we can see that:

1 joule (J) = 1 watt (W) for one second (s)
1000 joules (J) = 1 kilowatt (kW) for one second

In one hour there are 3600 seconds. Therefore:

3600 s x 1,000 J = 1 kW for one hour (kWh)

so:

1 kWh = 3.6 × 10⁶ J

The kilowatt hour is the unit used by the electrical supply companies to charge their customers for the supply of electrical energy. Have a look in your house. You will see that the electric meter is measuring in kWh. However, these are more often referred to as units by the time they appear on your bill!

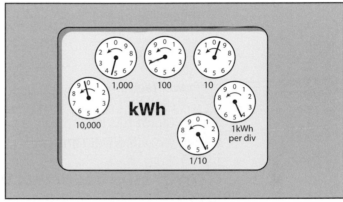

Figure 10.30 Typical electric meter dials

Efficiency

We looked at efficiency in chapter 4. The calculations and theory for efficiency applied to electrical circuits are very similar.

You already know that:

$$\text{Percentage efficiency} = \frac{\text{Output}}{\text{Input}} \times 100$$

Let us have a look at two examples.

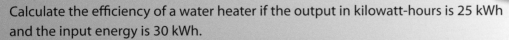

Example 1

Calculate the efficiency of a water heater if the output in kilowatt-hours is 25 kWh and the input energy is 30 kWh.

$$\text{Efficiency (\%)} = \frac{\text{Output}}{\text{Input}} \times 100 = \frac{25}{30} \times 100 = \textbf{83.33\%}$$

Example 2

The power output from a generator is 2700 W and the power required to drive it is 3500 W. Calculate the percentage efficiency of the generator.

$$\text{Efficiency (\%)} = \frac{\text{Output}}{\text{Input}} \times 100 = \frac{2700}{3500} \times 100 = \mathbf{77.1\%}$$

Power factor

When we are dealing with a.c. circuits, we are often looking at the way power is used with particular types of component within the circuit. We have already looked at power factor and phase angle in previous pages of this chapter.

Generally speaking, power factor is a number less than 1.0, which is used to represent the relationship between the apparent power of a circuit and the true power of that circuit. In other words:

$$\text{Power factor (PF)} = \frac{\text{True power (PT)}}{\text{Apparent Power (PA)}} \quad \text{or: } PF = \frac{PT}{PA}$$

In terms of units:

$$\text{Power factor (PF)} = \frac{\text{Power (W)}}{\text{Voltage} \times \text{Current (VA)}}$$

Power factor has no units, it is a number. It is also determined by the phase angle, which we will cover shortly.

Power in an a.c. circuit

If you were to try to push something against a resistance, you would get hot and bothered as you used up energy in completing the task. When current flows through a resistor, a similar thing happens in that power is the rate of using up energy – in other words, the amount of energy that was used in a certain time.

We already know that, if a resistor (R) has a current (I) flowing through it for a certain time (t), then the power (energy being used per second) given in watts can be calculated by the following formula.

$$P = I^2 \times R$$

This power will be dissipated in the resistor as heat and reflects the average power in terms of the r.m.s. values of voltage and current (see 'Alternating current theory' on pages 249–252 of this chapter).

This effectively means that the average power in a resistive circuit (one which is non-reactive, i.e. does not possess inductance or capacitance) can be found by the product of the readings of an ammeter and a voltmeter.

In other words, in the resistive circuit the power (energy used per second) is associated with that energy being transferred from the medium of electricity into another medium, such as light (filament lamp) or heat (electric fire/kettle). We call this type of power the active power. When we look at the capacitive circuit, we find that current flows to the capacitor but we have no power.

Look at this wave diagram for voltage (Figure 10.31).

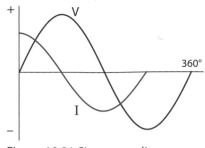

Figure 10.31 Sine wave diagram

During the first period of the cycle, the voltage is increasing and this provides the energy to charge the capacitor. However, in the second period of the cycle, the voltage is decreasing and therefore the capacitor discharges, returning its energy back to the circuit as it does so. The same is also true of the third and fourth periods.

This exchange of energy means that we have voltage and current, but no average power and therefore no heating effect. Thus the formula ($P = I^2 \times R$) is no longer useful.

We therefore say that the result of voltage and current in this type of circuit is called reactive power and we express this in reactive volt-amperes (VAr). Equally, we say that the current in a capacitive circuit, where there is no resistance and no dissipation of energy, is called reactive current.

In an inductive circuit there is a similar position. This time, as voltage increases during the first period of the cycle, the energy is stored as a magnetic field in the inductor. This energy will then be fed back into the circuit during the second period of the cycle as the voltage decreases and the magnetic field collapses. In other words, once again the exchange of energy produces no average power (energy used per second).

Circuits are likely to comprise combinations of resistance, inductance and capacitance. In a circuit, which has resistance and reactance, there will be a phase angle between the voltage and current. This relationship has relevance as power will only be expended in the resistive part of the circuit.

Let us now look at this relationship via a phasor diagram for a circuit containing resistance and capacitance, remembering that for this type of circuit the current will lead the voltage.

This circuit has two components (resistance and capacitance), so we have drawn two current phasors. In reality, as we already know, these are not currents that will actually flow, but their phasor sum will be the actual current in the circuit I_p (actual).

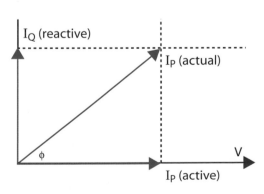

Figure 10.32 Phasor diagram

We also said that in a resistive circuit the voltage and current are in phase, and therefore this section of the current has been represented by the phasor I_p (active). This part of the current is in the active section and we therefore refer to this as the active current.

We then show that part of the current in the reactive section (capacitor) is leading the voltage by 90 degrees and this has been represented by the phasor I_Q (reactive). As stated previously, we refer to this as the **reactive current**.

We also know that for a resistive circuit, we calculate the power by multiplying together the r.m.s. values of voltage and current ($V \times I$).

Logically, as we know that no power is consumed in the reactive section of the circuit, we can therefore calculate the power in the circuit by multiplying together the r.m.s. value of voltage and the value of current which is in phase with it I_p (active). This would give us the formula for the apparent power:

$$P = V \times I_p \text{ (active)}$$

But as the actual current will be affected by the reactive current and therefore the phase angle ($\cos \varnothing$), our formula for true power becomes:

$$P = V \times I_p \text{ (active)} \times \cos \varnothing$$

If I_p (active) is zero then $\varnothing = 90°$ and $\cos \varnothing = 0$. Therefore $P = 0$.

We have now established that it is possible in an a.c. circuit for current to flow, but no power to exist.

We also say that the product of voltage and current is power given in watts. However, it would be fair to say that this is not the actual power of the circuit. The actual (true) power of the circuit has to take on board the effect of the phase angle ($\cos \varnothing$), the ratio of these two statements being the power factor. In other words:

$$\text{Power factor } (\cos \varnothing) = \frac{\text{True power (P)}}{\text{Apparent power (S)}}$$

$$= \frac{V \times I_p (\text{actual}) \times \cos \varnothing}{V \times I_p (\text{actual})} = \frac{\text{watts}}{\text{volt-amperes}}$$

To summarise, what we perceive to be the power of a circuit (the apparent power) can also be the true power, as long as we have a unity power factor (1.0). However, as long as there is a phase angle, there is a difference between apparent power and reality (true power). This difference is caused by the power factor (a value less than unity).

In reality we will use a wattmeter to measure the true power and a voltmeter and ammeter to measure the apparent power.

The power triangle

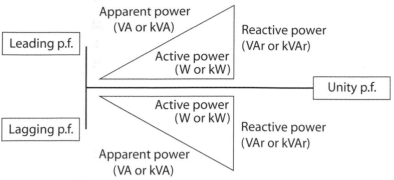

Figure 10.33 The power triangle

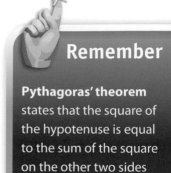
We can use Pythagoras' theorem to help us calculate the different power components within a circuit. We do so by using Pythagoras' formula as follows:

$$(VA)^2 = (W)^2 + (VAr)^2$$

This is then applied to Figure 10.33, where we have shown both inductive and capacitive conditions.

Example

A resistor of 15 Ω has been connected in series with a capacitor of reactance 30 Ω. If they are connected across a 230 V supply, establish both by calculation and by drawing a scaled power triangle the following:

(a) the apparent power   (b) the true power

(c) the reactive power   (d) the power factor.

In order to establish the elements of power, we must first find the current. To do this we need to find the impedance of the circuit. Therefore:

$$Z = \sqrt{R^2 + X_C^2} = \sqrt{15^2 + 30^2} = \sqrt{225 + 900} = \sqrt{1125}$$

Therefore $Z = 33.5\ \Omega$

$$I = \frac{V}{Z} = \frac{230}{35.5} = 6.9\ A$$

$$PF\ Cos\ \emptyset = \frac{R}{Z} = \frac{15}{33.5} = 0.45$$

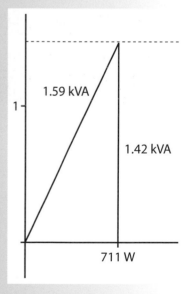

True power $= I \times V \times PF$

                            $= 6.9 \times 230 \times 0.45$

                            $= 714.15$ W or 0.714 kW

Apparent power $= I \times V$

                            $= 6.9 \times 230$

                            $= 1587$ VA or 1.587 kVA

Reactive power $= \sin \emptyset \times$ apparent power

However, as we do not know the value of $\sin \emptyset$, we must therefore convert $\cos \emptyset$ to an angle and then find the sine of that angle.

$\cos \emptyset = 0.45$ INV cos of $0.45 = 63.2$

sine of $63.2° = 0.89$

Therefore Reactive power $= \sin \emptyset \times$ apparent power

                                     $= 0.89 \times 1587$

                                     $= 1412.43$ VAr or 1.412 kVAr

Activity

You can check for the magnetic field surrounding a conductor by placing a compass near the conductor and switching the power on. This is the principle on which clamp-type ammeters work. If you have one, you could try measuring the current drawn by some equipment. Why won't this work if you place the ammeter around the cable?

FAQ

Q   **What is the link between electricity and magnetism?**

A   Electricity and magnetism are inextricably linked – you can't have one without the other! Whenever you pass a current through a conductor a magnetic field is produced around the conductor. Similarly if you pass a conductor through a magnetic field, electricity is produced.

Q   **What's the difference between a generator and an alternator?**

A   'Generator' is a term usually applied to a machine which consists of a set of conductors or coils rotating inside a magnetic field to produce a d.c. output. An 'alternator' usually consists of a magnet rotating inside a set of coils to produce an alternating current output.

Q   **Is it true that motors and generators are basically the same thing?**

A   Essentially yes. They both consist of two sets of windings, or one winding and one permanent magnet, one of which is fixed and the other free to rotate. With a motor you put electricity in and get mechanical power out. With an alternator you put mechanical power in and get electricity out.

Q   **How does a transformer step-up the voltage? Where does the extra voltage come from?**

A   A transformer steps the up voltage by having more turns on the secondary than on the primary. This means that more voltage is induced into the secondary than is produced by the primary, as the number of volts per turn is the same for the primary as it is for the secondary. But you don't get something for nothing! If you double the voltage, then you will only be able to draw half the available current.

Knowledge check

1. Explain the meaning of the term r.m.s. value.

2. If a sinusoidal a.c. supply has an r.m.s. voltage of 230 V, what would be the peak value of voltage?

3. Explain the purpose of the commutator in a d.c. machine.

4. Give two reasons why a.c. has become the preferred generating method rather than d.c.

5. Briefly describe the operation of a fluorescent luminaire.

6. What is the name of the principle on which a conventional core-type, double-wound transformer operates?

7. Explain how eddy currents may be reduced in a transformer core.

8. Explain the relationship between voltage and current in the following circuits, in terms of:

    *   resistance

    *   inductance

    *   capacitance.

9. Explain why the impedance of a long run of cable would increase as the frequency increased.

10. Draw a labelled diagram of the power triangle.

chapter 11

Polyphase systems

Unit 3 Outcome 4

So far we have looked at electrical supplies within a normal installation. But what we have not covered is how electricity is transmitted. Polyphase systems are the primary means for distributing power. The most common example is the **three-phase power** system.

Polyphase systems have two or more energized conductors carrying alternating current. The polyphase system must have a clear direction of phase rotation. It is this system that is most commonly used for electricity within domestic installations.

On completion of this chapter the candidate will be able to:

- describe the production, operation, transmission and distribution of energy in a polyphase system

- differentiate between voltages and currents in a balanced star and delta connected three-phase supply

- state the reason for balancing single-phase loads across a three-phase supply.

Production, operation, transmission and distribution of energy

On completion of this subject area the candidate will be able to describe the production, operation, transmission and distribution of energy in a polyphase system.

Production

Electricity is generated in power stations. The shaft of a three-phase alternator (a.c. generator) is turned, in the majority of cases, by using steam. Most electricity in the UK is produced by this method. Water is heated until it becomes high-pressure steam, which is forced on to the vanes of a steam turbine, which in turn rotates the alternator. A variety of energy sources can be used to heat the water in the first place. The more popular ones are coal, gas, oil and nuclear power. The basic components of the production system are shown in Figure 11.01.

Transmission

In Figure 11.01 electricity goes from the alternator to a transformer. This is because the output of most alternators is about 25,000 V (25 kV) and it must be transformed to:

- 400 kV and 275 kV for the super grid
- 132 kV for the original national grid
- 66 kV and 33 kV for secondary transmission
- 11 kV for local sub-station distribution and industry
- 400 V for commercial consumer supplies
- 230 V for domestic consumer supplies.

Electricity is transmitted at very high voltage values to compensate for the power losses that occur in the power lines. For example, we know that for a current to flow through a cable, power is used. The result is the production of heat – which is why we have electric heating. For an a.c. circuit the power can be found from $P = VI\cos\theta$. It therefore follows that an increase in voltage will result in a reduction of current for a given value of power.

The effect of decreasing the current in the transmission lines (volt drop) can be demonstrated with the following basic calculation. Volt drop can be calculated using the formula:

$$Vd = IR$$

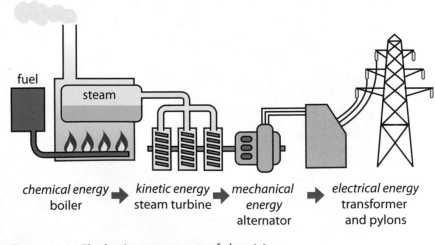

fuel
steam

chemical energy
boiler
→
kinetic energy
steam turbine
→
mechanical
energy
alternator
→
electrical energy
transformer
and pylons

Figure 11.01 The basic components of electricity generation systems

So, if a supply cable was carrying a current of 1000 amps and the resistance of the cable was 2 Ω, then 2000 volts would be dropped over the length of the supply cable. However, if the supply cable was carrying 100 amps then the volt drop would be reduced to 200 volts.

Transmission at low voltage values with high currents would necessitate the installation of very large cables and switchgear indeed.

From the super grid the electricity is fed into the National Grid system.

The National Grid

The National Grid is a network of nearly 5000 miles of overhead and underground power lines that link power stations together and are interconnected throughout the country. The concept is that, should a fault develop in any one of the contributing power stations or transmission lines, then electricity can be requested from another station on the system.

Electricity is transmitted around the grid, mainly via steel-cored aluminium conductors which are suspended from steel pylons. This is done for three main reasons.

- The cost of installing cables underground is excessive.

- Air is a very cheap and readily available insulator.

- Air also acts as a coolant for the heat being generated in the conductors.

Electricity is then 'taken' from the National Grid via a series of appropriately located substations. These will eventually transform the grid supply back down to 11 kV and then distribute electricity at this level to a series of local substations. It is their job to take the 11 kV supply, transform it down to 400 V and then distribute this via a network of underground radial circuits to the customer. However, in rural areas this distribution sometimes takes place using overhead lines.

It is also at this point that we see the introduction of the neutral conductor. This is normally done by connecting the secondary winding of the transformer in star and then connecting the star point to earth via an earth electrode beneath the substation.

Distribution to the customer

Once the electricity has left the local substation, it will eventually arrive at the customer. This is called the main intake position.

There are many different sizes of installation, but generally speaking we will find certain items at every main intake position. These items, which belong to the supply company, are:

- a sealed overcurrent device that protects the supply company's cable

- an energy metering system to determine the customer's electricity usage.

It is after this point that we say we have reached the consumer's installation.

Did you know?

Substations are dotted throughout UK cities. These small brick buildings are normally connected together on a ring circuit basis

The consumer's installation must be controlled by a main switch, which must be located as close as possible to the supply company equipment and be capable of isolating all phase conductors. In the average domestic installation this device is merged with the means of distributing and protecting the final circuits in what we know as the consumer unit.

Three-phase supplies

On completion of this topic area the candidate will be able to differentiate between voltages and currents in balanced star and delta connected three-phase supplies and state the reason for balancing single-phase loads across a three-phase supply.

Three-phase connections

So far, everything that we have looked at has revolved around a single-phase circuit. In such a circuit we normally use two conductors, where one delivers current and one returns it. You could logically assume that we would therefore need six conductors for a three-phase system, with two being used per phase. However, in reality we only use three or four conductors depending upon the type of connection of load that is being used.

We call these connections either **star** or **delta**.

Current flows along one conductor and returns along another called the neutral.

But what if there were no neutral? And what exactly is the neutral conductor for?

Keeping things simple, when we generate an e.m.f. we do so by spinning a loop of wire inside a magnetic field. To get three phases, we just spin three loops inside the magnetic field. Each loop will be mounted on the same rotating shaft but the loops will be 120 degrees apart, as shown in Figure 11.02.

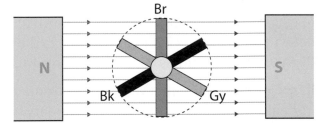

Figure 11.02 Three-phase generation

In the diagram, each loop will create an identical sinusoidal waveform, or in other words, three identical voltages, each 120 degrees apart, which provides us with a three-phase supply.

Whether or not we need a neutral is dependent upon the load. If we connect our three-phase output to three identical (i.e. balanced) loads then the current in each line will be the same but 120 degrees apart. By phasor addition, the resultant current will be zero, so there is no requirement for a neutral line. However, if the loads are not balanced then we do need a fourth cable to carry it.

Delta connection

We tend to use the delta connection when we have a balanced load. This is because there is no need for a neutral connection and therefore only three wires are needed. Thus we find that this configuration is used for power transmission from power stations or to connect the windings of a three-phase motor.

Figure 11.03 shows a three-phase load which has been delta connected. You can see that each leg of the load is connected across two of the lines, e.g. Br–Gy, Gy–Bk and Bk–Br. We refer to the connection between phases as being the line voltage and have shown this on the drawing as V_L.

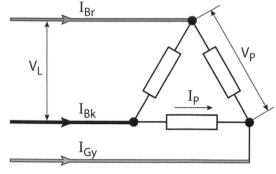

Figure 11.03 Delta connection

Equally, if each line voltage is pushing current along, we refer to these currents as being line currents, which are represented on the drawing as I_{Br}, I_{Gy} or I_{Bk}. These line currents are calculated as being the phasor sum of two phase currents, which are shown on the drawing as I_P and represent the current in each leg of the load. Similarly, the voltage across each leg of the load is referred to as the phase voltage (V_P).

In a delta connected balanced three-phase load, we are then able to state the following formulae:

$$V_L = V_P \qquad I_L = \sqrt{3} \times I_P$$

Note that a load connected in delta would draw three times the line current and consequently three times as much power as the same load connected in star. For this reason, induction motors are sometimes connected in **star–delta**. This means they start off in a star connection (with a reduced starting current) and are then switched to delta. In doing the heat that would otherwise be generated in the windings is reduced.

Star connection

Although we can have a balanced load connected in star as shown in Figure 11.04, we tend to use the star connection when we have an **unbalanced load**. All three star-connected loops are connected to a central point and it is from this point that we take our neutral connection, which in turn is connected to Earth. This is the three-phase four-wire system.

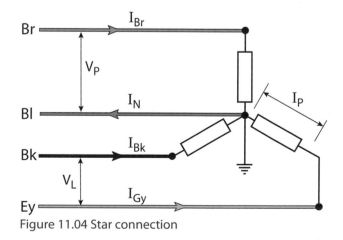

Figure 11.04 Star connection

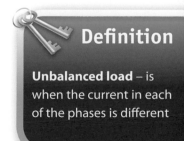

Definition

Unbalanced load – is when the current in each of the phases is different

Figure 11.04 shows a three-phase load that has been star-connected. As with delta, we refer to the connection made between phases as the line voltage and have shown this on the drawing as V_L. However, unlike delta, the **phase voltage** exists between any phase conductor and the neutral conductor and we have shown this as V_P. Our line currents have been represented by I_{Br}, I_{Gy} and I_{Bk} with the phase currents being represented by I_P.

In a star-connected load, the line currents and phase currents are the same, but the line voltage (400 V) is greater than the phase voltage (230 V).

In a star connected load, we are therefore able to state the following formulae:

$$I_L = I_P \qquad V_L = \sqrt{3} \times V_P$$

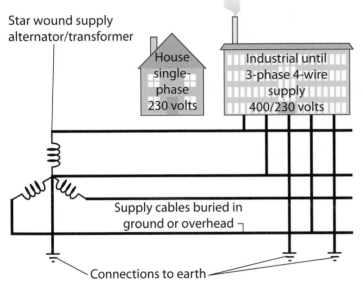

Figure 11.05 A three-phase four-wire distribution system

Using the star connected load we have access to a 230 V supply, which we use in most domestic and low load situations.

So what kind of load is an unbalanced load? If we were to look at a three-phase four-wire supply to a large building we may get some idea.

In the building there may be many single-phase circuits, all of different sizes, drawing different values of current. Some will be connected to the brown line, some to the grey line and some to the black line – although the electrician would have tried to spread the loads evenly across the three phases.

If we were to measure the current in any of the phase conductors of the supply cable, they would probably be quite high. If we were then to measure the current in the neutral it would be low because the neutral only carries the out of balance current.

Another advantage of the star-connected system is that it allows there to be two voltages – one to connect between any two phases (400 V) and another to connect between any phase and neutral (230 V). You should note that there will also be 230 V between any phase and earth.

Example 1

A three-phase star connected supply feeds a delta-connected load as shown in Figure 11.06 below.

If the star-connected phase voltage is 230 V and the phase current is 20 A, calculate the following:

- the line voltages and line currents in the star connection

- the line and phase voltages and currents in the delta connection.

Star connection

In a star system the line current (I_L) is equal to the phase current (I_P). Therefore if we have been given I_P as 20 A, then I_L must also be 20 A.

We find line voltage in a star connection using the formula:

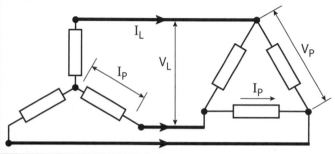

Figure 11.06 Three-phase star-connected supply feeding delta connected load

$$V_L = \sqrt{3} \times V_P$$

$\sqrt{3}$, the square root of 3, is a constant having the value 1.732. Therefore if we substitute our values, we get:

$$V_L = 1.732 \times 230 = 398 \text{ V}$$

Delta connection

In a delta system, the line current (I_L) is 1.732 times greater than the phase current (I_P). We calculate this using the formula: $I_L = \sqrt{3} \times I_P$

However, we know that I_L is 20 A, so if we transpose our formula and substitute our values we get:

$$I_P = \frac{I_L}{\sqrt{3}} = \frac{20}{1.732} = 11.5 \text{ A}$$

We know that for a delta connection, line voltage and phase voltage have the same values.

Therefore: $V_L = V_P = \textbf{398 V}$

Example 2

Three identical loads of 30 Ω resistance are connected to a 400 V three-phase supply. Calculate the phase and line currents if the loads were connected:

(a) in star

(b) in delta.

Star connection

First, we need to establish the phase voltage. If $V_L = \sqrt{3} \times V_P$ and $\sqrt{3} = 1.732$ then by transposition:

$$V_P = \frac{V_L}{\sqrt{3}} = \frac{400}{1.732} = 230.9 \text{ V}$$

Using Ohm's law:

$$I_P = \frac{V_P}{Z} = \frac{230.9}{30} = 7.7 \text{ A}$$

However, in a star-connected load, $I_P = I_L$, therefore I_L will also = 7.7 A.

Delta connection

In a delta connection $V_L = V_P$ and therefore we know the phase voltage will be 400 V.

Using Ohm's law:

$$I_P = \frac{V_P}{Z} = \frac{400}{30} = 13.33 \text{ A}$$

However, the line current equals:

$$I_L = \sqrt{3} \times I_P = 1.732 \times 7.7 = 23.09 \text{ A}$$

As can be seen from this example, the current drawn from a delta-connected load (23.09 A) is three times that of a star-connected load (7.7 A).

Load balancing

The UK's regional electricity companies require load balancing as a condition of their electricity supply because it is important to try to achieve balanced currents in the mains distribution system.

In order to design a three-phase four-wire electrical installation for both efficiency and economy, the load needs to be subdivided into load categories. By doing this the maximum demand can be assessed and items of equipment can be spread over all three phases of the supply, to achieve a balanced system. The designer needs to make a careful assessment of the various installed loads, which in turn leads to the proper sizing of the main cable and associated switchgear.

Standard circuit arrangements exist for many final circuits operating at 230 volts. For example, a ring final circuit is rated at 30 A, a lighting circuit at 5 A, and a cooking appliance at 30–45 A. Where more than one standard circuit arrangement is present, such as three ring final circuits and/or two cooking appliances, then a diversity allowance can be applied.

Once the designer has made these allowances for diversity, the single-phase loads can be evenly spread over all three phases of the supply so that each phase takes approximately the same amount of current. If this is done carefully, minimum current will flow along the neutral conductor, the sizes of cables and switchgear can be kept to a minimum, thus reducing costs, and the system is therefore said to be reasonably well balanced.

Activity

If you have access to some three-phase motors (preferably spare ones), and can do so safely, have a look at the rating plates and terminals to see how they would be connected to an electrical supply. Alternatively, if no motors are available, have a look at some manufacturers' catalogues.

FAQ

Q Why do we need ugly pylons stretching across the countryside?

A Simply because it would cost too much to make a 400 kV cable and bury it for miles underground.

Q Why are the three-phase colours brown, black and grey? It was much easier to tell them apart when they were red, yellow and blue.

A Well yes, but these are the colours that have been agreed and standardised across Europe. As for telling them apart, well it may be a problem with your eyesight – in which case a visit to an optician would seem wise. If there is insufficient light to distinguish which one is which, then you need to bring in more light!

Q One of our motors doesn't have a neutral connection. How does it work then?

A In a three-system, if all of the loads are balanced, i.e. the same, then the current flowing to the load up one phase, returns to the source via the other two phases not via the neutral. The neutral is only used when the loads are unbalanced, so with a balanced load, the neutral can be removed.

Knowledge check

1. List the standard voltages for generation, transmission, distribution and utilisation of UK electricity supplies.

2. Draw a labelled diagram of a three-phase load connected in:

    (a) star

    (b) delta.

3. In a delta-connected circuit, the line voltage is equal to what?

4. Explain why it is important to have a balanced load in a three-phase system.

chapter 12

Overcurrent, short-circuit and earth fault protection

Unit 3 Outcome 5

Where there is electricity, there also exists the potential for danger. As we have seen throughout this book, the danger from electric shock can be lethal. One of the crucial tasks in any installation is to reduce, as far as humanly possible, this hazard.

To do this we install protection devices into systems. Protective devices and simple fuses are designed to operate when they detect large currents due to excess temperature. In the case of a short circuit fault, high levels of fault current can develop causing high temperatures and breakdown of insulation. Such faults, if allowed to develop, can cause fires or lead to electric shock.

This chapter will build upon the topics we have dealt with earlier in this book.

On completion of this chapter the candidate will be able to:

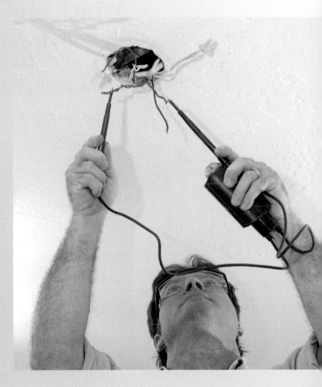

- state the need for protective devices
- state the factors on which the selection of protective devices depend
- list the exposed and extraneous conductive parts within a building
- list the essential requirements for a device designed to protect against overcurrent
- state the action of a fuse under fault conditions
- state the need for correct 'discrimination' of devices
- state the causes of current flowing to earth
- state the need to maintain a low impedance path
- state the path taken to earth in the event of a fault
- state the need for an RCD or RCBO.

Protection devices

On completion of this topic area the candidate will be able to state the need for protective devices, the factors that dictate the selection of devices.

Why protective devices?

Protective devices operate in the event of a fault occurring on the circuit or equipment that the device is protecting. Typical faults include:

- short circuits between live conductors – line to neutral single-phase, line to phase three-phase

- earth faults – overcurrents between any live conductor and earth.

Which protective device?

When selecting a protection device, it is crucial to remember all the factors that will have an effect on your final choice. These factors are:

- prospective short circuit current (PSCC)

- design current (I_b)

- rating of the protective device (I_n)

- disconnection time limitation.

Prospective short circuit current (PSCC)

Regulation 434.1 states that: 'The prospective fault current shall be determined at every relevant point in the installation.' This prospective short circuit fault current must be taken into account when selecting the type of overcurrent device to be installed.

The effects of short circuit current are:

- the thermal effect, which can cause melting of conductors/insulation, fire, alteration of the properties of materials etc.

- the mechanical effects of large magnetic fields that can build up when short circuit currents are flowing, resulting in conductor distortion, breaking of supports/insulators etc. This is the worst case of short circuit fault and rapid disconnection of the supply is essential to prevent damage.

Possible causes of the occurrence of a short circuit fault are:

- contact between two poles of the supply due to incorrect connection

- equipment failure

- ingress of moisture

- accidental damage.

The causes of a fault are many and varied, e.g. it may result from the rupturing of a lamp filament or a JCB cutting an underground cable.

A short circuit is said to be a connection of two live conductors. 'Live conductors' means all those carrying current under normal conditions; **this includes the neutral conductor**.

A short circuit can occur between conductors connected to different lines or, indeed, any line to neutral.

If a short circuit is two conductors touching, then it can be assumed that the resistance or impedance of that connection would be so low that it could be neglected.

If a fault has negligible impedance, then the only restriction to the amount of current that will flow in the circuit is that of all the conductors. As conductors are of low resistance, then this total will be low and the current that flows can be very high.

Consequently, if for example a line to neutral short circuit fault occurs within a final circuit, as shown in Figure 12.01, the final circuit protective device should operate first.

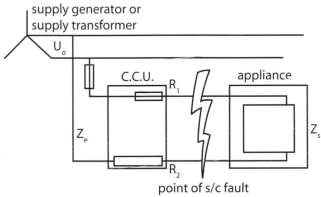

Figure 12.01 Short circuit loop impedance

- C.C.U. = customer consumer unit
- U_o = nominal supply voltage
- Z_e = loop impedance external to the installation
- R_1 = the resistance of the installation phase conductor to the fault point
- R_2 = the resistance of the installation return conductor to the fault point
- Z_s = total fault loop impedance.

$$Z_s = Z_e + (R_1 + R_2)$$

Note the return can be via either the neutral or another line depending on the supply arrangement.

To reduce this hazard we would use a residual current device (RCD). These provide extra protection to reduce the risk of electric shock. We will look at RCDs later in this chapter (pages 307–308).

Example

Calculate the prospective short circuit current (I_{sc}) flowing in the circuit shown in Figure 12.02 at:

(a) the customer consumer unit supply terminals

(b) the terminals of the load given the conditions shown.

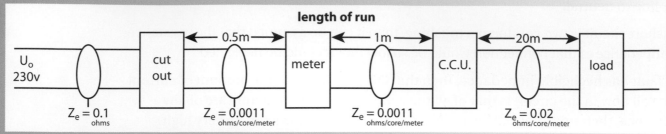

length of run

U_o 230v — cut out ←0.5m→ meter ←1m→ C.C.U. ←20m→ load

$Z_e = 0.1$ ohms $\quad Z_e = 0.0011$ ohms/core/meter $\quad Z_e = 0.0011$ ohms/core/meter $\quad Z_e = 0.02$ ohms/core/meter

Figure 12.02 Prospective short circuit current

(a) $Z_s = Z_e + (R_1 + R_2)$

$Z_s = 0.1 + (0.0011 \times 0.5 \times 2) + (0.0011 \times 1 \times 2) = 0.1033 \ \Omega$

$I_p = \dfrac{U_o}{Z_s} = \dfrac{230}{0.1033} = 2226.53 \ A$

(b) $Z_s = 0.1 + (0.0011 \times 0.5 \times 2) + (0.0011 \times 1 \times 2) + (0.02 \times 20 \times 2)$

$Z_s = 0.1 + (0.0011 \times 0.5 \times 2) + (0.0011 \times 1 \times 2) + (0.02 \times 20 \times 2) = 0.9033 \ \Omega$

$I_{sc} = \dfrac{U_o}{Z_s} = \dfrac{230}{0.9033} = 254.6 \ A$

Design current (I_b)

The cable current carrying capacity will have an effect on your choice of protection device. You will have to calculate this; it is the normal resistive load current in the circuit. The following formulae apply to single- and three-phase supplies:

Single-phase supplies:

$U_o = 230V$

$I_b = \dfrac{Power}{U_o}$

where U_o is the phase voltage to earth (supply voltage).

Three-phase supplies:

$U_o = 400V$

$I_b = \dfrac{Power}{\sqrt{3} \times U_o}$

In a.c. circuits, the effects of either highly inductive or highly capacitive loads can produce a poor power factor (PF). You will have to allow for this. To find the design current you may need to use the following equations:

Single-phase circuits:

$$I_b = \frac{\text{Power}}{U_o \times PF}$$

Three-phase circuits:

$$I_b = \frac{\text{Power}}{\sqrt{3} \times U_o \times PF}$$

where PF is the power factor of the circuit concerned.

Rating of the protective device (I_n)

When you have worked out the design current of the circuit (I_b), you must next work out the current rating or setting (I_n) of the protective device.

Regulation 433.1.1, on page 57 in the *IEE Wiring Regulations* says that current rating (I_n) must be no less than the design current (I_b) of the circuit.

The reason for this is that the protective device must be able to pass enough current for the circuit to operate at full load, but without the protective device operating and disconnecting the circuit.

Protective devices come in standard values – for example, 13 A for plug tops. You can find these values in the *IEE Wiring Regulations*, Tables 41.2, 41.3 or in 41.4.

The table you use to select the value of your protective device will depend on the type of equipment or circuit to be supplied and the requirements for disconnection times.

Disconnection time limitation

Protective devices have two basic methods through which overcurrent can operate or 'trip' the latch.

- **Thermal tripping** – the load passes through a small heater coil wrapped around a bi-metal strip. When the current is too high, the strip rises and trips the latch.
- **Magnetic tripping** – a magnetic field is set up around a flexible strip. When the current is too high the latch is operated.

The thermal tripping naturally takes a longer time, as it relays on a build up of heat. The magnetic tripping, however, can occur almost immediately. The selection of which tripping device to use will depend on the type of circuit.

For some circuits, immediate tripping will be required, such as when the overload is large or with a short circuit. Some systems will only feature small overloads, where a sudden shutdown of power may cause more damage. For these thermal tripping is more advisable.

Remember

Not every overload or short circuit is the same – you will need a different solution for every problem you might face

Exposed and extraneous conductive parts

We covered these earlier in Chapter 6 page 141. Please refer back to this section for more information.

Protective device requirements

On completion of this topic area the candidate should be able to list the essential requirements of a device fitted into a circuit to protect it from overcurrent.

Operation of overload and fault current devices

When a fault is noticed it is generally because a circuit or piece of equipment has stopped working and this is usually because the protective device has done its job and operated. The rating of a protective device should be greater than, or at least equal to, the rating of the circuit or equipment it is protecting, e.g. 10 × 100 watt lamps equate to a total current use of 4.35 amperes. Therefore a device rated at 5 or 6 amperes could protect this circuit.

A portable domestic appliance which has a label rating of 2.7 kW equates to a total current of 11.74 amperes. Therefore a fuse rated at 13 amperes should be fitted in the plug.

Protective devices are designed to operate when an excess of current (greater than the design current of the circuit) passes through it. The fault current's excess heat can cause a fuse element to rupture or the device mechanism to trip, dependent on which type of device is installed. These currents may not necessarily be circuit faults but short-lived overloads specific to a piece of equipment or outlet. The Regulations categorise these as **overload current** and **overcurrent**.

For conductors, the rated value is the current-carrying capacity. Most excess currents are, however, due to faults, either **earth faults** or **short circuit** type, which cause excessive currents.

Whichever type of fault occurs the designer should take account of its effect on the installation wiring and choose a suitable device to disconnect the fault quickly and safely. The fundamental effect of any fault is a rise in current and therefore a rise in temperature.

High temperature destroys the properties of insulation, which in itself could lead to a short circuit. High currents damage equipment, and earth-fault currents are associated with the risk of electric shock. Some examples of each are given in the following text.

Definition

Overload current – an overcurrent occurring in a circuit which is electrically sound

Overcurrent – a current exceeding the rated value

Earth-fault current – a fault current which flows to earth

Short-circuit current – an overcurrent resulting from a fault of negligible impedance between live conductors

Overload faults

Examples of these include:

- adaptors used in socket outlets exceeding the rated load of the circuit
- extra load being added to an existing circuit or installation
- not accounting for starting current on a motor circuit.

Short circuit faults

Examples of these include:

- insulation breakdown
- severing of live circuit conductors
- wrong termination of conductors energised before being tested.

Earth faults

Examples of these include:

- insulation breakdown
- incorrect polarity
- poor termination of conductors.

Wrong type of starter in fluorescent tube has melted due to excess power demand

Fuse holder has melted due to overloading

Fuses

On completion of this topic area the candidate will be able to state the action of a fuse under fault conditions.

BS 3036 rewirable fuses

Early rewirable fuses had a very low short-circuit capacity and were very dangerous when operating under fault conditions. This was because the fuse element melted and splashed the melted copper around and could cause fires. Later rewirable fuses incorporated asbestos to protect the fuse holder during the fusing period, thus reducing the risk of fire from scattering hot metal when rupturing.

The rewirable fuse consists of a fuse, holder, a fuse element and a fuse carrier. The holder and carrier are made of porcelain or bakelite. The circuits for which this type of fuse is designed have a colour code marked on the fuse holder as follows:

Poor termination of circuit protective conductors (cpc)

BS 3036 rewirable fuse

- 5 A white
- 15 A blue
- 20 A yellow
- 30 A red
- 45 A green.

This type of fuse was very popular in domestic installations. However, with the exception of old installations, it is not normally used because it has some serious disadvantages.

Disadvantages of rewirable fuses	Advantages of rewirable fuses
• Easily abused when the wrong size of fuse wire is fitted • **Fusing factor** of around 1.8–2.0 means they are not guaranteed to operate until up to twice the rated current is flowing; as a result cables protected by them must have a larger current carrying capacity • Precise conditions for operation cannot be easily predicted • Do not cope well with high short-circuit currents • Fuse wire can deteriorate over time • Danger from hot scattering metal if the fuse carrier is inserted into the base when the circuit is faulty.	• Low initial cost • Can easily see when the fuse has blown • Low element replacement cost • No mechanical moving parts • Easy storage of spare fuse wire

Table 12.01 Advantages and disadvantages of rewirable fuses

Nominal current of fuse wire (A)	Nominal diameter of wire (mm)
3	0.15
5	0.20
10	0.35
20	0.50
20	0.60
25	0.75
30	0.85
45	1.25
60	1.53
80	1.80
100	2.00

Table 12.02 Size of tinned copper wire for use in semi-enclosed fuses

BS 1361/1362 cartridge fuses

These cartridge fuses consist of a porcelain tube with metal end caps to which the element is attached. The tube is then filled with granulated silica.

BS 1361 and 1362 fuses

The BS 1362 fuse is generally found in domestic plug tops used with 13 A BS 1363 domestic socket outlets. There are two common fuse ratings available: the 3 A, which is for use with appliances up to 720 watts (e.g. radios, table lamps, electric blankets) and the 13 A fuse, which is used for appliances rated over 720 watts (e.g. irons, kettles, fan heaters, electric fires, lawn mowers, toasters, refrigerators, washing machines and vacuum cleaners). There are, however, other sizes available (which are more difficult to come across). These are 1, 5, 7 and 10 amp sizes.

The BS 1361 fuse is normally found in distribution boards and at main intake positions.

Disadvantages of cartridge fuses	Advantages of cartridge fuses
• More expensive to replace than rewirable fuses • Can be replaced with an incorrect size fuse (plug top type only) • The cartridge can be shorted out with wire or silver foil in extreme cases of bad practice • Not possible to see if the fuse has blown • Require a stock of spare fuses to be kept	• No mechanical moving parts • Declared rating is accurate • The element does not weaken with age • Small physical size and no external arcing, which permits their use in plug tops and small fuse carriers • Low fusing factor; around 1.6–1.8 • Easy to replace

Table 7.9 Advantages and disadvantages of cartridge fuses

BS 88 high breaking capacity (HBC) fuses

The HBC fuse is a sophisticated variation of the cartridge fuse and is normally found protecting motor circuits and industrial installations. It consists of a porcelain body filled with silica, a silver element and lug-type end caps. Another feature is the indicating bead, which shows when the fuse element has blown. It is a very fast acting fuse and can discriminate between a starting surge and an overload.

A typical BS 88 HBC fuse

These types of fuses would be used when an abnormally high prospective short-circuit current exists.

Disadvantages of BS 88 fuses	Advantages of BS 88 fuses
• Very expensive to replace • Stocks of these spares are costly and take up space • Care must be taken when replacing them, to ensure that the replacement fuse has the same rating and also the same characteristics as the fuse being replaced	• No mechanical moving parts • The element does not weaken with age • Operation is very rapid under fault conditions • It is difficult to interchange the cartridge, since different ratings are made to different physical sizes

Table 7.03 Advantages and disadvantages of BS 88 fuses

Type 'D' and Neozed fuses

Both these fuses are manufactured in Germany and have been developed to European testing regulations where all European testing authorities have approved them. The Neozed is the successor to the D-type fuse. You may in the course of your work come into contact with either type of fuse.

Neozed fuse

Discrimination

On completion of this topic area the candidate will be able to state the need for correct discrimination of devices when a number of devices are fitted between the supply and the load.

In both large and small installations there is usually a series of fuses and/or circuit breakers between the incoming supply and the electrical outlets. The relative rating of the protective devices used will decrease the nearer they are located to the current-using equipment.

Ideally, when a fault occurs only the device nearest the fault operates, thus ensuring minimum disruption to other circuits not associated with the fault. Discrimination is said to have taken place when the smaller rated local device operates before the larger device.

Discrimination is generally only a problem when a system uses a mixture of devices. Obviously a particular type of fuse will discriminate against a similar type of fuse if it is of larger rating.

Discrimination is also known as the co-ordination between fuses.

Figure 12.03 relates to a 15 A semi-enclosed rewireable fuse serving a final sub-circuit, backed up by a 30 A MCB. When the fault current approaches 200 amps, the MCB will operate quicker than the fuse. Here discrimination has failed.

Simply because protective devices have different ratings, it cannot be assumed that discrimination will be achieved. This is especially the case where a mixture of different types of device is used. However, as a general rule when fuses are used in series in an installation a 2:1 ratio with the lower-rated devices will be satisfactory.

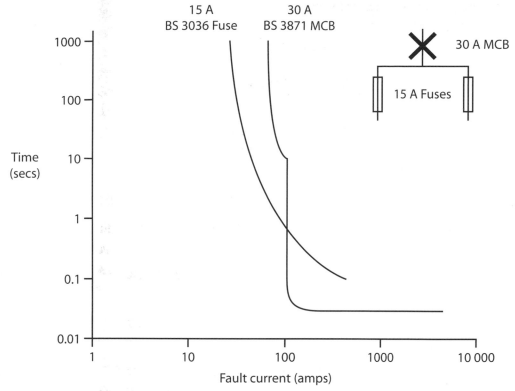

Figure 12.03 Discrimination

Earthing

On completion of this topic area the candidate will be able to: state the causes of current flowing to earth; the need to maintain a low impedance path to ensure that overcurrent and earth-fault protective devices will operate within design parameters; and the path taken by an earth-fault current within the installation.

The purpose of earthing

We covered this in detail in chapter 6 page 140. Please refer back to this section for more information.

The earth-fault loop path

On a full earth-fault, the overcurrent protective device will operate, causing the automatic disconnection of the supply. On a low-level earth leakage fault, the overcurrent protective device will not detect the fault. With a faulty or disconnected earth-fault loop path, metalwork may become live under fault conditions.

A residual current device (RCD) will detect any imbalance between line and neutral conductors, as any difference is assumed to be flowing to earth. This earth 'leakage' will cause the RCD to operate at its rated residual operating current. RCDs are discussed in more detail in the next section.

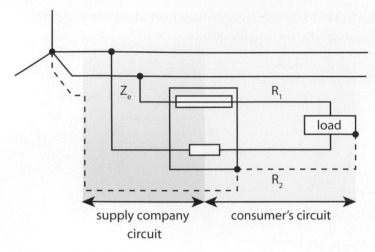

Figure 12.04 Earth-fault loop path

Z_s is the sum of Z_e (the supply earth loop impedance) plus R_1 (the resistance of the circuit's line conductor) and R_2 (the resistance of the circuit protective conductor), i.e.:

$$Z_s = Z_e + (R_1 + R_2)$$

Fault current $I_f = \dfrac{U_o}{Z_s}$

where U_o is the nominal supply voltage.

Note: This is also known as the prospective short circuit current (PSSC) I_{sc}.
In situations where the value of Z_s is too high to achieve rapid disconnection on a fault then:

• cable conductor size may be increased to reduce the Z_s value

• RCDs may be installed in combination with overcurrent protection devices and will provide a safe solution.

Earth-fault leakage currents will then be detected and the supply automatically disconnected.

The value of Z_s governs the flow of earth-fault loop current, the operation of earth-fault protective devices and their circuit disconnection times.

From Tables 41.2, 41.3 and 41.4 and Appendix 3 Figures 3.3B, 3.5 and 3.6 of BS 7671, it can be seen that:

• a BS 88 40 A fuse installed in a circuit of Z_s = 0.86 ohms will operate (at a fault current of 280 A) in 0.4 seconds on a 230 volt system

• a BS EN 60898 40 A Type C MCB installed in a circuit of Z_s = 0.6 ohms will operate (at a fault current of 400 A) in 0.4 seconds on a 230 volt system

• a BS EN 60898 40 A Type D MCB installed in a circuit of Z_s = 0.3 ohms will operate (at a fault current of 800 A) in 0.4 seconds on a 230 volt system.

Regulations state that the protective device protecting a circuit must operate within a time of:

- 0.4 seconds for portable equipment supplied from socket outlet circuits
- 5 seconds for fixed equipment.

Supplementary protection

On completion of this topic area the candidate will be able to state the need for supplementary protection against electric shock by the use of a RCD or RCBO.

Residual current devices (RCD)

Residual current devices (RCDs) are a group of devices providing a modern approach to the enhancement of safety in electrical systems. They provide extra protection to people and livestock by reducing the risk of electric shock. Although RCDs operate on small currents, there are circumstances where the combination of operating current and high earth-fault loop impedance could result in the earthed metalwork rising to a dangerously high potential.

The Regulations draw attention to the fact that if the product of operating current (A) and earth-fault loop impedance (Ω) exceeds 50 V, the potential of the earthed metalwork will be more than 50 V above earth potential and, hence, dangerous. This situation must not be allowed to occur.

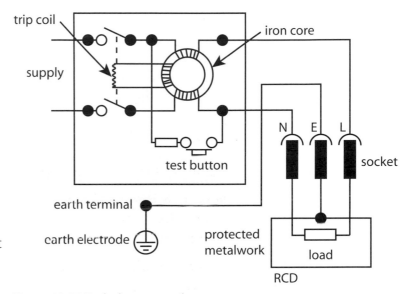

Figure 12.05 Fault detector coil

Method of operation under fault conditions

The current taken by the load is fed through two equal and opposing coils wound on to a common iron core. Under normal safe working conditions, the current flowing in the line conductor into the load will be the same as that returning via the neutral conductor from the same load. When this happens, the line and neutral conductors produce equal and opposing magnetic fluxes in the iron core contained within the RCD, resulting in no voltage being induced into the trip coil.

If more current flows in the phase side than the neutral side (or less returns via the neutral because the current has 'leaked' to earth within the installation) an out-of-balance flux will be produced which will induce a voltage into the fault detector coil. The fault detector coil then energises the trip coil and opens the double pole (DP) switch, due to the residual current produced by the induced voltage in the trip coil.

Functional testing of an RCD should be carried out by operating the test button at regular intervals. Initial and periodic inspection and testing procedures additionally require the RCD to be tested using the appropriate instrument. RCDs are available for single and three-phase applications.

Practical application of RCDs (Regulation 531.2) include the following.

- RCDs must disconnect all live conductors at the same time (this includes the neutral).

- All live conductors of the protected circuit are to be passed through the RCD's magnetic circuit (this includes the neutral).

- An RCD should be selected so that any expected earth leakage current, which may occur during normal operation of the connected loads (e.g. some computers), will be unlikely to cause unnecessary tripping of the device.

- An RCD must be located outside stray magnetic fields.

- RCDs are to be used to protect equipment being used outside the earthed equipotential zone, e.g. lawn mowers, hedge cutters, water pumps. This implies that a combination of earthed equipotential bonding, overcurrent protection devices and residual current devices in the design of an electrical installation should offer total electric shock protection.

Sensitivity-rated residual operating current	Level of protection	Applications
10 mA	Personnel	High risk areas: schools, colleges, workshops, laboratories; areas where liquid spillage may occur
30 mA	Personnel	Domestic, commercial and industrial
100 mA	Personnel fire	Only limited personnel protection Excellent fire protection
300 mA	Fire	Commercial and industrial

Table 12.03 Overcurrent protection sensitivity

Residual current circuit breaker with overload protection (RCBO)

An RCBO is essentially a marriage between a miniature circuit breaker (MCB) and an RCD, as it comprises both of these components. Consequently, such a device offers protection against the effects of earth leakage, overload and short-circuit currents, while reducing the number of outgoing ways required within a distribution board.

If, groups of circuits are protected by an RCD, all circuits would be interrupted under fault conditions, which can cause inconvenience. This device has the advantage of allowing earth-fault protection to be restricted to a single circuit, thus ensuring that only the circuit with the fault is interrupted.

A standard range of 10 A/30 mA, 16 A/30 mA, 20 A/30 mA, 32 A/30 mA and 40 A/30 mA are available from most manufacturers.

Activity

Obtain a copy of BS 7671 and using the time/current characteristics tables in Appendix 3, find out how much current it takes to blow a 5 A, 15 A, 20 A and 30 A BS 1361 fuse in 0.1s and 5s.

FAQ

Q **What does the rating written on the fuse mean?**

A It is a common misconception that the rating on the fuse is the current that will cause the fuse to blow. It isn't – it is the amount of current that the fuse will carry continuously without blowing.

Q **So how much current does it take to blow a fuse?**

A Well that depends on how quickly you want it to blow! Take, for example, a 30 A BS 1361 cartridge fuse – to get it to blow in one hour would require about 58 A; if you want it to blow more quickly, say 0.1 s, then you'd need 280 A.

Q **Why do fuses need so much current to get them to blow quickly?**

A This apparent crudeness in a fuse's operation can be a good thing. Imagine a 30 A ring circuit that has two 13 A applicances already plugged into it. If you then switched the kettle on to make a cup of tea (another 13 A), if the fuse was too accurate, it would blow. As long as the overcurrent doesn't stay there for too long (i.e. 58 A for one hour), then the fuse will not blow (and we'll get a cup of tea!).

Knowledge check

1. What are the two potentially dangerous effects a short-circuit current can cause?

2. State the formula for calculating Z_s and explain the meaning of the terms used.

3. In a miniature circuit breaker, which mechanism is most likely to cause the device to trip in the case of:

   • a short-circuit

   • an overload current.

4. Explain why BS 3036 (rewirable) fuses are no longer preferred.

5. Explain the principle of 'discrimination' when applied to protective devices. Why is it a good thing?

6. When would it be a good idea to use an RCBO?

chapter 13

Statutory regulations and codes of practice

Unit 4 Outcome 1

Electricity, when properly installed, used and maintained, is a very safe and useful commodity. However, if we 'break the rules' electricity can deliver death and destruction with lightning speed – often there will be no second chance.

Although each installation is different, you will notice that they all contain the same elements; fuses, switches, cables, trunking etc. All are installed in a similar way, no matter what the situation. This is because we have the electrical Regulations. These must be followed to ensure a certain amount of standardisation and, most importantly, safety.

In this chapter we will look at the Regulations and how they affect the way we do things everyday.

On completion of this chapter the candidate will be able to:

- state that the electrical Regulations concern all aspects of electrical systems
- state how site-based responsibility may make a person a designated duty holder
- state the need for statutory regulations of installations
- state the purpose and function of a device identified by a BS or BS EN number.

Electrical Regulations

For the UK electrical industry (and throughout this book) the term 'Regulation(s)' by itself is understood to refer to BS 7671, the *IEE Wiring Regulations*; though there are many other regulations that electricians must concern themselves with, particularly the Electricity at Work Regulations (EAW). On completion of this topic area the candidate will understand the role and structure of BS 7671 Regulations and the EAW; and also how site-based responsibility may make a person a designated duty holder.

The history and role of the Regulations

When the first electrical Regulations were laid down, virtually no one had electricity in their homes or workplaces; now everyone has access to electricity. The Regulations have moved on since 1882 and we are now up to the *17th Edition 2008*. Over the years the Regulations have taken account of the new types of electrical equipment available and its usage, whether that be computers, lighting, overcurrent protection etc. and the effects on all aspects of the installation process.

The Regulations are designed to protect persons, property, and livestock from electric shock, fire and burns and injury from mechanical movement of electrically operated equipment. They are not designed to instruct untrained persons, take the place of a detailed specification or to provide information for every circumstance.

The use of other British Standards is needed to supplement the information contained in these Regulations, such as BS 6551 which deals with protection of structures against lightning.

In 1992 the 16th Edition of the *IEE Wiring Regulations* became a British Standard, BS 7671. These Regulations are not a statutory document (i.e. not legislated by an Act of Parliament). However, compliance with the Regulations ensures compliance with such legislation as *Electricity at Work Regulations 1989* and the Health and Safety at Work Act 1974, which are statutory documents and are enforceable by law.

Many other British Standards are referred to throughout these Regulations and in some cases these other standards have a BS EN number. This refers to European harmonisation of standards, which will then be applicable throughout Europe. These harmonised standards are co-ordinated by representatives from all the countries in the European Union via an organisation known as CENELEC.

The IEE has published eight 'Guidance Notes' that simplify BS 7671 requirements:

1. *Selection and Erection of Equipment*, 4th Edition (IEE Publication)
2. *Isolation and Switching*, 4th Edition (IEE Publication)
3. *Inspection and Testing*, 4th Edition (IEE Publication)
4. *Protection against Fire*, 4th Edition (IEE Publication)
5. *Protection against Electric Shock*, 4th Edition (IEE Publication)
6. *Protection against Overcurrent*, 4th Edition (IEE Publication)
7. *Special Locations*, 2nd Edition (IEE Publication)
8. *Earthing & Bonding*, 1st Edition (IEE Publication)

Plan and style of the Regulations

The Regulations are divided into seven parts with fifteen appendices, and each of the Regulations is numbered in a unique way to identify which part of the Regulations it refers to.

For example, for Regulation 412.2.2.4:

- 1st digit, 4, refers to the part of the Regulations, in this case part 4.
- 2nd digit, 1, when attached to the 1st digit, refers to the chapter within that part of the Regulations, which in this case is chapter 41.
- 3rd digit, 2, when attached to the previous two digits, refers to the section within the chapter, which in this case is section 412.
- 4th digit, when attached to the previous ones, normally refers to a group of associated regulations.
- 5th and 6th digits refer to individual regulations.

The parts of BS 7671 IEE Regulations

Part 1 Scope, object and fundamental principles

There are three chapters within this part and numerous sections within each chapter, the first chapter (Chapter 11) dealing with what the Regulations actually cover and what they do not.

The Regulations apply to the design, selection, erection and inspection and testing of electrical installations such as:

- residential premises
- commercial premises
- public premises
- industrial premises

Remember

A useful book is the *IEE On Site Guide*, which is an abbreviated version of BS 7671 which also gives practical explanations, and in some cases drawings, of the more commonly used regulations as well as useful cable tables etc. Another useful reference is produced by the union Amicus

Did you know?

The IEE has joined with other institutions to form the Institution of Engineering & Technology (IET); but published documents will continue to bear the IEE imprint until republished at some future date

- agricultural and horticultural premises
- prefabricated buildings
- caravans, caravan parks and similar sites
- construction sites, exhibitions, fairs and other installations in temporary buildings
- highway power supplies, street furniture and outdoor lighting
- marinas
- external lighting and similar installations
- medical locations
- mobile or transportable units
- photovoltaic systems
- low voltage generating sets.

The Regulations are crucial to electrical installations

They do not cover the following:

- distributors' equipment
- railway traction equipment
- equipment of motor vehicles
- equipment on board ships
- equipment of mobile and fixed offshore installations
- equipment of aircraft
- mines and quarries
- radio interference equipment (unless it affects safety of electrical installations)
- lightning protection of buildings covered by BS 6651
- lift installations covered by BS 5655
- electrical equipment of machines.

Voltage ranges covered are up to 1000 volts a.c. or 1500 volts d.c. between conductors or 600 volts a.c. or 900 volts d.c. between conductors and earth.

Fundamental principles (Chapter 13) deals with what the Regulations are designed to protect: 'The regulations are designed to protect persons, property and livestock from dangers and damage which may arise in the reasonable use of electrical installations.'

Part 2 Definitions

In this section of BS 7671 the terms used throughout the Regulations are given a specific meaning, so that when one person talks about a circuit breaker, for example, then everyone else knows what they mean by that word.

Part 3 Assessment of general characteristics

This part is about assessing the general characteristics of the supply and the type of installation proposed, and takes into account the following:

Remember

Think of Part 2 of the Regulations as a type of dictionary relating to electrical words

- what the installation is to be used for, the supply used (single phase or three phase), type of earthing (TNS, TNC-S, TT etc.); these are in Chapter 31

- the external influences to which the installation will be exposed such as temperature, water, corrosion etc.; these external influences are listed in Appendix 5 of the Regulations

- the compatibility of the equipment to be used with reference to such things as other electrical equipment or services, and the supply; the effects are things like fluctuating loads, starting currents and so on; details of other effects are in Chapter 33

- the maintainability of the installation, how often it will need maintenance, how easily periodic inspections can be carried out, and reliability of the equipment.

- The requirements for safety services and continuity of service.

Part 4 Protection for safety

This part of the Regulations deals with the protective measures to be taken to prevent electrical shock risk and other dangers arising from the use of electricity.

If the risk of electric shock is increased because of the location of the installation, e.g. a bathroom or construction site, then the requirements of Part 7 of the Regulations should be used as well.

Chapters 41–44 deal with what the requirements are for protection, such as protection against shock (41), burns and overheating (42), overcurrent (43), as well as Chapter 44 dealing with the requirements for protection against voltage and electromagnetic disturbances.

Part 5 Selection and erection of equipment

This part deals with the selection of equipment and its erection to provide compliance with the following:

- measures of protection for safety (for example, correct voltage and current rating)

- proper functioning for the intended use of the installation (for example, isolators and switches)

- appropriate requirements for the likely (and foreseen) external influences for example, presence of water or corrosive substances).

Generally speaking, any item of equipment used must have a relevant British or European Standard, which confirms compliance with the use for which it is intended. Any item which does not have either of these standards can be specified by a designer provided that the equipment provides the same level of conformance with the Regulations.

Part 6 Inspection and testing

Every installation during erection and on completion and before being put into use must be inspected and tested to verify that the requirements of the Regulations have been met. While this process is being carried out precautions must be taken to prevent danger to persons, property and the installed equipment.

Any alteration or addition to an existing installation requires the relevant inspections and tests to be carried out to ensure compliance with the Regulations and also to ensure that the existing installation has not been made less safe.

Installations should also be tested on a regular basis; this is known as Periodic Inspection and Testing, and the time interval varies depending on what type of installation it is.

The initial verification (visual inspection) covers such things as identification of conductors, routing of cables, connection of conductors and so on.

The test requirements themselves are not listed in an easy-to-read format within this part of the Regulations, nor does it detail how to carry out these tests. Those details, along with topics such as periodic test intervals, are in the *Guidance Note 3* booklet published by the IEE, or alternatively in the *Amicus Guide* and the *IEE On-Site Guide* (Section 9).

Reference must, of course, be made to the relevant Regulations to ensure that the results obtained are within the limits set. Electrical Installation Certificates are based around the model shown in Appendix 6 of the Regulations.

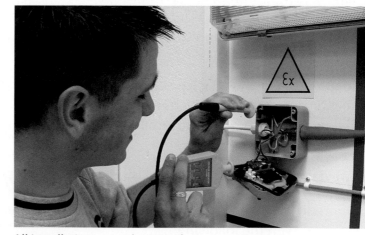

All installations must be tested

Part 7 Special installations or locations

Certain locations are deemed to be more hazardous to persons and livestock because of the environment in which the equipment is used. Additional (or in some cases replacement) regulations are needed to provide greater safety.

The special locations identified by the Regulations are as follows:

- **locations containing a bath or shower**
- swimming pools
- hot air saunas
- **construction site installations**
- **agricultural and horticultural premises**
- conducting locations with restricted movement
- **caravan parks**
- marinas
- exhibitions
- solar photovoltaic systems
- mobile units
- caravans and motor caravans
- temporary installations in amusement parks
- floor and ceiling heating systems.

We will be looking at key areas, listed above and on page 316 in bold type, in Chapter 14. For more information on the topics covered in *Guidance Note 7*, please refer to BS 7671.

Appendices

Appendix 1 lists the appropriate British and European Standards referred to throughout the Regulations.

Appendix 2 is about the statutory regulations that have to be complied with in Great Britain, such as Building Standards, Electricity at Work Regulations 1989 and Electricity Safety, Quality and Continuity Regulations 2002. These are all Acts or Regulations passed by Parliament and are therefore law.

Appendix 3 deals with the time current characteristics of overcurrent protective devices, in other words how quickly (time) a fuse or circuit breaker will interrupt the supply when a fault current flows. The information contained in this part will be used extensively when you do cable calculations.

Appendix 4 deals with the current carrying capacity and voltage drops of cables and flexible cords. It also lists all the different methods of installation, such as embedded in plaster, in trunking, conduit etc.

This information will also be use extensively when you do cable calculations.

Appendix 5 deals with all the external influences that can apply to the use of electrical equipment, things such as water, corrosion etc. The lists in this part are used to ensure correct selection of equipment for the environment in which it is expected to operate.

Appendix 6 contains examples of the various certification forms that are used when inspecting and testing electrical installations. These include certificates for new installations, minor works and periodic inspections.

Appendix 7 deals with the new cable colours that came into effect on 1 April 2004. The colours used for fixed and flexible cables are now the same as the rest of Europe that is part of the CENELEC Agreement. You will come across many installations where the old colours are still in use, and where a mixture of the two will exist; it is important that you can distinguish between the two sets of colour standards, so learn them well!

Appendix 8 provides the current-carrying capacities of trunking and powertrack systems

Appendix 9 contains the definitions for multiple source and other systems

Appendix 10 deals with the protection of conductors in parallel

Appendix 11 provides correction factors for the effects of harmonic currents

Appendix 12 provides voltage drop values

Appendix 13 provides methods for measuring the insulation resistance to floors and wells

Appendix 14 considers effects of temperature when measuring fault loop impedance

Appendix 15 Ring and Radial circuits

The Electricity at Work (EAW) Regulations 1989

The Electricity at Work Regulations were passed through Parliament in 1989 and came into force on 1 April 1990. Their purpose is to require precautions to be taken against the risk of death or personal injury from electricity in work activities.

The Regulations were made under the Health and Safety at Work Act 1974, which imposes duties on employers, employees and the self-employed. These Regulations (EAW) are more specific and concentrate on work activities at or near electrical equipment and make one person primarily responsible to ensure compliance in respect of systems, electrical equipment and conductors; this person is referred to as the 'duty holder'.

They were also designed to include all the systems etc. that BS 7671 does not cover, such as voltages above 1000 volts a.c.

There are 33 Regulations and three appendices within EAW 1989. However, not all of them apply to all situations. This chapter contains an overall look at the relevant EAW Regulations and what they mean to give you an appreciation of what they are about. For detailed information read *The Memorandum of Guidance on the Electricity at Work Regulations 1989*.

EAW Regulation 1 only states that these EAW Regulations came into force on 1 April 1990.

EAW Regulation 2 contains definitions of what is meant by certain words or phrases.

EAW Regulation 3 deals with **duty holders** and the requirements imposed on them by these Regulations. There are three categories of duty holder, namely:

- employers
- employees
- self-employed persons.

The duty holder is the person who has a duty to comply with these EAW Regulations because they are relevant to circumstances within their control. Such a person must be competent.

EAW Regulation 4 has four parts. The first deals with the construction of the electrical systems (all parts), in that the equipment should be suitable for its intended use so that it does not give rise to danger 'as far as is reasonably practicable'.

Remember

Whenever a Regulation does use the phrase **'as far as is reasonably practicable'** this means that the duty holder must assess the magnitude of the risks against the costs in terms of physical difficulty, time, trouble and expense involved in minimising that risk(s). The onus is on the duty holder to prove in a court of law that he or she took all steps, as far as is reasonably practicable

The second part deals with the maintenance of systems, whereby all systems should be maintained to prevent danger (this includes portable appliances) 'as far as is reasonably practicable'. Records of maintenance including test results should be kept.

The third part deals with ensuring safe work activities near a system including operation, use and maintenance 'as far as is reasonably practicable'. This could include non-electrical activities such as excavation near underground cables and erecting scaffolding near overhead lines. Safe work activities include things such as:

- company health and safety policy
- permit to work systems
- clear communication

The Regulations ensure that equipment and conditions are safe

- use of competent people

- personnel attitudes.

The fourth part deals with provision of protective equipment, such as insulated tools, test probes, insulating gloves, rubber mats, etc. which must be suitable for use, maintained in that condition and properly used. This is an **absolute** duty.

EAW Regulation 5 has four parts. This Regulation states that 'no electrical equipment must be used where its strength and capability may be exceeded and give rise to danger' – for example, switchgear should be capable of handling fault currents as well as normal load currents, correct size cable etc.

EAW Regulation 6 deals with the siting and/or selection of electrical equipment and whether it would be exposed to, or could you foreseeably predict it being exposed to, adverse or hazardous environments.

The following list provides what needs to be considered when siting and/or selecting electrical equipment, not just for the present situation but for what you could reasonably expect could be the situation in the future 'as far as is reasonably practicable':

- protection against mechanical damage

- effects of weather, natural hazards, temperature or pressure

- effects of wet, dirty, dusty or corrosive conditions

- any flammable or explosive substances, including dusts, vapours or gases.

EAW Regulation 7 is concerned with conductors in a system and whether they present a danger to persons. All conductors must either be suitably covered with insulating material and protected or, if not insulated (such as overhead power lines), placed out of reach 'as far as is reasonably practicable'. The definition of placed out of reach is in part 2 of BS 7671.

EAW Regulation 8 deals with the requirements for earthing or other such suitable precautions that are needed to reduce the risk of electric shock when a conductor (other than a circuit conductor) becomes live under fault conditions. This consists of such things as earthing the outer conductive parts of electrical equipment that can be touched, and other conductive metalwork in the vicinity such as water and gas pipes. Other methods could be reduced voltage systems, double insulated equipment and RCDs. This is an **absolute** Regulation.

EAW Regulation 9 is about maintaining the integrity of referenced conductors. In simple terms this means that the neutral conductor must not have a fuse or switch placed in it. The only exception is that a switch may be placed in the neutral conductor if that switch is interlocked to break the line conductor(s) at the same time. This is an **absolute** Regulation.

EAW Regulation 10 requires that all joints and connections in a system must be mechanically and electrically suitable for its use. For example, things like taped joints on extension leads are not allowed. This is an **absolute** Regulation.

EAW Regulation 11 states that every part of a system must be protected from excess current that may give rise to danger. This means that suitably rated fuses, circuit breakers etc. must be installed so that, in a fault situation, they will interrupt the supply and prevent a dangerous situation happening. This is an **absolute** Regulation.

EAW Regulation 12 deals with the need for switching off and isolating electrical equipment and, where appropriate, identifying circuits. Isolation means cutting off from every source of electrical energy in such a way that it cannot be switched back on accidentally, in other words a means of 'locking off' the switch securely. This is an **absolute** Regulation.

EAW Regulation 13 refers to the precautions required when work is taking place at or near electrical equipment which is to be worked on, whether it be electrical or non-electrical work that is taking place. The electrical equipment must be isolated, locked off and tested for absence of live parts before the work takes place and must remain so until all persons have completed their work. A safe isolation procedure or written Permit to Work scheme should be used. This is an **absolute** Regulation.

EAW Regulation 14 refers to the precautions needed when working on or near live conductors. An **absolute** duty is imposed that the conductors must be isolated unless certain conditions are met. Precautions considered appropriate are:

- properly trained and competent staff
- provision of adequate information regarding nature of the work and system
- use of appropriate insulated tools, equipment, instruments, test probes and protective clothing
- use of insulated barriers
- accompaniment of another person
- effective control of the work area.

This is an **absolute** Regulation.

EAW Regulation 15 concerns itself with matters relating to work being carried out at or near electrical equipment whereby to prevent danger, adequate working space, adequate means of access and adequate lighting must be provided. This is an **absolute** Regulation.

EAW Regulation 16 deals with the competency of persons working on electrical equipment to prevent danger or injury. To comply with this regulation a person should conform to the following or be under such a degree of supervision as appropriate given the type of work to be carried out:

- adequate understanding and practical experience of the system to be worked on
- an understanding of the hazards that may arise
- the ability to recognise at all times whether it is safe to continue.

The Regulation tries to ensure that no one places themselves or anyone else at risk due to their lack of technical knowledge or practical experience. This is an **absolute** Regulation.

EAW Regulations 17 to 28 apply only to mines and quarries.

EAW Regulation 29 is what is known as the 'Defence' Regulation. If an offence is committed by the duty holder under these Regulations (the absolute ones) and criminal proceedings are brought by the HSE (Health and Safety Executive), then if the duty holder can prove that they took all reasonable steps and exercised due diligence to avoid committing that offence, they will not be found guilty.

EAW Regulation 30. A duty holder can apply to the HSE for exemption from these Regulations for the items listed:

- any person
- any premises
- any electrical equipment
- any electrical system
- any electrical process
- any activity.

Exemptions will only be granted by the HSE provided they do not prejudice the health and safety of any persons.

EAW Regulation 31 deals with work activities and premises outside of Great Britain. If the activity or premises is covered by sections 1 to 59 and sections 80 to 82 of the Health and Safety at Work Act 1974, then these regulations apply.

EAW Regulation 32 details what these regulations do not apply to:

- sea going ships
- aircraft or hovercraft moving under their own power.

EAW Regulation 33 deals with changes and modifications to these regulations since they were brought in.

EAW Appendix 1 lists the HSE guidance publications available for help in understanding and applying the regulations.

EAW Appendix 2 lists various other codes of practice and British Standards that could help in the understanding and application of these regulations.

EAW Appendix 3 deals with the legislation concerning working space and access regulations.

Hazardous installations

On completion of this topic area the candidate will be able to state the need for statutory regulation of given hazardous installations.

Definition of hazardous areas

A hazardous area can be defined as: 'An area in which explosive gas/air mixtures are, or may be expected to be, present in quantities such as to require special precautions for the construction and use of electrical apparatus.'

Zoning

Hazardous areas are defined in the Dangerous Substances and Explosive Atmospheres Regulations 2002 (DSEAR) as: 'Any place in which an explosive atmosphere may occur in quantities such as to require special precautions to protect the safety of workers'. In this context, 'special precautions' is best taken as relating to the construction, installation and use of apparatus, as given in BS EN 70079 -10.

Area classification is a method of analysing and classifying the environment where explosive gas atmospheres may occur. The main purpose is to facilitate the proper selection and installation of apparatus to be used safely in that environment, taking into account the properties of the flammable materials that will be present. DSEAR specifically extends the original scope of this analysis, to take into account non-electrical sources of ignition, and mobile equipment that creates an ignition risk.

Hazardous areas are classified into zones based on an assessment of the frequency of the occurrence and duration of an explosive gas atmosphere, as follows.

- **Zone 0**: An area in which an explosive gas atmosphere is present continuously or for long periods.
- **Zone 1**: An area in which an explosive gas atmosphere is likely to occur in normal operation.
- **Zone 2**: An area in which an explosive gas atmosphere is not likely to occur in normal operation and, if it occurs, will only exist for a short time.

Various sources have tried to place time limits on to these zones, but none have been officially adopted. The most common values used are listed below.

- **Zone 0**: Explosive atmosphere for more than 1000 hours per year.
- **Zone 1**: Explosive atmosphere for more than 10, but fewer than 1000 hours per year.
- **Zone 2**: Explosive atmosphere for fewer than 10 hours per year, but still sufficiently likely as to require controls over ignition sources.

Where people wish to quantify the zone definitions, these values are the most appropriate, but for the majority of situations a purely qualitative approach is adequate.

Remember

Ignition temperature is not the same as flash point, so don't confuse them!

When the hazardous areas of a plant have been classified, the remainder will be defined as non-hazardous, sometimes referred to as 'safe areas'.

Term	Definition
Explosive limits	The upper and lower percentages of a gas in a given volume of gas/air mixture at normal atmospheric temperature and pressure that will burn if ignited.
Lower explosive limit (LEL)	The concentration below which the gas atmosphere is not explosive.
Upper explosive limit (UEL)	The concentration of gas above which the gas atmosphere is not explosive.
Ignition energy	The spark energy that will ignite the most easily ignited gas/air mixture of the test gas at atmospheric pressure; Hydrogen ignites very easily, whereas Butane or Methane require about ten times the energy.
Flash point	The minimum temperature at which a material gives off sufficient vapour to form an explosive atmosphere.
Ignition temperature or auto ignition temperature of a material	The minimum temperature at which the material will ignite and sustain combustion when mixed with air at normal pressure, without the ignition being caused by any spark or flame.

Table 13.01 Definitions of key terms

Selection of equipment

DSEAR sets out the link between a zone and the equipment that may be installed in that zone. This applies to new or newly modified installations. The equipment categories are defined by the ATEX equipment directive, set out in UK law as the Equipment and Protective Systems for Use in Potentially Explosive Atmospheres Regulations 1996.

Standards set out different protection concepts, with further subdivisions for some types of equipment according to gas group and temperature classification. Most of the electrical standards have been developed over many years and are now set at international level, while standards for non-electrical equipment are only just becoming available from CEN.

The DSEAR ACOP describes the provisions concerning existing equipment. There are different technical means (protection concepts) of building equipment to the different categories. The standard current and the letter giving the type of protection are listed in the text that follows.

Correct selection of electrical equipment for hazardous areas requires the following information:

- temperature class or ignition temperature of the gas or vapour involved according to Table 13.02

- classification of the hazardous area (as in zones shown in Table 13.03).

Temperature classification	Maximum surface temperature, °C	Ignition temperature of gas or vapour °C
T1	450	>450
T2	300	>300
T3	200	>200
T4	135	>135
T5	100	>100
T6	85	>85

Table 13.02 Ignition temperatures

Zone 0	Zone 1	Zone 2
Category 1	Category 2	Category 3
'ia' intrinsically safe EN 50020, 2002	'd' flameproof enclosure EN 50018 2000	Electrical Type 'n' EN 50021 1999 Non electrical EN 13463-1, 2001
Ex s – Special protection if specially certified for Zone 0	'p' pressurised EN 50016 2002	
	'q' powder filling EN 50017, 1998	
	'o' oil immersion EN 50017, 1998	
	'e' increased safety EN 50019, 2000	
	'ib' Intrinsic safety EN 50020, 2002	
	'm' encapsulation EN 50028, 1987	
	's' special protection	

Table 13.03 Classification of hazardous areas

If several different flammable materials may be present within a particular area, the material that gives the highest classification dictates the overall area classification. The IP code considers specifically the issue of hydrogen containing process streams as commonly found on refinery plants.

Consideration should be shown for flammable material that may be generated due to interaction between chemical species.

Installations in potentially explosive areas

Within hazardous areas there exists the risk of explosions and/or fires occurring due to electrical equipment 'igniting' the gas, dust or flammable liquid.

These areas are not included in BS 7671 but are instead covered by IEC Standard BS EN 60079 as follows:

- BS EN 60079 Part 10 – Classification of hazardous areas
- BS EN 60079 Part 14 – Electrical apparatus for explosive gas atmospheres
- BS EN 60079 Part 17 – Inspection/maintenance of electrical installations in hazardous areas.

The BS EN 60079 has been in place since 1988, replacing the old BS 5345. However, many installations obviously still exist that were completed in accordance with BS 5345 and new European Directives (ATEX) address safety where there is a danger from potentially explosive atmospheres.

Other statutory regulations such as the Petroleum Regulation Acts 1928 and 1936 and local licensing laws govern storage of petroleum.

Ignition sources: identification and control

Ignition sources may be:

- flames
- direct fired space and process heating
- use of cigarettes/matches etc.
- cutting and welding flames
- hot surfaces
- heated process vessels such as dryers and furnaces
- hot process vessels
- space heating equipment
- mechanical machinery
- electrical equipment and lights
- spontaneous heating
- friction heating or sparks
- impact sparks
- sparks from electrical equipment
- stray currents from electrical equipment
- electrostatic discharge sparks

- lightning strikes
- electromagnetic radiation of different wavelengths
- vehicles (unless specially designed or modified are likely to contain a range of potential ignition sources).

Sources of ignition should be effectively controlled in all hazardous areas by a combination of design measures and systems of work:

- using electrical equipment and instrumentation classified for the zone in which it is located; new mechanical equipment will need to be selected in the same way (see earlier)
- earthing of all plant/equipment (see *Technical Measures Document on Earthing*)
- elimination of surfaces above auto-ignition temperatures of flammable materials being handled/stored (see above)
- provision of lightning protection
- correct selection of vehicles/internal combustion engines that have to work in the zoned areas (see *Technical Measures Document on Permit to Work Systems*)
- correct selection of equipment to avoid high intensity electromagnetic radiation sources, e.g. limitations on the power input to fibre optic systems, avoidance of high intensity lasers or sources of infrared radiation
- prohibition of smoking/use of matches/lighters
- controls over the use of normal vehicles
- controls over activities that create intermittent hazardous areas, e.g. tanker loading/unloading
- control of maintenance activities that may cause sparks/hot surfaces/naked flames through a Permit to Work System.

Petrol filling stations

The primary legislation controlling the storage and use of petrol is the Petroleum (Consolidation) Act 1928 (PCA). This requires anyone who keeps petrol to obtain a licence from the local Petroleum Licensing Authority (PLA). The licence may be, and usually is, issued subject to a number of licence conditions. The PLA set the licence conditions, but they must be related to the safe keeping of petrol.

The Local Authority Co-ordinating Body on Food and Trading Standards (LACOTS) has issued a set of standard licence conditions which most, if not all, PLAs apply to their sites.

Installations within petrol filling stations are effectively also covered by BS EN 60079 Parts 10, 14 and 17.

Additionally there is industry-developed guidance for this sector in the form of the electrical section of IP/APEA's Guidance for the Design, Construction, Modification and Maintenance of Petrol Filling Stations (Institute of Petroleum and the Association for Petroleum and Explosives Administration). This guidance replaced most of HS(G)41 and was published in 1999 by IP/APEA with input from HSE.

Lightning Protection Systems (LPS)

Many larger buildings or those of historical interest are fitted with Lightning Protection Systems (LPS). These are designed to protect the building and persons within the building from the harmful effects of a lighting strike by creating a 'Faraday's Cage' around the building.

A typical LPS system has three main parts:

- An air termination system which is designed to "catch" the lightning strike before it hits the building structure.

- A system of down-conductors to conduct the current away from the building and down to earth.

- An earth termination system which provides a solid and reliable connection to earth to disperse the current into the ground

Installation of these systems is often carried out by specialist contractors working to BS 6651:1999 Code of Practice for Protection of Structures Against Lightning. This BS is shortly to be replaced by BSEN 62305 parts 1-4.

BS 7671 has a requirement for the LPS to be connected to the main earth terminal (MET) of the electrical installation via a main equipotential bonding conductor. The designer of the LPS should be consulted when deciding on the best position for the main bonding connection.

Activity

Find out if your company has a policy document (company procedures) telling you how to work safely on or near electrical systems. Some companies also have risk assessment and permit to work forms specifically for electrical work.

FAQ

Q In the Electricity at Work Regulations, what is an absolute regulation?

A An absolute regulation is one which must be complied with 'regardless of cost or any other consideration'.

Q My washing machine has wiring of different colours in it. Does this contravene the Regulations?

A No. BS 7671 does not apply to the manufacture of items of equipment, only how they are installed.

Q My company never inspects and tests anything; we know we do a good job.

A Well, this is in direct contravention of BS 7671, which says that every addition or alteration to an installation must be inspected and tested before being put into service. This means that it will also be in breach of the Electricity at Work Regulations (a criminal offence) and possibly (depending on the type of work) against Part P of the Building Regulations.

Knowledge check

1. State two reasons for having electrical regulations in force.

2. What do the Electricity at Work Regulations say about working on or near live conductors?

3. Under EAWR, what is the definition of a 'competent person'?

4. What are the voltage ranges covered by the EAWR and BS 7671?

5. Define the terms 'explosive limit', 'ignition energy', and 'flash point'.

chapter 14

Electrical installations and systems

Unit 4 Outcomes 2–4

There are several different types of installation that you will be involved in as an electrician. It is important to be familiar with all of these in order to complete all the work you may need to carry out to the highest possible standard. We have covered some of these installation types and the regulations that govern their operation earlier in this book.

Wiring systems and enclosures have special factors that need to be considered according to circumstance. The same system will not apply for two different installations. When working on installations at all times, it is important to remember safe working practices.

This chapter will cover all the material from Outcomes 2, 3 and 4 from Unit 4 that have not been covered earlier in this book. The majority of this material comes from Unit 3. A page reference grid will be given for Units 2 and 4 at the end of the chapter, to allow the candidate to refer to them.

On completion of this chapter the candidate should be able to:

- **state types of electrical installations, components and related functions**

- **state wiring systems and wiring enclosures, factors determining choice of system and applications/limitations of specialised types of cable**

- **state factors affecting selection of conductor size**

- **carry out calculations to determine voltage drop, circuit current and protective devices.**

Types of electrical installation and components

On completion of this topic area the candidate will be able to state types of electrical installation, components and related functions, and describe the function, operation and wiring of components for electrical systems.

There are several different types of electrical installation that you will undertake in your career as an electrician. Many of these have been covered earlier in this book. We looked at the various types of installation in chapter 2 pages 39–41, so you should be familiar with the different varieties of job you may be called upon to do within the electro-technical industry.

Lighting

The switching of lighting circuits was covered fully in chapter 6 pages 124–132. In that chapter we saw that wiring can be undertaken in two different ways:

- conduit and trunking – this uses PVC single-core insulated cables

- multicore or composite cables – this uses a sheathed multi-core twin and earth or a three cores and earth. A 'loop in' or 'joint box' method may be employed with this type of installation. More information about this can be found on pages 131–132.

The primary role of lighting circuits is to provide functional levels of illumination in buildings, allowing the people in the building to get around in the dark. Lighting circuits may also be used outside, such as streetlights, or used to direct people, such as on road signs and traffic lights, or floodlights at stadiums. Emergency lighting is covered separately below, as there are special conditions relating to its installation and operation.

Operation of lighting systems

The operation of lighting systems was covered in chapter 6 pages 125–131. The other major area in lighting is the operation of lamps.

Incandescent lamps

In this method of creating light, a fine filament of wire is connected across an electrical supply. This makes the filament wire heat up until it is white-hot and gives out light. The filament wire reaches a temperature of 2500–2900°C. These lamps are very inefficient and only a small proportion of the available electricity is converted into light; most of the electricity is converted into heat as infrared energy. The light output of this type of lamp is mainly found at the red end of the visual spectrum, which gives an overall warm appearance.

Remember

The filament wire in an incandescent lamp reaches a temperature of about 2500–2900°C

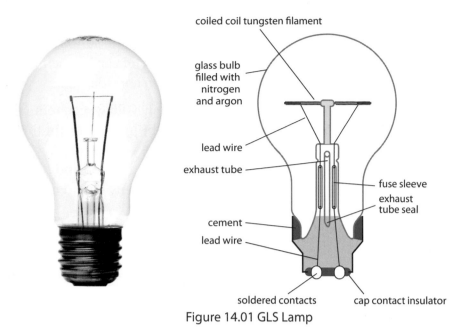

coiled coil tungsten filament

glass bulb filled with nitrogen and argon

lead wire

exhaust tube

fuse sleeve

exhaust tube seal

cement

lead wire

soldered contacts   cap contact insulator

Figure 14.01 GLS Lamp

Operation of GLS lamps

The General Lighting Service (GLS) lamp is one type of incandescent lamp and is commonly referred to as the light bulb. It has at its core a very thin tungsten wire that is formed into a small coil and then coiled again.

A current is passed through the tungsten filament, which causes it to reach a temperature of 2500°C or more so that it glows brightly. At these temperatures, the oxygen in the atmosphere would combine with the filament to cause failure, so all the air is removed from the glass bulb and replaced by gases such as nitrogen and argon. Nitrogen is used to minimise the risk of arcing and argon is used to reduce the evaporation process. On low-power lamps, such as 15 and 25 watt, the area inside the bulb remains a vacuum. The efficiency of a lamp is known as the efficacy. It is expressed in lumen per watt (lm/w). For this type of lamp the efficacy is between 10 and 18 lumens per watt. This is low compared with other types of lamp, and its use is limited. However, it is the most familiar type of light source used and has many advantages including:

- comparatively low initial costs
- immediate light when switched on
- no control gear
- it can easily be dimmed.

When a bulb filament finally fails it can cause a very high current to flow for a fraction of a second – often sufficient enough to operate a 5 or 6 amp miniature circuit breaker which protects the lighting circuit. High-wattage lamps, however, are provided with a tiny integral fuse within the body of the lamp to prevent damage occurring when the filament fails.

Did you know?

The first lamp that was developed for indoor use was the carbon-filament lamp. Although this was a dim lamp by modern standards it was cleaner and far less dangerous than the exposed 'arc lamp'

Did you know?

The average life of this type of lamp is 1000 hours, after which the filament will rupture

Did you know?

The ancient Romans were the first to invent central heating. They installed ducts in the floors and walls, into which hot air from a fire was fed

If the lamp is run at a lower voltage than that of its rating, the light output of the lamp is reduced at a greater rate than the electricity used by the lamp, and the lamp's efficacy is poor. This reduction in voltage, however, increases the lifespan and can be useful where lamps are difficult to replace or light output is not the main consideration.

It has been calculated that an increase in 5 per cent of the supply voltage can reduce the lamp life by half. However, if the input voltage is increased by just 1 per cent this will produce an increase of 3.5 per cent in lamp output (lumens). When you consider that the Electricity Distributor is allowed to vary its voltage up to and including 10 per cent it is easy to see that if this was carried on for any length of time the lamps would not last very long.

Heating and environmental control

Central heating is a system available in nearly every building you will find yourself in – whether at home or in the place where you work. Heating, or climate control as it is sometimes known, aims at keeping the building it is in at a stable and comfortable temperature. Central heating systems operate by heating water, and then using a pump to transfer either the hot water, air or steam around pipes and ducts throughout the building.

The water is usually held at a central point in the building. The main components of a heating system are as follows:

- a boiler – this contains the water that the system heats through powering a furnace
- a pump – used to transfer the heat into the pipework
- piping or ductwork – this transmits the heated water, steam or air around the building
- heat emitters – to transfer the heat from the pipework to the room.

Rooms can also be heated through space heaters. These are electric radiators that convert electricity into heat. The electrical current flowing through the heater is converted into heat.

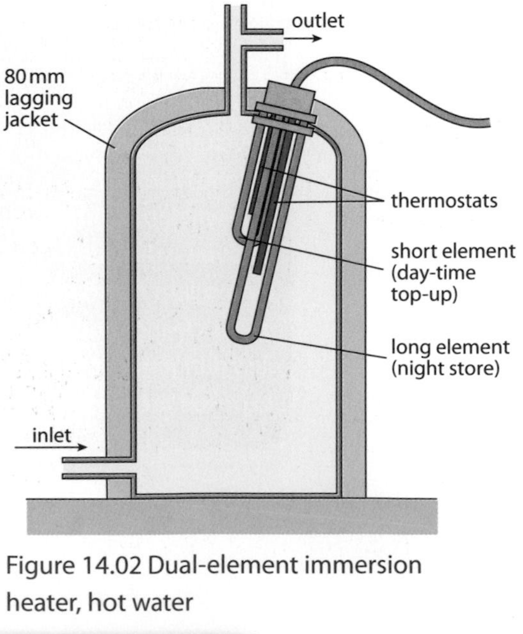

Water heating

There are two main methods of heating water electrically: either heating a large quantity stored in a tank or heating only what is required when it is needed. With both of these types of heater it is important to ensure that the exposed and extraneous conductive parts are adequately bonded to earth: water and electricity do not mix well! It is also important to ensure that the cables selected are of the correct size for full load current, since no diversity is allowed for water heaters.

Heating large tanks of stored water (typically 137+ litres) is done using an immersion heater (see Figure 14.02) fitted into a large water tank. When it is on or off is controlled via either a timer switch or an on/off switch.

outlet
80 mm lagging jacket
thermostats
short element (day-time top-up)
long element (night store)
inlet

Figure 14.02 Dual-element immersion heater, hot water

The temperature of the water is controlled by a stem-type thermostat which is incorporated within the housing of the heating element. This type of heater is used in domestic situations, although larger multiple immersion heaters can be used in industrial situations. The heater in a domestic situation must be fed from its own fuse/MCB in the consumer unit and have a double pole isolator fitted next to the storage tank. The final connection to the heating element must be with heatproof flexible cable due to the high ambient temperatures where the water tank is normally located.

This type of system sometimes has two elements. One is controlled via a separate supply which operates only at night time (Economy 7 or white-meter supply) when cheap electricity is available, thus heating a full tank of water ready for use the next day. The other is switched on as and when needed during the day to boost the amount of hot water available.

The rest of this section will describe the different types of water system:

- cistern type
- instantaneous.
- non-pressure

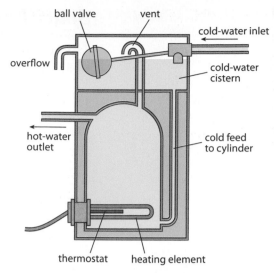

Figure 14.03 Cistern-type water heater

Cistern-type

Where larger volumes of hot water are needed, for example in a large guest house, then a cistern-type water heater (9 kW+) is used which is capable of supplying enough hot water to several outlets at the same time.

Non-pressure

Non-pressure water heaters, which are typically rated at less than 3 kW and contain less than 15 litres of water, heat the water ready for use and are usually situated directly over the sink, such as in a small shop or hairdressers' salon.

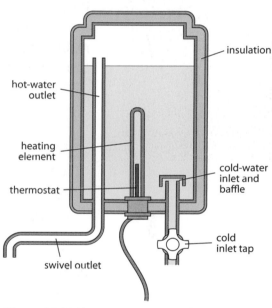

Figure 14.04 Non-pressure water heater

Instantaneous

Instantaneous water heaters heat only the water that is needed. This is done by controlling the flow of water through a small internal water tank which has heating elements inside it; the more restricted the flow of water, the hotter the water becomes.

The temperature of the water can therefore be continuously altered or stabilised locally at whatever temperature is selected. This is how an electric shower works, and showers in excess of 10 kW are currently available. The shower-type water heater must be supplied via its own fuse/MCB in the consumer unit and have a double pole isolator located near the shower.

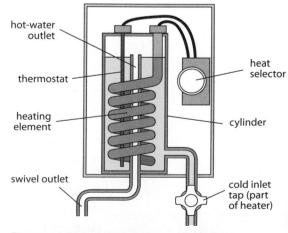

Figure 14.05 Instantaneous water heater

Space heating

The type of electric heating available falls into two main categories: direct acting heaters and thermal storage devices.

Direct acting heaters

Direct acting heaters are usually just switched on and off when needed; some of them can be thermostatically controlled. Direct heaters fall into two categories: radiant and convection.

Radiant heaters

The radiant-type heaters reflect heat and come in a variety of shapes, sizes and construction as follows.

- **Traditional electric fire:** has a heating element supported on insulated blocks with a highly polished reflective surface behind it; these range in size from about 750 W to 3 kW.

- **Infrared heater:** consists of an iconel-sheathed element or a nickel-chrome spiral element housed in a glass silica tube which is mounted in front of a highly polished surface. Sizes vary from about 500 W to 3 kW; the smaller versions are usually suitable for use in bathrooms and may be incorporated with a bulb to form a combined heating and lighting unit.

- **Oil-filled radiator:** consists of a pressed steel casing in which are housed heating elements; the whole unit is filled with oil. Oil is used because it has a lower specific gravity than water and so heats up and cools down more quickly. Surface temperature reaches about 70°C; power sizes range from about 500 W to 2 kW.

- **Tubular heater:** low-temperature unit designed to supplement the main heating in the building. Consists of a mild steel or aluminium tube of about 50 mm diameter in which is mounted a heater element. The elements themselves are rated at 200 W to 260 W per metre length and can range in length from about 300mm up to 4.5m. The surface temperature is approximately 88°C.

- **Under-floor heater:** consists of heating elements embedded under the floor which heat up the tiles attached to the floor surface. The floor then becomes a large low-temperature radiant heater. A room thermostat controls the temperature within the room and the floor temperature does not normally exceed 24°C. The elements have conductors made from a variety of materials such as chromium, copper, aluminium, silicon or manganese alloys. The insulating materials used are also made from a variety of materials such as asbestos, PVC, silicon rubber and nylon.

Convection heaters

Convection heaters consist of a heating element housed inside a metal cabinet that is insulated both thermally and electrically from the case so that the heat produced warms the surrounding air inside the cabinet. Cool air enters the bottom of the cabinet and warm air is passed out at the top of the unit at a temperature of between 80°C and 90°C. A thermostatic control is usually fitted to this type of heater.

Fan heater

Operates in the same way as a convector heater but uses a fan for expelling the warm air into the room. Fan heaters usually have a two-speed fan incorporated into the casing and up to 3 kW of heating elements.

Thermal storage devices

Thermal storage devices heat up thermal blocks within the unit during off-peak times to enable use of cheap-rate electricity. The heat stored is then released during the day when it is needed.

A thermal storage unit consists of several heating elements mounted inside firebricks, which in turn are surrounded by thermal insulation such as fibreglass, all housed inside a metal cabinet. The firebricks are made from clay, olivine, chrome and magnesite, which have very good heat-retaining properties. The bricks are heated up during off-peak hours (usually less than half the normal price per electrical unit) and the heat is stored within the bricks until the outlet vent is opened the following day and allows the warm air to escape and hence heat up the room.

Cooker thermostats and controllers

Simmerstat (energy regulator)

This device is used to control the temperature of electric cooking plates. It uses a bi-metal strip as its main principle of operation; it is not controlled by the temperature of the hotplates. Operation is by the opening and closing of a switch at short definite time intervals by the heating up (via an internal heating coil) of a bi-metal strip. The length of time that the switch is opened or closed is determined by the control knob mounted on the front of the device, and hence the length of time that the hotplate has power is varied. The control knob is normally calibrated from either 0 to 10 or 0 to 5, with the highest number being the hottest temperature that the hotplate will reach.

There are two basic ways that these devices are arranged. One is with a shunt-connected (parallel) heating coil, and the other is with a series-connected heater coil. The heater coil in each case responds to the current flowing through it and hence determines the control of the bi-metal strip.

The shunt-connected thermal regulator (Figure 14.06) consists of a two-part bi-metal strip block where one strip 'A' has a small-gauge heater wire wrapped around it; this is in turn connected in parallel with the hotplate element. The second part of the bi-metal strip block 'B' is in mechanical contact with the cam of the control knob. Both of these strips are connected together at one end and pivoted on a fulcrum point 'C'. When the control knob is in the 'off' position, the cam is pushing against the bi-metal strip and hence keeps the contacts at 'F' open, so that no current flows and therefore the hotplate does not heat up.

When the control knob is moved to one of the 'on' positions, the cam moves and pressure on the bi-metal strip is reduced, thus allowing the contacts at 'F' to close and start to heat up the hotplate. As this happens, current flows

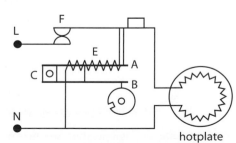

Figure 14.06 Shunt-wired regulator

through the heater coil as well as the hotplate and causes the bi-metal strip to bend, causing the contacts at 'F' to open. When the heater coil and bi-metal strip cool down the bi-metal strip bends and allows the contacts at 'F' to close again, thus repeating the cycle.

The hotplate therefore has power switched on and off rapidly and hence stays at a constant temperature. This technique is known as simmering, hence the name of the device 'simmerstat'. At low-temperature settings the contacts will be open for longer, and in the fully 'on' position the contacts at 'F' will be closed and no regulation occurs.

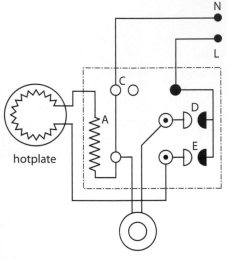

Figure 14.07 Series-wired regulator

In a series type of simmerstat, as shown in Figure 14.07, 'A' is the bi-metallic heater coil which is connected in series with the hotplate. On the right of the diagram there are two sets of contacts, 'D' and 'E'. When the control knob is operated as if to turn the hotplate on, both contacts will close. One of them, 'D', will bring a pilot lamp into the circuit to show that the hotplate is being heated the other 'E' is the main contact that will energise the hotplate.

As the heater coil 'A' transmits heat it causes the bi-metal strip to bend, and eventually this will cause the contacts at 'C' to open, thus breaking the neutral to the coil and the pilot lamp. Consequently the pilot lamp will go out and the hotplate is switched 'off'.

When this happens the heater coil begins to cool and the bi-metal strip returns to its original position, allowing the contacts at 'C' to close again, and thus the cycle is repeated. Each time the contacts open, the pilot light goes off and gives a visual indication that the hotplate is up to temperature.

Oven thermostats

There are two basic types of oven thermostat, both of which work in similar ways. A capillary type has a capillary tube (typically about 800mm long) filled with liquid which, when the phial containing it heats up, expands. The liquid then pushes against the

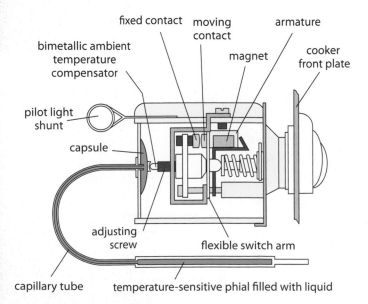

Figure 14.08 Capilliary type oven thermostat

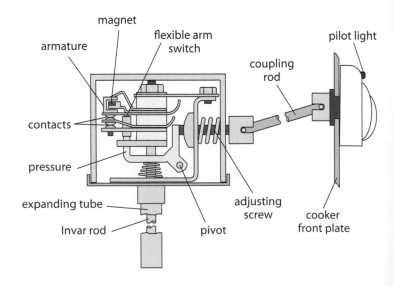

Figure 14.09 Stem type oven thermostat

capsule in the control housing, which in turn pushes against the plunger in the pressure block and causes the contacts to open and hence switch off the oven-heating elements.

The other type of oven housing is the stem type. These come in various lengths to suit different types of oven. The operation of the stem type relies on two dissimilar metals expanding at different rates. It consists of a copper tube with an **Invar** rod inside. When the temperature increases, the copper tube expands faster than the Invar rod and pushes against the pressure rod; this opens the contacts and the oven element is switched off. When it cools down, the copper tube 'shrinks' and allows the Invar plunger rod to move down and close the contact for the oven element. This type of thermostat is also used in water-heating systems.

Power

We covered the mathematics behind the creation of electrical power in chapter 4. We looked at the principles and operation behind motors in Chapter 10. In Chapter 11, we covered how power is generated and transmitted around the country.

Without power generation, we would not have the electricity we need to run electrical machines.

The major components of electrical systems that we have looked at elsewhere in this book are:

- ring and radial circuits and sockets (see chapter 6 pages 132–135)
- motors (see chapter 10).

Emergency management systems

The electrical supply in this country is very reliable and secure. However, as with all systems there are occasional interruptions, which for some installations would be dangerous as well as inconvenient. Hospitals, air-traffic control and the petro-chemical industry are just a few installations that could not tolerate an interruption to the mains supply, so a standby system needs to be available.

Smaller establishments such as small offices cannot afford complex standby generation systems, but nevertheless they may have computer systems that cannot afford to be off or, worse still, risk losing data. In this situation standby power systems known as Uninterruptible Power Supplies (UPS) are used, which consist of a battery supply that is charged up via the mains when not in use. When the mains supply is lost the UPS automatically cuts in and, via the electronics contained in it, converts the d.c. battery supply to a mains supply capable of powering several computers.

Emergency power systems are installed in order to provide a back-up source of power in the event of a power or system failure. These will often supply power to certain lights and generators. They are found in a range of buildings and structures – you might find them both in a nursing home and a military base!

Remember

Power is measured in joules per second or J/s known as watts (W)

Remember

Large installations need a standby generating system, whereby a large combustion engine cuts in automatically and drives a generator capable of supplying the load needed to continue working safely

Did you know?

Power systems were used in the Navy during the Second World War. If a ship's steam engine was put out of action, it had diesel engine to take its place

'Normal' power can be lost in a building for a variety of reasons, such as weather conditions. Most modern emergency power systems rely on a diesel engine driven generator. In the event of the main power supply being knocked out, an automatic transfer switch will connect the back-up generator to the supply. This switch is operated by a solenoid (see chapter 5, page 113). At the same time, a starter switch activates the back-up generator. Once this is complete, the emergency power is activated.

Equipment on emergency power can include:

- lighting
- fire fighting equipment
- life-support machines in hospitals
- technical equipment in theatres
- computers and machines that hold electronic records.

Emergency lighting

Emergency lighting is not required in private homes because the occupants are familiar with their surroundings. However, in public buildings, people are in unfamiliar surroundings and, in an emergency, they will require a well-illuminated and easily identified exit route.

Emergency lighting should be planned, installed and maintained to the highest standards of reliability and integrity, so that it will operate satisfactorily when called into action. It must be installed in accordance with the British Standard Specification BS 5266: Part 1: 1999 – Code of Practice for Emergency Lighting.

Emergency-lighting terminology

For the purposes of the European Standard EN 1838, emergency lighting is regarded as a general term. There are actually several types, as shown in Figure 14.10.

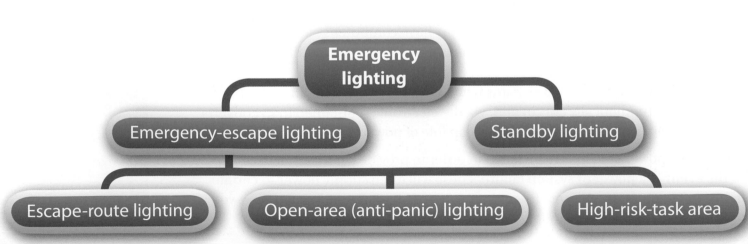

Figure 14.10 Specific forms of emergency lighting

- **Emergency-escape lighting:** provided to enable safe exit in the event of failure of the normal supply.

- **Standby lighting:** provided to enable normal activities to continue in the event of failure of the normal mains supply.

- **Escape-route lighting:** provided to enable safe exit for occupants by providing appropriate visual conditions and direction-finding on escape routes and in special areas/locations, and to ensure that fire-fighting and safety equipment can be readily located and used.

- **Open area (or anti-panic area) lighting:** provided to reduce the likelihood of panic and to enable safe movement of occupants towards escape routes by providing appropriate visual conditions and direction-finding.

- **High-risk-task area lighting:** provided to ensure the safety of people involved in a potentially dangerous process or situation and to enable proper shutdown procedures to be carried out for the safety of other occupants of the premises.

Types

Emergency lighting comes in two main formats: individual, self-contained systems with their own emergency battery power source, and centralised battery-backup systems. In using both these formats there are then three types available.

Maintained

The same lamp is used by both the mains and the emergency backup system and therefore operates continuously. The lamp is supplied by an alternative supply when the mains supply fails.

The advantage of this system is that the lamp is continuously lit and therefore we can see whether a lamp needs replacing. The disadvantage is that, although the lamp is lit, we do not know whether it is being powered by the mains supply or the batteries. It is therefore common to find a buzzer and indicator lamp that show which supply is being used. Emergency lighting should be of the maintained type in areas in which the normal lighting can be dimmed, e.g. theatres or cinemas, or where alcohol is served.

Non-maintained

The emergency lighting lamp only operates when the normal mains lighting fails. Failure of the mains supply connects the emergency lamps to the battery supply. The disadvantage of this system is that a broken lamp will not be detected until it is required to operate. It is therefore common to find an emergency-lighting test switch available that disconnects the mains supply for test purposes.

Sustained

An additional lamp housed in the mains luminaire is used only when the mains fails.

The duration of the emergency lighting is normally three hours in places of entertainment and for sleeping risk, or where evacuation is not immediate, but one hour's duration may be acceptable in some premises if evacuation is immediate and re-occupation is delayed until the system has recharged.

Siting of luminaires

BS 5266 and IS 3217 provide detailed guidance on where luminaires should be installed and what minimum levels of illuminance should be achieved on escape routes and in open areas. It also specifies what minimum period of duration should be achieved after failure of the normal mains lighting.

Local and national statutory authorities, using legislative powers, usually require escape lighting. Escape-lighting schemes should be planned so that identifiable features and obstructions are visible in the lower levels of illumination that will occur during an emergency.

Current UK regulations require the provision of a horizontal illuminance at floor level, on the centre line of a defined escape route, of not less than 0.2 lux (similar to the brightness of a full moon). In addition, for escape routes of up to 2m wide, 50 per cent of the route width should be lit to a minimum of 0.1 lux. Wider escape routes can be treated as a number of 2m wide bands.

Emergency-escape lighting should:

- indicate the escape routes clearly, allowing for changes of direction or of level
- provide illumination along escape routes to allow safe movement towards the final exits
- ensure that fire-alarm call points and fire-fighting equipment can be readily located.

Standby lighting is required in, for example, hospital operating theatres and in industry, where an operation or process, once started, must continue even if the mains lighting fails. Additional emergency lighting should also be provided in:

- lift cars – potential for the public to be trapped
- toilet facilities – particularly disabled toilets – and open tiled areas over 8m²
- escalators – to enable users to get off them safely
- motor generator, control or plant rooms – these require battery-supplied emergency lighting to help any maintenance or operating personnel
- covered car parks along the normal pedestrian routes.

Illuminance levels for open areas

Emergency lighting is required for areas larger than 60m² or open areas with an escape route passing through. Illuminance BS 5266 requires 1 lux average over the floor area. The European standard EN 1838 requires 0.5 lux minimum anywhere on the floor level excluding the shadowing effects of contents. The core area excludes the 0.5m next to the perimeter of the area.

Did you know?

Cashpoints in commercial buildings need to be illuminated at all times to discourage acts of theft occurring during a mains failure

High-risk-task area lighting

BS 5266 requires that higher levels of emergency lighting are provided in areas of particular risk, although no values are defined. The European standard EN 1838 states that the average horizontal illuminance on the reference plane (note that this is not necessarily the floor) should be as high as the task demands in areas of high risk. It should not be less than 10 per cent of the normal illuminance, or 15 lux, whichever is the greater. It should be provided within 0.5 seconds and continue for as long as the hazard exists. This can normally only be achieved by a tungsten or permanently illuminated and maintained fluorescent lamp source. The required illuminance can often be achieved by careful location of emergency luminaires at the hazard, and may not require additional fittings.

Maintenance

Essential servicing should be defined to ensure that the system remains at full operational status. This would normally be performed as part of the testing routine, but for consumable items, such as replacement lamps, spares should be provided for immediate use.

Fire-alarm systems

A correctly installed fire-alarm system installation is of paramount importance compared to any other electrical undertaking, as life could be lost and property damaged as a result of carelessly or incorrectly connected fire-detection and alarm equipment. The subject is detailed, and therefore this section sets out only to give an overview of requirements.

BS 5839 Part 1 classifies fire-alarm systems, perhaps better described as fire-detection and alarm systems, into the following general types.

- **Type M:** break-glass contacts operating sounders for protection of life; no automatic detection.

- **Type L:** automatic detection systems for the protection of life.

- **Type P:** automatic detection systems for the protection of property.

It is essential that the installation of fire-alarm systems is carried out in compliance with the requirements of BS 5839 Part 1, BS 7671 and manufacturers' instructions, but remember: local government can enforce even stricter requirements in the interests of public safety. BS 5839 and BS 7671 (528-01-04) state that fire-alarm circuits must be segregated from other circuits and, in order to comply with BS 7671, a dedicated circuit must be installed to supply mains power to the fire-alarm control panel.

Fire-alarm systems can be designed and installed for one of two reasons:

- property protection

- life protection.

Did you know?

The Building Regulations require the installation of mains-fed smoke detectors in new-build domestic installations

A fire-alarm system might have prevented this

Property protection

A satisfactory fire-alarm system for the protection of property will automatically detect a fire at an early stage, indicate its location and raise an effective alarm in time to summon fire-fighting forces (both resident staff and the fire service). The general attendance time of the fire service should be less than 10 minutes. Therefore an automatic direct link to the fire service is a normal part of such a system.

Protection for property is classed as either P1 or P2.

- **P1:** All areas of the building must be covered with detectors with the exception of lavatories, water closets and voids less than 800mm in height, such that spread of fire cannot take place in them prior to detection by detectors outside the void.

- **P2:** Only defined areas of high risk are covered by detectors. A fire-resisting construction should separate unprotected areas.

Life protection

A satisfactory fire-alarm system for the protection of life can be relied upon to sound a fire alarm in sufficient time to enable the occupants to escape. Life protection is classed as M, L1, L2 or L3.

- **M:** the most basic and minimum requirement for life protection. It relies upon manual operation of call points and therefore requires people to activate the system. Such a system can be enhanced to provide greater cover by integrating any, or a combination, of L1, L2 and L3.

- **L1:** same as P1 above.

- **L2:** only provides detection in specified areas where a fire could lead to a high risk to life, e.g. sleeping areas, kitchens, day accommodation etc., and places where the occupants are especially vulnerable owing to age or illness or are unfamiliar with the building. An L2 system always includes L3 coverage.

- **L3:** protection of escape routes. The following areas should therefore be included:

  (i) corridors, passageways and circulation areas

  (ii) all rooms opening on to escape routes

  (iii) stairwells

  (iv) landing ceilings

  (v) the top of vertical risers, e.g. lift shafts

  (vi) at each level within 1.5m of access to lift shafts or other vertical risers.

Types of fire-alarm system

All fire-alarm systems operate on the same general principle, i.e. if a detector detects smoke or heat or if a person operates a break-glass contact, then the alarm will sound. We will look at the devices that may be incorporated into the system later. That said most fire-alarm systems belong to one of the following categories.

Conventional

In this type of system, a number of devices (break-glass contacts/detectors) are wired as a radial circuit from the control panel to form a zone (e.g. one floor of a building). The control panel would have lamps on the front to indicate each zone and, if a device operates, then the relevant zone lamp would light up on the control panel. However, the actual device that has operated is not indicated.

Identifying accurately where the fire has started would therefore depend on having a number of zones and knowing where in the building each zone is. Such systems are therefore normally found in smaller buildings or where a cheap, simple system is required.

Addressable

The basic principle here is the same as for a conventional system – the difference being that, by using modern technology, the control panel can identify exactly which device initiated the alarm.

These systems have their detection circuits wired as loops, with each device then having an 'address' built in. Such systems therefore help fire location by identifying the precise location of an initiation, and thus allow the fire services to get to the source of a fire more quickly.

Radio addressable

These are the same as addressable systems, but have the advantage of being wireless and can thus reduce installation time.

Analogue

Sometimes known as intelligent systems, analogue systems incorporate more features than either conventional or addressable systems. The detectors may include their own mini-computer, and this evaluates the environment around the detector and is therefore able to let the control panel know whether there is a fire, a change in circumstance likely to lead to a fire, a fault, or even if the detector head needs cleaning. Consequently these systems are useful in preventing the occurrence of false alarms.

Fire-prevention systems

Although still incorporating fire-detection systems, one recent innovation has been the introduction of the fire-reduction system. This type of system is still under development, but works by reducing levels of one of the main components in the fire triangle – oxygen – and thus seeks to create a 'fire-free' area. Although not without problems, usage of these systems could be appropriate in critical areas such as historical archives or identified unmanned areas such as chemical storage.

Zones

To ensure a fast and unambiguous identification of the source of fire, the protected area should be divided into zones. Although less essential in analogue addressable systems, the following guidelines relate to zones as follows.

- If the floor area of each building is not greater than 300m² then the building only needs one zone, no matter how many floors it has. This covers most domestic installations.

- The total floor area for one zone should not exceed 2000m².

- The search distance should not exceed 30m. This means the distance that has to be travelled by a searcher inside a zone to determine visually the position of a fire should not be more than 30m. The use of remote indicator lamps outside of doors may reduce the number of zones required.

- Where stairwells or similar structures extend beyond one floor but are in one fire compartment, the stairwell should be a separate zone.

- If the zone covers more than one fire compartment, then the zone boundaries should follow compartment boundaries.

- If the building is split into several occupancies, no zone should be split between two occupancies.

System devices

The control panel

This is the heart of any system, as it monitors the detection devices and their wiring for faults and operation. If a device operates, the panel operates the sounders as well as any other related equipment and gives an indication of the area in which the alarm originated.

Break-glass contacts (manual call points)

The break-glass call point is a device to enable personnel to raise the alarm in the event of a fire, by simply breaking a fragile glass cover (housed in a thin plastic membrane to protect the operative from injury sustained by broken or splintered glass). A sturdy thumb pressure is all that is required to rupture the glass and activate the alarm. The following guidance relates to the correct siting and positioning of break-glass call points.

- They should be located on exit routes and in particular on the floor landings of staircases and at all exits to the open air.

- They should be located so that no person need travel more than 30m from any position within the premises to raise the alarm.

- Generally, call points should be fixed at a height of 1.4m above the floor, at easily accessible, well-illuminated and conspicuous positions free from obstruction.
- The method of operation of all manual call points in an installation should be identical unless there is a special reason for differentiation.
- Manual and automatic devices may be installed on the same system, although it may be advisable to install the manual call points on separate zones for speed of identification.

Automatic detectors

When choosing the type of detector to be used in a particular area it is important to remember that the detector has to discriminate between fire and the normal environment existing within the building – for example, smoking in hotel bedrooms, fumes from fork-lift trucks in warehouses, or steam from kitchens and bathrooms. There are several automatic detectors available, as described below.

Heat detectors (fixed-temperature type)

The fixed-temperature heat detector is a simple device designed to activate the alarm circuit once a predetermined temperature is reached. Usually a choice of two operational temperatures is available: either 60°C or 90°C. This type of detector is suitable for monitoring boiler-rooms or kitchens where fluctuations in ambient temperature are commonplace.

Heat detector (rate-of-rise type)

This type of detector responds to rapid rises of temperature by sampling the temperature difference between two heat-sensitive thermocouples or thermistors mounted in a single housing (a thermistor is a device whose resistance quickly changes with a change in temperature).

Smoke detectors

May be either of the 'ionisation' or 'optical' type. Smoke detectors are not normally installed in kitchens, as burning toast and so on could activate the alarm.

The ionisation detector is very sensitive to smoke with fine particles such as that from burning paper or spirit, whereas the optical detector is sensitive to 'optically dense' smoke with large particles such as that from burning plastics.

The optical smoke detector, sometimes known as the photoelectric smoke detector, operates by means of the light-scattering principle. A pulsed infrared light is targeted at a photo-receiver but separated by an angled non-reflective baffle positioned across the inner chamber. When smoke and combustion particles enter the chamber, light is scattered and reflected on to the sensitive photo receiver, triggering the alarm.

Detector heads for fire-alarm systems should only be fitted after all trades have completed work, as their work could create dust, which impairs the detector operation. Strict rules exist regarding the location of smoke detectors.

Break-glass call point

Remember

It is wise not to install a rate-of-rise heat detector unit in a boiler-room or kitchen, where fluctuations in ambient temperature occur regularly. This will help to avoid nuisance alarms

Smoke detector

Alarm bell

Alarm sounders

These are normally either a bell or an electronic sounder, which must be audible throughout the building to alert (and/or evacuate) the occupants of the building. The following gives guidance for the correct use of alarm sounders.

- A minimum level of either 65 dBA, or 5 dBA above any background noise likely to persist for a period longer that 30 seconds, should be produced by the sounders at any occupiable point in the building.

- If the alarm system is to be used in premises such as hotels, boarding houses etc. where it is required to wake sleeping persons, then the sound level should be a minimum of 75 dBA at the bedhead.

- All audible warning devices used in the same system should have a similar sound.

- A large number of quieter sounders rather than a few very loud sounders may be preferable. At least one sounder will be required per fire compartment.

- The level of sound should not be so high as to cause permanent damage to hearing.

Wiring systems for fire alarms

BS 5839 Part 1 recommends eleven types of cable that may be used where prolonged operation of the system in a fire is not required. However, only two types of cable may be used where prolonged operation in a fire is required.

It is obvious that the cabling for sounders and any other device intended to operate once a fire has been detected must be fireproof. However, detection wiring can be treated differently, as it can be argued that such wiring is only necessary to detect the fire and sound the alarm.

In reality, fire-resistant cabling tends to be used throughout a fire-alarm installation for both detection and alarm wiring. Consequently, as an example, MICC cable used throughout the system is considered by many as the most appropriate form of wiring, but there are alternatives, such as Fire-tuf.

Irrespective of the cable type and the circuit arrangements of the system, all wiring must be installed in accordance with BS 7671. Where possible, cables should be routed through areas of low fire risk and, where there is risk of mechanical damage, they should be protected accordingly.

Because of the importance of the fire-alarm system, it is wise to leave the wiring of the system until most of the constructional work has been completed. This will help prevent accidental damage occurring to the cables. Similarly, keep the control panel and activation devices in their packing cartons, and only remove them when building work has been completed in the area where they are to be mounted, thus preventing possible damage to the units.

Standby back-up for fire-alarm systems

The standby supply, which is usually a battery, must be capable of powering the system in full normal operation for at least 24 hours and, at the end of that time period, must still have sufficient capacity to trigger the alarm sounders in all zones for a further 30 minutes.

Typical maintenance checks for a fire-alarm system

BS 5839 Part 1 makes the following recommendations:

Daily inspection	Annual test
• Check that the control panel indicates normal operation. Report any fault indicators or sounders not operating to the designated responsible person.	• Repeat the quarterly test. • Check all call points and detectors for correct operation. • Enter details of test in logbook.
Weekly test	**Every two to three years**
• Check panel key operation and reset button. • Test fire alarm from a call point (different one each week) and check sounders. • Reset fire-alarm panel. • Check all call points and detectors for obstruction. • Enter details of test in logbook.	• Clean smoke detectors using specialist equipment. • Enter details of maintenance in logbook.
Quarterly test	**Every five years**
• Check all logbook entries and make sure any remedial actions have been carried out. • Examine battery and battery connections. • Operate a call point and detector in each zone. • Check that all sounders are operating. • Check that all functions of the control panel are operating by simulating a fault. • Check sounders operate on battery only. • Enter details of test in logbook.	• Replace battery (see manufacturer's information).

Table 14.01 BS 5839 recommendations

Intruder alarms

Intruder alarm systems are increasingly seen as standard equipment in a house or office. They act as a deterrent to some intruders but will never stop the more determined ones. People feel more secure when they have an alarm installed, and in most cases it will reduce their insurance premiums. There are basically two ways to protect a property: one is called perimeter protection and the other is space detection.

Perimeter protection detects a potential intruder before they gain entry to the premises, whereas space detection only detects when the intruder is already on the premises. Sometimes both types are used together for extra security.

Typical systems

In this section we will look at some the component parts of an alarm system and some of the more common types of detection devices available and what they do:

- proximity switches
- inertia switches
- passive infrared
- ultrasonic devices
- control panels
- audible and visual warning devices.

Proximity switch

This is a two-part device: one part is a magnet and the other contains a reed switch. The two parts are fixed side by side (usually less than 6mm apart) on a door or window, and when the door or window is opened the reed switch opens (because the magnet no longer holds it closed) and activates the alarm panel. The switch can be surface-mounted or can be recessed into the door or window frame. This device is generally used for perimeter protection and does not rely on a power supply to operate.

Inertia switch

This type of switch detects the vibration created when a door or window is forced open. This then sends a signal to the alarm panel and activates the sounder. The sensitivity of these devices can be adjusted, and they are used for perimeter protection. These need a 12 volt d.c. supply to operate.

Passive infrared

These devices are used to protect large areas of space and are only activated when the intruder has already gained entry. The device monitors infrared so detects the movement of body heat across its viewing range; this in turn sends a signal to the panel and activates the sounder. These can be adjusted for range and, by fitting different lenses, the angle of detection can also be adjusted. These need a 12 volt d.c. supply to operate.

Ultrasonic devices

These devices send out sound waves and receive back the same waves when noone is in the building. However, when an intruder enters the detection range, the sound waves change (because of deflection) and trigger the alarm panel. These devices also require a 12 volt d.c. supply for operation and are used for space-detection systems.

Control panels

Control panels are the 'brains' of the system to which all the parts of the system are connected. They used to be key-operated but nowadays they virtually all use a digital keypad, either on the panel itself or mounted remotely elsewhere in the building, for switching the system on or off.

The panels are all programmable whereby entry- and exit-route zone delays can be adjusted, new codes selected for switching on/off, automatic telephone diallers set to ring any phone selected etc.

Control panels have a mains supply installed, which is reduced (via a transformer) down to 12 volts d.c. for operation of all the component parts that need it. A rechargeable battery backup is provided in case of mains failure.

Audible and visual warning devices

When an alarm condition occurs a means of attracting attention is obviously needed, either audibly or visually or sometimes both. The most common audible sounder is the electronic horn (I'm sure you've all heard them before!), which will sound for 20 minutes (the maximum allowed by law) before being switched off by the panel automatically. The panel then re-arms itself and monitors the system again.

To help identify which alarm has sounded (especially when there are several in the same area) a visual warning is usually fitted to the sounder box, which activates at the same time. This is a xenon light (strobe light) and can be obtained in a variety of colours. This light usually remains on after the alarm has automatically been reset to warn the occupant upon their return that an alarm condition has occurred. It is only reset when the control panel itself is reset by the occupant.

Closed circuit television (CCTV)

There are many different types of CCTV systems in use today, ranging from those suitable for domestic properties through to sophisticated multi-camera/multi-screen monitoring for large commercial and industrial premises.

Typical systems

In this section we will look at the component parts that make up a typical system. The following topics will be looked at:

- wireless CCTV
- wired CCTV
- cameras
- light levels
- monitoring and recording
- other systems.

Wireless CCTV

These systems do not require cabling back to a monitor or video recorder, as they have an in-built transmitter which transmits the image seen back to these pieces of equipment. Typically they can transmit 100m outdoors and 30m indoors. They do, however, still require a power supply (usually 9–12 V d.c.), which is usually obtained via a small power supply transformer connected to the mains. These systems are useful where it is difficult to install video cable back to the monitor or video recorder but they can suffer from interference problems.

Wired CCTV

These systems do require cabling back to the monitor or video recorder but can be positioned many hundreds of metres away from them. Usually the same cable will provide power and the video signal back to the recording device, so all the power supplies for many cameras can be located at one central control point.

Cameras

There are many different types of camera available, ranging from very cheap (less than £100) to those costing many thousands of pounds. There are two common types: CMOS and CCD. The CMOS type is the cheapest but the images produced are not very clear or sharp. The CCD camera, on the other hand, produces very clear and sharp images from which people are easily identifiable.

Most cameras are installed outside and therefore virtually all cameras available are weatherproof; if they are not they will need to be fitted into a weatherproof housing. Virtually all cameras have the lens integrated into the camera and are sealed to prevent moisture getting in; thus they do not need a heater built in to keep the lens dry. With the lens being sealed into the camera, the former cannot be adjusted, so only one field of view is possible.

Colour and monochrome types of camera are available, with colour cameras being the more expensive. Colour cameras can only transmit colour if the light level is high, so generally speaking they will not transmit colour images at night. Monochrome cameras, on the other hand, can incorporate infrared (IR) sensitivity, allowing for clearer images where discreet IR illumination is available.

Light levels

Light levels available where the camera is to be used are an important consideration. Table 14.02 shows some typical light levels.

When choosing a suitable camera for a particular environment, it is best to select one that is specified at approximately ten times the minimum light level for the environment. One that is specified at the same level of light will not produce the clear images needed, because the camera will not have enough light to 'see'.

Environment	Typical light level
Summer sunlight	50,000 lux
Dull daylight	10,000 lux
Shop/office	500 lux
Main street lighting	30 lux
Dawn/dusk	1–10 lux
Side street lighting	3 lux

Table 14.02 Typical light levels

Monitoring and recording

Most CCTV systems use several cameras, each relaying images back to a central control where they are either viewed or recorded. Three methods for recording or viewing these images are by using a video switcher, a quad processor or a multiplexer.

Switcher

A CCTV switcher, as the name suggests, is a device that switches between camera images one at a time. The image can either be viewed or recorded on to a video recorder; only one image at a time can be accessed.

Quad processor

This device enables four camera images to be viewed on one screen at the same time, or one image or all four to be recorded at the same time. The quality of the image when recording all four is not as good.

Multiplexer

This device allows simultaneous recording of multiple full-sized images on to one VCR, or can allow more than one camera image to be displayed at the same time without losing picture quality.

For recording purposes, a slower-moving tape can record the images for long periods of time. The time lapse can be set for either 24, 240 or 960 hours of recording on standard tapes.

Other systems

PC-based systems

By adding a video capture card and surveillance software to a PC, a powerful digital system can be created. Some of the advantages are:

- it is easy to expand the system
- it is easy to record (via hard drive)
- images can be emailed
- text alerting is possible
- software allows many configurations for monitoring
- remote viewing is possible.

Motion detectors

The camera and recording facilities are only activated when movement is detected within the camera's range. Typically this is activated by the use of passive infrared sensors (PIRs) similar to those used on security lighting and alarm systems.

Wiring systems and enclosures

On completion of this topic area the candidate will be able to state wiring systems and enclosures, and state the applications and limitations of more specialised cables.

Throughout this book we have looked at the wiring systems and enclosures that are used in installations. We have already looked at single and multicore PVC insulated cables (armoured and non-armoured) (chapter 7, pages 159–163). Many of these topics will be covered in more depth at Level 3.

MICC cable

Mineral-insulated copper cables (MICC) consist of high-conductivity copper conductors insulated by a highly compressed white powder (magnesium oxide). A seamless copper sheath encapsulates the conductors and powder.

This type of cable originated in France and was introduced into the UK in 1936. The first company to market these cables in the UK was Pyrotenax and from this name came the term 'pyro', which is still sometimes used when referring to this cable. The cable is made by placing solid copper bars in a hollow copper tube. The magnesium oxide powder is then compacted into the tube and finally the whole tube, powder and copper bars are drawn out by pulling and rolling. This reduces the overall size while further compressing the powder.

Properties of MICC cables

MICC cable

Mineral-insulated cables have very good fire-resisting properties: copper can withstand 1000°C and magnesium 2800°C. The limiting factor of the whole cable system is the seal and, where a high working temperature is required, special seals must be used. MI cables have the following qualities.

- The cable is very robust and can be bent or twisted within reasonable limits, hence its use in emergency lighting and fire-alarm systems.

- For a given cross-sectional area, MI cables have a very high current-carrying capacity.

- Relative spacing between the conductors and sheath is maintained when the cable is flattened, hence maintaining the cable's insulation properties.

- MI cables are non-ageing (many cables installed in the 1930s are still in operation today).

- The cable is completely waterproof, although where it is to be run underground or in ducts a PVC oversheath must be used.

- Bare copper unsheathed MI cables do not emit smoke or toxic gases in fires.

- Where PVC oversheath is used, the reduced volume of PVC in comparison with PVC-insulated cables keeps down smoke output. Special oversheaths are also available where a further reduction of flame propagation is required.

- The copper sheath can be used for earth continuity, saving the need for a separate protective conductor.

- These cables come into their own in areas such as boiler houses where the ambient temperature can become high and there is moisture present.

Steel conduit installations

Annealed mild steel tubing, known as **conduit**, is widely used as a commercial and industrial wiring system. PVC-insulated (non-sheathed) cables are run inside the steel tubing. Conduit can be bent without splitting, breaking or kinking, provided the correct methods are employed. Available with this system is a very extensive range of accessories to enable the installer to carry out whole installations without terminating the conduit. It offers excellent mechanical protection to the wiring and in certain conditions may also provide the means of earth continuity.

The British Standard covering steel conduit and fittings is BS 4568. The two types of commonly used steel conduit are known as black enamel conduit, which is used indoors where there is no likelihood of dampness, and galvanised conduit, which is used in damp situations or outdoors.

Plastic conduit (PVC)

Plastic conduit is made from polyvinyl chloride (PVC), which is produced in both flexible and rigid forms. It is impervious to acids, alkalis, oil, aggressive soils, fungi and bacteria, and is unaffected by sea, air and atmospheric conditions. It withstands all pests and does not attract rodents. PVC conduit is preferable for use in areas such as farm milking parlours. PVC conduit may be buried in lime, concrete or plaster without harmful effects.

Trunking

Trunking is a fabricated casing for cables, normally of rectangular cross-section, one side of which may be removed or hinged back to permit access. It is used where a number of cables follow the same route, or in circumstances where it would otherwise be expensive to install a large number of separate conduits or runs of mineral-insulated cable. Trunking is commonly installed, for example in factories, where the introduction of new equipment and the relocation of existing equipment may involve frequent modification of the installation.

PVC trunking

Cable tray

Cable tray, ladder and basket

On large industrial and commercial installations, where several cables take the same route, **cable tray** is frequently used. This provides a cost-effective means of supporting groups of cables. A wide range of cable tray and accessories is available to match any cabling requirement, from lightweight instrumentation cable through to the heaviest multicore power cable. In situations where heavy multicore cables are required to cross long, unsupported spans, cable ladders should be used.

Cable ladders (or **ladder racking**) are an effective method of transporting cables across long unsupported spans or where the number of supports is to be reduced. They can be used in the most adverse site conditions and can withstand high winds, heavy snow, sand or dust settlement or high humidity.

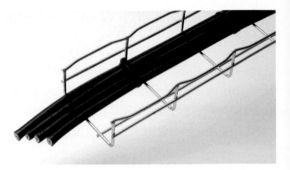

Cable ladder

Cable ladder design allows the maximum airflow around the cables and so prevents possible derating of power cables. They may be mounted in virtually any direction.

Cable basket is similar to cable ladder. It is made from wire steel basket and it requires similar installation techniques. Cutting of the basket to form bends or tees is normally achieved using bolt cutters. Any cuts then need to be made smooth, as with tray or ladder systems.

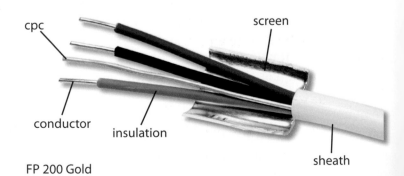

Cable basket

FP 200 cable

Normally used for fire alarms and fire detection systems, there are two types of FP 200 cable. FP 200 Gold and FP 200 Flex. FP 200 Gold has solid conductors, FP 200 Flex uses stranded conductors. We will only look here at FP 200 Gold.

The solid copper conductors are covered with a fire- and damage-resistant insulation (Insudite). An electrostatic screen is provided by a laminated aluminium tape, which is applied longitudinally and folded around the cores to give an overlap. The aluminium tape is applied metal side down and in contact with

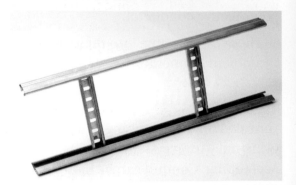

cpc

screen

conductor

insulation

sheath

FP 200 Gold

the uninsulated circuit protective conductor (cpc. The sheath is a robust thermoplastic low-smoke, zero-halogen sheath, which is an excellent moisture barrier.

FP200 Gold has two, three or four cores as standard with others available on request. It is less robust and durable than MICC cables.

Communication cables

A wide range of cable types is available for communications and computing. The commonest are fibre optic for high speed links and Category 5 (Cat5) cable to supply high signal integrity over shorter distances (typically less than 100m). Coaxial cable is still found but rapidly dropping out of use.

Cat5 cable

This usually contains four UTP (universal twisted pair) cables within a cable jacket. The basic cable is intended for data transmission up to 100 MHz. It is being replaced by Cat5E (enhanced), recommended for new installations and capable of transmission speeds up to one gigabit/second.

Cables of this type are often used in computer networks, such as an internet or network cable. A common example of Category 5 cable (Cat 5), which is specifically designed to supply high signal integrity. Many of these cables are unshielded. The primary limitation of this cable are length restrictions. Cables are still limited to 100m in length.

Fibre-optic cable

This cable is used for digital transmissions by equipment such as telephones or computers. They are made from optical-quality plastic (the same as spectacles) where digital pulses of laser light are passed along the cable from one end to another with no loss or interference from mains cables. They look like steel wire armoured (SWA) cables but are much lighter and contain either one core or many dozens of cores. Tight radius bends in this type of cable should be avoided, as should 'kinks', as the cable will break. Jointing of these cables requires specialist tools and equipment. Never look into the ends of the cable as the laser light could damage your eyes.

The applications of optical fibre communications have increased at a rapid rate since the first commercial installation of a fibre-optic system in 1977. Telephone companies quickly began replacing their old copper-wire systems with optical-fibre lines. Today's telephone companies use optical fibre throughout their system as the backbone architecture and as the long-distance connection between city phone systems.

Fibre-optic cable

Light rays, modulated into digital pulses with a laser or a light-emitting diode, move along the core without penetrating the cladding. The light stays confined to the core because the cladding has a lower refractive index (a measure of its ability to bend light).

Cable television companies have also begun integrating fibre-optics into their cable systems. The trunk lines that connect central offices have generally been replaced with optical fibre. Some providers have begun experimenting with fibre to the curb using a fibre/coaxial hybrid. Such a hybrid allows for the integration of fibre and coaxial at a neighbourhood location. This location, called a node, would provide the optical receiver that converts the light impulses back to electronic signals. The signals could then be fed to individual homes via coaxial cable.

Fibre-optic cables are also used in Local Area Networks (LAN). These collective groups of computers, or computer systems, connected to each other, allow for shared programme software or databases. Colleges, universities, office buildings and industrial plants, just to name a few, all make use of fibre-optic cables within their LAN systems.

Power companies are emerging as big users of fibre optics in their communication systems. Most power utilities already have fibre-optic communication systems in use for monitoring their power grid systems.

Fibre-optic cables are limited by the bending radius restrictions mentioned above. They also require special techniques and equipment to be terminated.

Armoured/braided cables

These are multicore cables in which the cores are surrounded by a braided tubular sheath. This metal sheath is placed around the cable to provide mechanical and electrical protection. Heavy duty cables are refered to as steel wire armoured (SWA). On lighter cables the braiding is formed by small wires wrapped round the cable. These are more resistant to vibration than larger wires and are usually oil resistant.

Choosing wiring systems and conductors

On completion of this topic area the candidate will be able to state factors determining the choice of wiring systems and conductor sizes.

Wiring and conductor selection

We covered the factors that determine the choice of wiring selection in chapters 6 and 7. The key factors involved in choosing wiring are:

- temperature
- effect of moisture
- corrosive substances
- UV damage/sunlight
- damage by animals
- mechanical stress
- aesthetic considerations.

More information on all these factors can be found earlier in this book.

Similarly we also looked at the factors used in deciding upon conductor size (see chapter 6). The key factors are:

- design current
- regulations concerning thermal constraints and shock protection
- voltage drop.

You will need to remember all these factors when deciding upon the components you will use during any installations you carry out.

Voltage drop

Cables in a circuit are similar to resistors, in that the longer the conductor, the higher its resistance becomes and thus the greater the voltage drop.

Applying Ohm's law (using the circuit current and the conductor resistance), it is possible to determine the actual voltage drop. To determine voltage drop quickly in circuit cables, BS 7671 and cable manufacturer data include tables of voltage drop in cable conductors. The tables list the voltage drop in terms of (mV/A/m) and are listed as conductor feed and return, e.g. for two single core cables or one two-core cable.

Regulation 525 states that the voltage drop between the origin of the installation (usually the supply terminals) and a socket-outlet (or the terminals of the fixed current using equipment) shall not exceed three per cent (3%) of the nominal voltage of the supply for lighting and 5% for other circuits.

In our example the voltage drop was smaller than the four per cent allowed by BS 7671. If the volt drop was larger than four per cent it may be necessary to change the cross sectional area (csa) of the cable, thus reducing the resistance and so lowering the volt drop.

Other options are to reduce the length of cable where possible, or you may reduce the load. If necessary, all alternatives must be considered so that you satisfy the requirements of BS 7671.

Special arrangements (Part 7)

On completion of this section the candidate will be able to state the special arrangements that need to be considered for bathrooms, external installations, flammable/explosive situations, temporary and construction site installations, agricultural and horticultural installations.

As we have mentioned certain locations are deemed to be more hazardous and additional (or in some cases replacement) Regulations are therefore needed to provide greater safety.

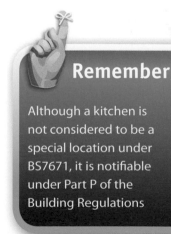

Remember

Although a kitchen is not considered to be a special location under BS7671, it is notifiable under Part P of the Building Regulations

Example

A low-voltage radial circuit supplying fixed equipment is arranged as shown in Figure 14.11. It is wired throughout with 50 mm² copper cable, for which the voltage drop is given as 0.95 mV/A/m.

We want to calculate:

(a) the total current drawn from the supply

(b) the total voltage drop

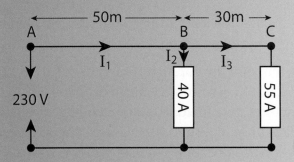

Figure 14.11 Low-voltage radial circuit

(c) whether the voltage drop meets Regulation 525-01-02.

Step 1: Total current flowing $= I_1 = I_2 + I_3 = 40 + 55 =$ **95 A**

Step 2: Calculate the voltage drop in section A–B.

We know the voltage drop (in mV) for each metre of our 50mm² copper cable is 0.95 for each amp of current flowing through it.

Therefore the total voltage drop in section A–B $= 0.95 \times I_1 \times$ length of A–B

$$= 0.95 \times 95 \times 50$$

$$= 4512.5 \text{ mV}$$

$$= 4.5125 \text{ V}$$

Step 3: Calculate the voltage drop in cable section B–C

We are using the same cable type. We can see from the diagram that the length is now 30m and the current flowing through is I_3.

Therefore total voltage drop in section B–C $= 0.95 \times I_3 \times$ length of B–C

$$= 0.95 \times 55 \times 30$$

$$= 1567.5 \text{ mV}$$

$$= 1.5675 \text{ V}$$

Step 4: Total voltage drop is therefore the sum of 4.5125 + 1.5675 = **6.08 V**

Step 5: Regulation 525 allows us a voltage drop of 5% of the nominal supply voltage, in this case 230 V. 5% of 230 V = 11.5 V but we are only dropping 6.08 V.

Hence we meet the requirement.

Locations containing a bath or shower

This section on special locations will concentrate on the locations (where electrical equipment may be installed) that contain a bath or shower, which present a greater risk of electric shock than in dry locations. This section discusses the regulations governing such installations.

Scope (Regulation 701.1)

The requirement of these Regulations apply to locations containing baths, showers and cabinets containing a shower and/or a bath and the surrounding zones. Baths and showers used as emergency facilities in industrial areas, or locations containing baths or showers for medical treatment and for the use of disabled people, are not covered in this book.

Assessment of general characteristics (Regulation 701.30)

The bathroom and rooms containing a shower or bath are now classified into zones. There are three zones: Zone 0, Zone 1 and Zone 2. These zones are determined by taking into account the presence of walls, doors, fixed partitions, ceilings and floors. The zones described in Part 7, section 701, of BS 7671 tell us what type of electrical equipment etc. may be installed (or not) within the different zones. Refer to Figures 14.12 to 14.20 below when reading this section.

Zone 0

This is the interior of the bath tub or shower basin. In a location containing a shower without a basin, Zone 0 is limited by the floor and by the plane 0.10m above the floor, for the same horizontal distance as zone 1.

Zone 1

This zone is limited by the following:

(i) the finished floor level and the horizontal plane of 2.25m above the floor, or the highest fixed showerhead, whichever is the greater.

(ii) (a) by the vertical plane circumscribing the bath tub or shower basin, which includes the space below the bath tub or shower basin where that space is accessible without the use of a tool, or

(b) for a shower without a basin and with a demountable showerhead able to be moved around in use, Zone 1 is limited by the vertical plane at a radius of 1.2m from the water outlet at the wall, or

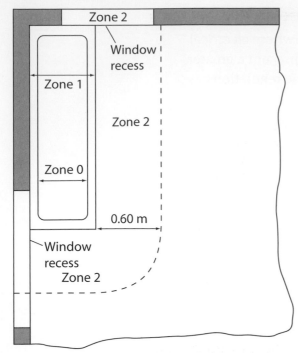

Figure 14.12 Bath tub

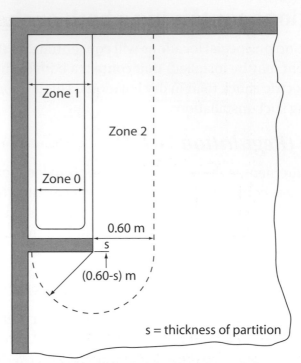

Figure 14.13 Bath tub with permanently fixed partition

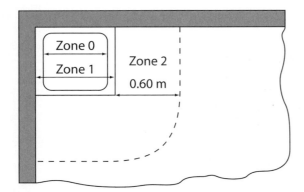

Figure 14.14 Shower basin

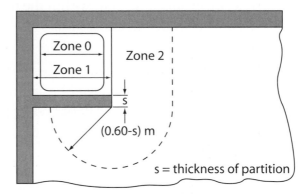

Figure 14.15 Shower basin with permanently fixed position

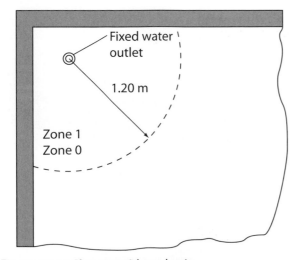

Figure 14.16 Shower, without basin

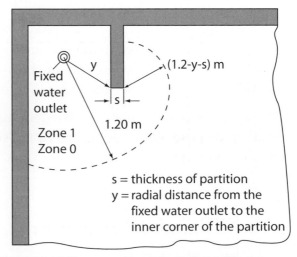

Figure 14.17 Shower, without basin but with permanently fixed partition – fixed water outlet not demountable

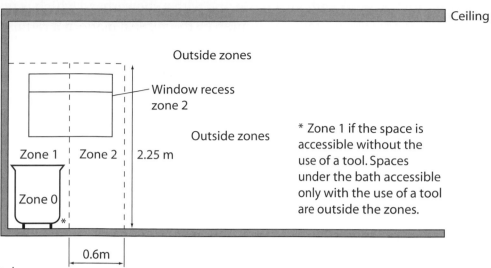

Ceiling

Outside zones

Window recess
zone 2

Outside zones

Zone 1 | Zone 2 | 2.25 m

Zone 0

0.6m

* Zone 1 if the space is
accessible without the
use of a tool. Spaces
under the bath accessible
only with the use of a tool
are outside the zones.

Figure 14.18 Bath tub

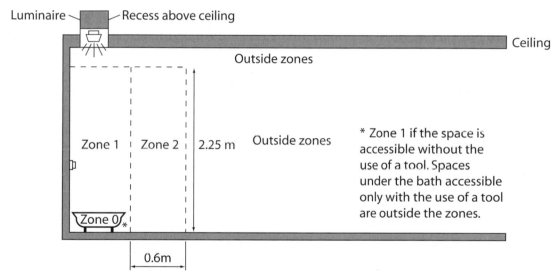

Luminaire — Recess above ceiling

Ceiling

Outside zones

Zone 1 | Zone 2 | 2.25 m | Outside zones

Zone 0

0.6m

* Zone 1 if the space is
accessible without the
use of a tool. Spaces
under the bath accessible
only with the use of a tool
are outside the zones.

Figure 14.19 Shower basin

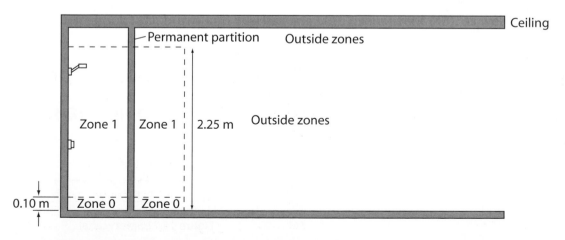

Ceiling

Permanent partition   Outside zones

Zone 1 | Zone 1 | 2.25 m | Outside zones

0.10 m | Zone 0 | Zone 0

Figure 14.20 Shower, without basin but with permanently fixed
partition – fixed water outlet not demountable

(c) for a shower without a basin and with a showerhead which is not demountable, Zone 1 is limited by the vertical plane at a radius 600mm from the showerhead.

Zone 2

This zone is limited by the following:

(i) the vertical plane external to Zone 1 and parallel vertical plane 600mm external to Zone 1

(ii) the floor and horizontal plane 2.25m above the floor or the highest fixed showerhead, whichever is the greater.

Protection for safety (Regulation 701.41)

The following productive measures are not permitted:

- obstacles
- placing out of reach
- non-conducting location
- earth-free local equipment bonding

Additional protection by RCD's (Regulation 701.411.3.3)

Every circuit in the location shall be provided with one or more RCD(s) with a rated residual operating current of not more than 30 mA. However, care must be taken to minimise inconvenience in the event of a fault or nuisance tripping of the RCD (Reg 314).

Supplementary equipotential bonding (Regulation 701.415.2)

Local supplementary equipotential bonding complying with Regulation 415.2 shall be provided connecting together the terminals of the protective conductors associated with Class I and Class II equipment in Zones 1 and 2 and extraneous conductive parts in these zones including the following:

Remember

All circuits in a bathroom must be RCD protected

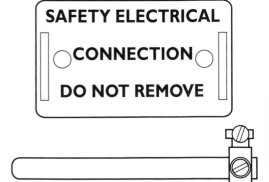

Figure 14.21 Safety label

(i)   metallic pipes supplying services and metallic waste pipes

(ii) metallic central heating pipes and air conditioning systems

(iii) accessible metallic structural parts of the building (not metallic doorframes or windows, unless connected to the metallic structure of the building)

(iv) metallic baths and metallic shower basins.

(v)   connections to pipes to be made with BS 951 clamps (complete with 'Safety Electrical Connection' label).

Where the bath or shower is in a building with a protective earthing and bonding system, the supplementary bonding may be omitted if the electrical equipment in the location is protected by an RCD not exceeding 30 mA.

External influences (Regulation 701.512.1)

Any external equipment should have the following degrees of protection:

(i)   in Zone 0, IPX7

(ii) in Zone 1 and 2, IPX4. Where water jets are likely to be used for cleaning purposes in communal baths or communal showers, IPX5

Switch gear and control gear (Regulation 701.512.3)

The following requirements do not apply to switches and controls which are incorporated in fixed current-using equipment suitable for use in that zone.

In Zone 0: switchgear or accessories shall not be installed.

In Zone 1: only switches of SELV circuits supplied at a nominal voltage not exceeding 12 volts rms a.c. or 30 volts ripple free d.c. shall be installed, the safety source being installed outside Zones 0, 1 and 2.

In Zone 2: switchgear, accessories incorporating switches or socket outlets shall not be installed with the exception of:

(i)   switches and socket outlets of SELV circuits, the safety source being outside the Zones 0,1 and 2

(ii) shaver supply units complying with BS EN 60742.

Socket outlets complying with section 414 may be installed in the location at a distance of more than 3m horizontally from the edge of Zone 1.

Did you know?

You do not have to bond the extraneous metalwork in a bathroom if the circuits are earthed and protected by an RCD.

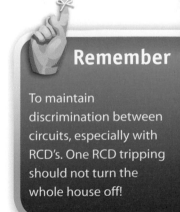

Remember

To maintain discrimination between circuits, especially with RCD's. One RCD tripping should not turn the whole house off!

Current using equipment (Regulation 701.55)

In Zone 0 only fixed current using equipment, can be installed but should be suitable for the conditions of this zone and be protected by SELV, operating at a voltage of less than 12 V a.c. or 30 V d.c.

The following fixed current using equipment may be installed in Zone 1 if it is suitable for that zone:

- water heaters
- shower pump
- SELV current-using equipment
- towel rails
- electric showers
- luminaires
- ventilation equipment
- whirlpool units

Electric heating units embedded in the floor and intended for heating the location may be installed below any zone provided that they are covered by an earthed metallic grid or by an earthed metallic sheath connected to the protective conductor of the supply circuit.

Construction site installations (Part 7: section 704)

This section on special locations will concentrate on the locations that are classed as construction sites, which present a greater risk of electric shock because of the conditions. This section sets out to look at these situations and discuss the Regulations governing such installations.

Scope (Regulation 704.1.1)

These Regulations cover the following types of construction sites:

- new building construction
- repair, alteration, extension or demolition of existing buildings

- engineering construction
- earthworks
- similar works.

The requirements of this section apply to both fixed and moveable installations.

They **do not** apply to construction site offices, cloakrooms, meeting rooms, canteens, restaurants, dormitories and toilets or to installations covered by IEC 60621 series 2 (mines and quarries).

Supplies (Regulation 704.313)

Equipment must be identifiable, colour coded and keyway interlocked in the case of plugs and socket outlets, and compatible with the particular supply from which it is energised (Regulation 704.313.3).

The following nominal voltages must not be exceeded (Regulation 704.313.4):

- **SELV** – to be used for portable hand-held lamps in confined or damp places
- **110 V, one-phase, centre point earthed** – used for reduced low voltage systems, portable hand lamps for general use, portable hand-held tools and local lighting up to 2 kW. 230 V hand held equipment up to 32 A must be protected by an RCD of 30 mA or less in addition to protection by automatic disconnection.
- **110 V, three-phase, star-point earthed** – used for reduced low voltage system, portable hand held tools and local lighting up to 2 kW, and small mobile plant to 3.75 kW
- **230 V, one-phase** – used for fixed floodlighting
- **400 V, three-phase** – used for fixed and moveable equipment above 3.75 kW.

The colour codes for plugs and socket outlets for the various voltages are:

- 400 V red
- 230 V blue
- 110 V yellow
- 50 V white
- 25 V violet.

Remember

230V hand held equipment on a construction sithe mus be protected by an RCD

Protection for safety (Regulation 704.4)

An IT supply system must not be used if there is an alternative type such as TNS, TT, as permitted by the Electrical Safety, Quality and Continuity regulations 2002. However, the distribution network operator (DNO) may not allow connection to a protective multiple earthed (PME) network because of the difficulty of installing and maintaining the main equipotential bonding.

Protecting against electric shock (Regulation 704.41)

The basic protective measures of using obstacles or placing out of reach are not allowed to be used on a construction site installation (704.410.3.5).

Each socket outlet and any permanently connected hand-held equipment up to and including 32 A must be protected by an RCD unless it is supplied using:

- reduced low voltage
- electrical separation
- SELV or PELV

Selection and erection of equipment (Regulation 704.5)

Every assembly used for the distribution of electricity on construction and demolition sites must comply with BS EN 60439-04. A plug or socket outlet with a rated current of more than 16 A must comply with BSEN 60309-2

Wiring systems (Regulation 704.52)

Cables must not be installed across a site road or walkway unless they are adequately protected against mechanical damage (Regulation 704.522.8.10).

For reduced low voltage systems the type of cable to be used must be low temperature, 300/500 V, thermoplastic (PVC) flexible cable or equivalent. For voltages above low voltage systems the flexible cable should be HO7 RN-F type or equivalent, having a rating of 400/750 V, and be resistant to abrasion and water (Regulation 704.522.8.11).

Isolation and switching devices (Regulation 704.53)

Every assembly for construction sites (ACS) must include devices for switching and isolating the incoming supply (Regulation 704.537.2.2).

Every circuit supplying current using equipment must be fed from an ACS which incorporates the following:

- overcurrent protection devices (fuses, MCBs)
- devices that protect against indirect contact (RCDs)
- socket outlets (if required).

Safety and standby supplies must be connected by means of devices (interlocking) arranged to prevent interconnection of diffcrent supplies.

Agricultural and horticultural premises (Part 7, section 705)

This section on special locations will concentrate on the locations that are classed as agricultural and horticultural premises, which present a greater risk of electric shock to people and livestock because of the conditions in which these premises exist. This section sets out to look at these situations and discuss the regulations governing such installations.

Scope (Regulation 705.1)

These regulations apply to all parts of fixed installations of agricultural and horticultural premises, both indoors and outdoors, and to locations where livestock are kept such as stables, chicken houses, piggeries, feed processing locations, loft areas for hay, straw and fertilisers.

They do not apply to dwellings intended solely for human habitation which are located on these types of premises.

Protection against electric shock (Regulation 705.41)

For basic protection against electric shock only barriers, enclosures and insulation of live parts shall be used. Protection by obstacles or placing out of reach are not to be used.

Remember

All circuits in within the location must be RCD protected

Protection by automatic disconnection of the supply (Regulation 705.411.1)

In locations in which livestock is intended to be kept, and where protection against indirect contact is provided by earthed equipotential bonding and automatic disconnection of supply, then the Regulations and tables in this section (Regulations 605-05 to 605-09) will apply and not those in Part 4 of BS 7671.

These include, for example, different disconnection times (faster) and different earth fault loop impedance values (lower).

All circuits shall be provided with an RCD to provide automatic disconnection regardless of the type of earthing system.

The RCD ratings must not exceed the following values:

- 30 mA for socket outlet circuits up to 32A

- 100 mA for socket outlet circuits of more than 32A

- 300 mA for all other circuits.

Supplementary equipotential bonding (Regulation 705.415.2.1)

In a location intended for livestock, supplementary bonding must connect all exposed and extraneous conductive parts which can be touched by livestock

Concrete reinforcement and other extraneous conductive parts in or on the floor shall be supplementary bonded.

Protection against fire and thermal effects (Regulation 705.422)

For protection against fire, a RCD having a rated current of not more than 300 mA must be installed for the supply to equipment other than that essential to the welfare of livestock (705.422.7).

Heating appliances must be kept at an appropriate distance from livestock and combustible material to minimise fire or risk of burns to livestock. For radiant heaters the clearance should not be less than 0.5m or as recommended by the manufacturer (Regulation 705.422.6).

Selection and erection of equipment

Electrical equipment for normal use must be protected to at least IP44; higher degrees should be provided, as appropriate, depending on the external influences (Regulation 705.512.2).

Switchgear and control gear

Each device for emergency switching (including emergency stopping) must be installed where it is inaccessible to livestock and will not be impeded by livestock. Account must be taken of conditions likely to arise in the event of panic by livestock.

Other equipment

Socket outlets of up to 20 A rated current may be of the standard BS 1363 type or should comply with BS 546 or BS 196.

Industrial plugs and sockets, complying with BS EN 60309-1 or BS EN 60309-2 may also be used.

To reduce the risk of fire any luminaire used must be suitable for mounting on normally flammable surfaces, e.g. an $\underline{\nabla F}$ rating.

For protection against the ingress of dust and moisture, $\underline{\nabla D}$ rated luminaires with a degree of protection of at least IP 54 are required. Any luminaire should comply with BSEN 60598.

Caravan parks (Part 7: section 708)

This section on special locations will concentrate on the locations that are classed as caravan parks, which present a greater risk of electric shock to members of the public and animals because of the conditions in which these premises exist.

Possible risks include open circuit faults of the PEN conductor in PME supplies (this raises the potential to true earth of all metalwork, including that of the caravan, to dangerous levels) and possible loss of earthing because of long cable runs, devices exposed to weather and flexible cord connections.

This section looks at these situations and discusses the Regulations governing such installations.

Scope

The Regulations apply to that portion of an electrical installation in caravan parks that provide facilities for the supply of electricity to, and connection of, leisure accommodation vehicles (caravans/motor caravans) or tents at nominal voltages not exceeding 230/400 V.

Protection against electric shock

Basic protection (708.410.3.5)

The following methods of protection shall not be used:

- protection by obstacles
- protection by placing out of reach.

Fault protection (708.410.3.6)

The following methods of protection shall not be used:

- non-conducting location
- earth free equipotential bonding.

Selection and erection of equipment

Wiring systems

Caravan pitch supply equipment shall preferably be connected by underground cables. Underground cables, unless provided with additional mechanical protection, shall be installed outside any caravan pitch or area where tent pegs or ground anchors may be driven.

All overhead conductors shall be:

- insulated
- at a height of not less than 6m in vehicle movement areas and 3.5m in all other areas.

Poles and other supports for overhead wiring shall be located or protected so that they are unlikely to be damaged by any reasonably foreseeable vehicle movements.

Switch gear and control gear

Caravan pitch supply equipment shall be located next to the pitch and no more than 20m from the pitch it is intended to serve.

Did you know?

All socket outlets in caravan parks must be individually RCD protected.

Plugs and socket outlets

Each socket outlet and its enclosure that forms part of the pitch supply equipment shall:

- comply with BS EN 60309-2 and meet the degree of protection IP44

- be placed at a height of between 0.5m and 1.5m from the ground to its lowest point

- have a current rating of no less than 16 A

- have at least one socket outlet provided for each pitch

- be individually protected by an overcurrent device

- be protected individually, by a RCD having the characteristics specified in Regulation 415.1 and which must not be bonded to the PME terminal.

Floor and ceiling heating systems (Part 7:section 753)

This new section covering floor and ceiling heating systems has been included because it is one of the new 'special locations' introduced by BS7671:2008 (IEE Wiring Regulations 17th Edition). These systems are becoming increasingly popular, particularly in kitchens, where wall space to hang conventional heat emitters may be limited, and in bathrooms for comfort and safety.

Scope

The requirements of this section apply to heating systems that are installed within the building fabric of either the floor or ceiling. They do not apply to wall-mounted heating systems or those intended for use outdoors.

Automatic disconnection of the supply

Automatic disconnection must be provided by an RCD with a rated residual operating current of 30 mA or less (Regulation 753.411.3.2).

If the manufacturer has not provided an earth terminal as part of the heating units, an earthed metallic grid shall be installed above the floor elements and below the ceiling elements.

Protection against burns

Precautions must be taken to limit the surface temperatures of under-floor heating systems, particularly in areas where contact with the skin is possible.

Protection against overheating

Floor and ceiling heating systems must be designed and installed to provide protection against overheating of the units themselves during normal operations (maximum 80°C) or under fault conditions.

The connection between the heating system and the cold tail must be a mechanically sound joint, for example a crimped connection.

Care must be taken to ensure that the heating system will not ignite nearby combustible materials. This can be achieved by providing a suitable gap (at least 10mm) between the heating system and the materials or by inserting a non-flammable barrier between the two.

External influences

The position of floor and ceiling heating systems within the building means that, both during and after installation, they are particularly at risk from mechanical damage. Once installed, a floor or ceiling heating system should be embedded into the building fabric as soon as possible. Heating units installed in ceilings should have a degree of protection of at least IP X1 and those embedded in a floor (in concrete, for example) should be rated at not less than IPX7.

Heating-free areas must be provided to allow for the installation and fixing of other electrical and non-electrical appliances and equipment and other contractors must be informed of their location (Regulation 753.522.4.3).

Identification and Notices

By their very nature, once installed, floor and ceiling heating systems "disappear" within the fabric of the building. The layout, area covered and power rating quickly become forgotten by the installer and, without notices, may never be known by the owner or user of the installation. Without knowing the details of the heating installation, any alteration to the building runs the risk of damaging the heating units or causing them to overheat.

The following information must be provided within or adjacent to the distribution board, detailing:

- Manufacturer and type
- Number of units
- Length/area of units
- Rated power
- Surface power density
- Layout drawing
- Position/depth of heating units
- Position of junction boxes

- Conductors and shields
- Heated area
- Rated voltage
- Rated resistance (cold)
- Rated current of over-current device
- RCD rating
- Insulation resistance
- Leakage capacitance

In addition, the user must be provided with information regarding the positioning of furniture, carpeting or other objects which may affect the operation or integrity of the heating system (Fig 753, BS7671:2008).

Remember

You must provide adequate information to the user of floor and ceiling heating systems so they can be used and maintained safely.

Other special locations

BS7671:2008 contains the following other special locations.

- Swimming pools
- Saunas
- Restricted movement
- Marinas
- Medical locations

- Exhibitions
- Solar photovoltaic
- Mobile units
- Caravans
- Amusement parks

These locations are less common and therefore more specialist than those already dealt with in this book, so they will not be covered in detail.

Electrical system arrangements

On completion of this topic area the candidate will be able to state earthing systems and the protection arrangements for electrical systems.

Earthing systems

TT system

The letter T when used in this way of describing wiring systems comes from the Latin word 'terra', which means earth. So the first letter T in this instance means that the supply is connected directly to earth at the source. This could be the generator, or transformer at one or more points.

The second letter T means that the exposed metalwork of the installation is connected to the earth by a separate earth electrode. The only connection between these two points is the general mass of earth (soil etc.) as shown.

When a fault to earth occurs on this system, the earth-fault current will flow around this circuit from the fault, through the earth and transformer windings, along the line conductor and back to the fault position.

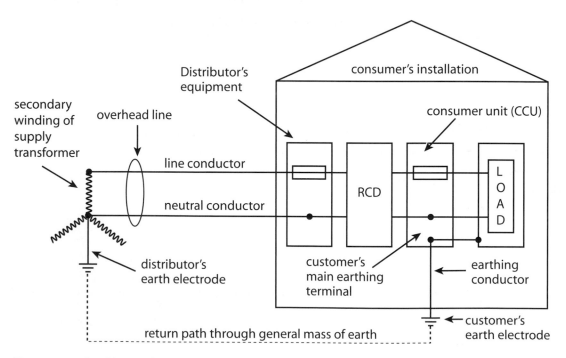

Figure 14.22 Earthing with customer's earth electrodes and ground earth return path

In this system the earth may have a very high value of impedance, for example 2300 ohms. From Ohm's law we can see that the current flow would be:

$$I = \frac{V}{R} = \frac{230}{2300} = 0.1A$$

This current flow of 0.1 A, that is 100 mA, is sufficient to be fatal to people, so protection against even very small currents is vital to prevent danger from electric shock. Regulation 411.5.2.states that a residual current device (RCD) or an overcurrent protective device or both should protect a TT system, with the RCD being preferred. Figure 14.23 illustrates the earth connection.

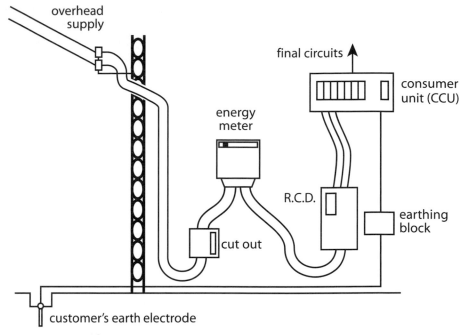

Figure 14.23 Intake earthing arrangements on a TT system

TN-S system

- T means that the supply is connected directly to the earth at one or more points.

- N means the exposed metalwork of the installation is connected directly to the earthing point of the supply.

- S means a separate conductor is used throughout the system from the supply transformer all the way to the final circuit to provide the earth connection.

Figure 14.24 illustrates this system.

This earth connection is usually through the sheath or armouring of the supply cable and then by a separate

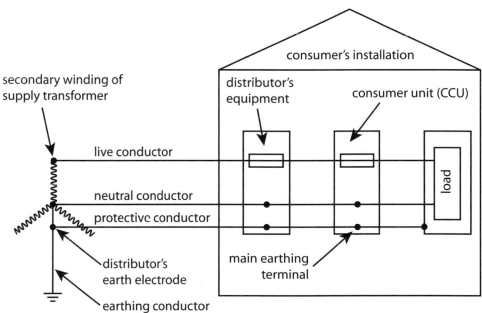

Figure 14.24 TN-S earthing system with metallic earth return path

conductor within the installation. As a conductor is used throughout the whole system to provide a return path for the earth-fault current, the return path should have a low value of impedance. Figure 14.25 illustrates the intake earthing arrangements for a TN-S system.

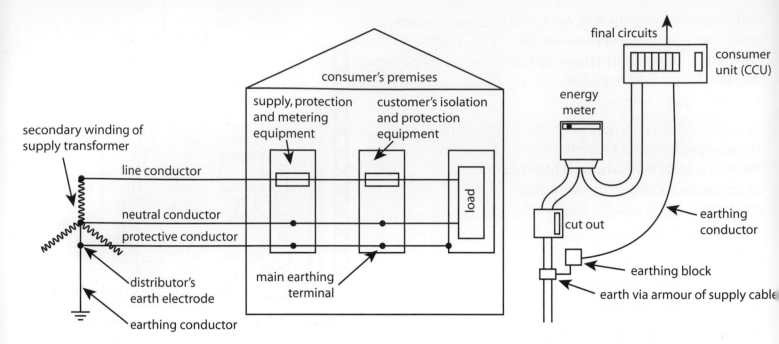

Figure 14.25 TN-S earth connection

TN-C-S system

- T means the supply is connected directly to earth at one or more points.

- N means the exposed metalwork of the installation is connected directly to the earthing point of the supply.

- C means that for some part of the system (generally in the supply section) the functions of neutral conductor and earth conductor are combined in a single common conductor.

- S means that for some part of the system generally in the installation, the functions of neutral and earth are performed by separate conductors.

Figures 14.25 and 14.26 illustrate this system.

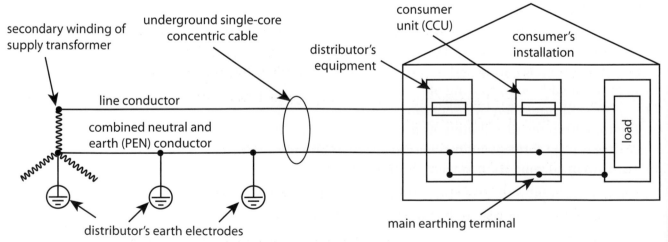

Figure 14.26 TN-C-S, protective multiple earthing (PME) system

In a TN-C-S system, the supply uses a common conductor for both the neutral and the earth. This combined conductor is commonly known as the Protective Earthed Neutral (PEN) or also sometimes as the Combined Neutral & Earth (CNE) conductor.

In such a system the supply PEN is required to be earthed at several points. This type of system is also known as Protective Multiple Earthing (PME) and this will be discussed more fully in the next section.

However, often this effectively means that the distribution system is TN-C and the consumer's installation is TN-S – this combination therefore giving us a TN-C-S system.

The intake earthing arrangements for a typical TN-C-S system are shown in Figure 14.27.

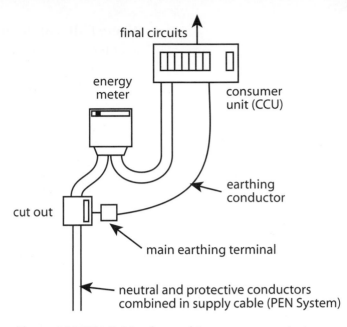

Figure 14.27 TN-C-S intake earthing arrangements

Protective Multiple Earthing (PME)

This system of earthing is used on TN-C-S systems. PME is an extremely reliable system of earthing and is becoming the most commonly used distribution system in the UK today.

With the PME system, the neutral of the incoming supply is used as the earth point and all cpcs connect all metal work, in the installation to be protected, to the consumer's earth terminal.

Consequently, all line to earth faults are converted into line to neutral faults, which ensures that under fault conditions a heavier current will flow, thus operating protective devices rapidly.

However this increase in fault current may produce two hazards.

- The increased fault current results in an enhanced fire risk during the time the protective device takes to operate.
- If the neutral conductor ever rose to a dangerous potential relative to earth, then the resultant shock risk would extend to all the protected metalwork on every installation that is connected to this particular supply distribution network.

Because of these possible hazards, certain conditions are laid down before a PME system is used. These include the following.

- PME can only be installed by the supply company if the supply system and the installations it will feed meet certain requirements.
- The neutral conductor must be earthed at a number of points along its length. It is this action that gives rise to the name 'multiple' earthing.
- The neutral conductor must have no fuse or link etc. that can break the neutral path.

- Where PME conditions apply, the main equipotential bonding conductor shall be selected in accordance with the neutral conductor of the supply and Table 54.8 of BS 7671.

Other systems

TN-C system

In this system the neutral and protective functions are combined in a single conductor throughout the system. This system is relatively uncommon and its use is restricted for specific situations. There must be no metallic connection between this system and supply company equipment.

IT system

This system must not be connected to the supply company's system. It is a special system used in quarries, telephone exchanges and some industrial processes etc. The system has no protective devices (e.g. fuses). The neutral point of the supply generator is bonded to earth via high value impedance. In the event of an earth leakage, the value of current that can flow is restricted. This also restricts the potential difference developed to earth.

The system is used in areas of production where it could be dangerous for machines to just stop in the event of a fault. Often it uses a system of indicating lights to show a fault exists, enabling machine operators to complete the process and repair the fault. Should a second fault occur before the first fault has been cleared, the machinery will shut down automatically.

Protective arrangements

On completion of this topic area the candidate will be able to state arrangements made for electrical systems with relevance to isolation, overcurrent protection and earth fault protection.

We have covered the protective arrangements for electrical systems throughout this book. Please refer to the following sections for more information:

- Isolation and switching (Chapter 6)
- Overcurrent protection (Chapter 6 pages 147–148)
- Earth fault protection (Chapter 6 pages 141–144).

Installation and testing

On completion of this topic area the candidate will be able to state the requirements for successful installation and electrical installation testing.

Installation

All installations have to meet three key criteria before the installation can be said to have been completed, as follows.

- The secure fixing of systems and components – this applies in particular to switching, cables and conduit, all of which must be adequate and suitable to the environment.

- Electrical continuity and maintenance of system integrity – this applies to all protective measures to ensure that the system is safe to use.

- Avoidance of damage to components and system – obviously, no one wants an installation to be damaged before it is even started!

In order to make sure that the work is carried out satisfactorily and that these conditions are met, the inspection and test procedure must be carefully planned and carried out, and the results correctly documented.

Inspection

Precise details of all equipment should be obtained from the manufacturers or suppliers in order to check that the required standards have been met, to ensure that satisfactory methods of installation have been used and to provide the information necessary to confirm correct operation. All this information must be included in the operation and maintenance manual prepared for the project. This should also include information on the operation and technical data for installations.

As well as doing functional testing on the protective equipment, the commissioning process is intended to confirm that the installation complies with the designer's requirements. As such, commissioning includes the functional testing of all equipment, isolation, switching, protective devices and circuit arrangements.

The results of all inspections and tests must be recorded and compared with the relevant design criteria (often the regulations). Any persons carrying out tests should be supplied with the necessary data in order to make a comparison. In the absence of such data the inspector should apply the requirements set out in BS 7671.

Initial verification procedures

We inspect and commission material after the completion of work to ensure:

- compliance with BS 7671
- continuity of protective conductors

Did you know?

The recording of inspection and test results is also a recommendation contained in *The Memorandum of Guidance on the Electricity at Work Regulations (EAWR)*. This states that records of all maintenance including test results should be kept throughout the life of an installation to enable the condition of equipment and the effectiveness of maintenance to be monitored

- ring final circuit
- polarity
- that it is safe to use.

Compliance with BS 7671

BS 7671 Part 7 states that every electrical installation shall, either during construction, on completion or both, be inspected and tested to verify, so far as is reasonably practicable, that the requirements of the Regulations have been met. In carrying out such inspection and test procedures, precautions must be taken to ensure no danger is caused to any person or livestock, and to avoid damage to property and installed equipment.

BS 7671 requires that the following information be provided to the person carrying out the inspection and test of an installation:

- the maximum demand of the installation expressed in amperes per phase
- the number and type of live conductors at the point of supply
- the type of earthing arrangements used by the installation, including details of equipotential bonding arrangements
- the type and composition of circuits, including points of utilisation, number and size of conductors, and types of cable installed (this should also include details of the 'reference installation method' used)
- the location and description of protective devices (fuses, circuit breakers etc.)
- details of the method selected to prevent danger from shock in the event of an earth fault, e.g. earthed equipotential bonding and automatic disconnection of supply
- the presence of any sensitive electronic devices.

It is important to remember that periodic inspection and testing must be carried out on installations to ensure that the installation has not deteriorated and still meets all requirements. Tests will also need to be carried out in the event of minor alterations or additions being made to existing installations.

Continuity of protective conductors

We also need to check that extraneous conductive parts have been correctly bonded with protective conductors. An extraneous conductive part is a conductive part that is liable to introduce a potential, generally earth potential, and not form part of an electrical installation. Examples of extraneous conductive parts are metal sink tops and metal water pipes etc. The purpose of the bonding is to ensure that all extraneous conductive parts which are simultaneously accessible are at the same potential.

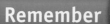

Remember

All electrical items *must* be tested before finally being put into service

Did you know?

For commercial or industrial installations, the requirements of the Electricity Safety, Quality and Continuity Regulations 2002 and the Electricity at Work Regulations 1989, both of which are statutory instruments, should also be taken into account

Ring final circuit

A test is required to verify the continuity of each conductor, including the circuit protective conductor (cpc), of every ring final circuit. The test results should establish that the ring is complete and has no interconnections. The test will also establish that the ring is not broken. Figure 14.28 shows a ring circuit illustrating these faults.

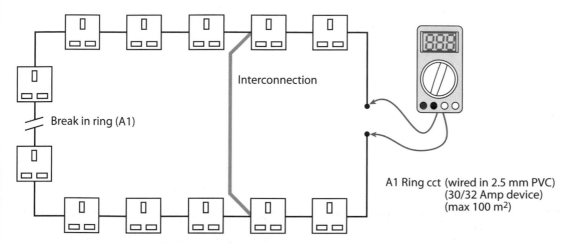

Interconnection

Break in ring (A1)

A1 Ring cct (wired in 2.5 mm PVC)
(30/32 Amp device)
(max 100 m²)

Figure 14.28 Test of continuity of ring final circuit conductors

Polarity

Polarity tests should be carried out to check that:

- polarity is correct at the intake position and the consumer unit or distribution board.

- single pole switches or control devices are connected in the line conductor only

- socket outlets and other accessories are connected correctly

- centre contact bayonet and Edison screw type lamp holders have their outer or screwed contact connected to the neutral conductor

- all multi-pole devices are correctly installed.

Safe to use

The final act of the commissioning process is to ensure the safe and correct operation of all circuits and equipment which have been installed, and that the customer's requirements have been met. This will also confirm that the installation works and, more importantly, will work under fault conditions. After all, it is under fault conditions that lives and property will be at risk.

Outcome 2 Specifications, drawings, instructions and other data involved with electrical installations

For information on the areas explored in this outcome but not covered in this chapter, please turn to the following pages in this book.

Outcome 4 Describe methods of undertaking installations of systems and components

For information on the areas explored in this outcome but not covered in this chapter, please turn to the following pages in this book.

Learning point	Sub points	Chapter	Page reference
State means of assessing and stating precautions	• Possible sources of hazards • Preparing the work area • Identifying PPE	Chapter 8 Chapter 1	 pages 9-16
Isolating and lock-off procedures		Chapter 1	pages 16–18
Liaising with contractors		Chapter 2	pages 45–49
Using work plans	• Identifying tasks • Developing specifications • Checking materials	Chapter 2 Chapter 9 Chapter 7	pages 45–49 page 228
Purpose and use of equipment	• Rules, tapes • Gauges • Levels and plumbs • Squares • Scribes	Chapter 7	
Measuring and marking out		Chapter 9	pages 240–241
Fixing enclosures and equipment to surfaces	• Woodscrews • Spring toggles • Girder clips • Crampets • Masonry bolts	Chapter 7	

Support and installation methods	• Application • Load-bearing capacity • Fabric of structure • Environmental conditions • Aesthetic considerations	Chapter 6	page 122 page 120 pages 138–139 pages 120–122
Use of tools in fixing and installing	• Cable cutters, wire strappers, knives, saws, drills • Files, reamers, spanners wrenches • Bending and forming machines • Adhesives	Chapter 7	
Safe use and storage of equipment	• Use of low voltage for power tools	Chapter 7	
Disposal of waste		Chapter 8	pages 223–224

Activity

1. Have a look around the properties you live and work in and see if you can find instances where:

    •   The supplementary equipotential bonding is missing.

    •   The supplementary equipotential bonding has been installed unnecessarily.

2. Look at the properties you are working in to find examples of TT, TN-S and TN-C-S systems

FAQ

Q Why is the hot water outlet taken from the top of the cylinder?

A Because hot water rises, it makes sense to fill the cylinder from the bottom and draw off the hot water from the top, where it will be hotter.

Q Why do room thermostats often have a neutral connection, when they are just ON/OFF switches?

A The neutral connection may be there for one of two reasons. In a conventional room thermostat, the neutral is connected to a small resistor called an anticipator. This increases the responsiveness of the thermostat. Some modern electronic room thermostats need a neutral connection because they draw their power from the mains supply.

Q I still don't understand the term 'maintained' when dealing with emergency lighting; surely all equipment has to be maintained?

A When talking about emergency lighting, the term 'maintained' has a specific meaning. It means that the same lamps are used for normal mains operation and emergency operation. They are normally fed from the mains but if the power fails they are fed from an emergency backup system – the same level of illumination is maintained using the same lamps.

Q Does every building need a fire alarm system?

A No, but most public buildings do have a fire alarm system and an evacuation procedure. Domestic properties are normally fitted with smoke alarms in accordance with the Building Regulations.

Q I've seen a supplementary equipotential bonding conductor connected between a wall heater and a pipe. Is this correct – I thought they just connected pipe-work together?

A Yes, depending on the layout it probably is correct, the Regulations ask us to connect together exposed and extraneous conductive parts within the zones.

Knowledge check

1   Discuss the relative advantages and disadvantages of incandescent and discharge lamps.

2   What would be the best type of water heater to use in the following locations and why:

*   A large guest house

*   A hairdressers salon

3   Why is oil used as the heat transfer medium in oil-filled radiators?

4   Explain the operation of the following type of thermostat:

*   Capillary type

*   Stem type

5   Explain the difference between a maintained and a non-maintained emergency lighting system.

6   Why is it necessary to divide a larger building up into fire alarm zones?

7   Which type of sensors would be best suited to perimeter protection of a property?

8   What special precautions would you need to take when installing fibre-optic cables?

9   In which zones of a bathroom would it be possible to install a convector heater? Are there any special requirements for each zone?

10   At what voltage should a 750 W hand-held portable tool for use on a construction site be supplied?

11   List the type of fixed current using equipment that may be installed (if it is suitable) in zone 2 of a bathroom.

12   In zone 0 of a bathroom, what type of switchgear and accessories may be fitted?

13   What is the maximum height that zone 1 of a bathroom extends to?

14   What voltage should a hand held portable tool on a construction site be supplied at?

Index

A

a.c. motors 233
ABC of resuscitation 27–8
a.c. (alternating current)
    capacitance 271–2
    frequency and period 250
    generator 248–9, 254–6
    impedance 273–4
    inductance 270–1
    motor 256
    power in a.c. circuit 278–80
    resistance 269–70
    r.m.s. method 250–2
    theory 247, 249–52
    values of alternating waveform 249–50
    see also relays; transformers
acceleration 71
access equipment 321
    see also height
access to site 217–18
accidents 200–1
    prevention 24–5, 212–16
    procedures 25–6
    reporting 25–6
    risks 200
    safety procedures 212–16
    see also risk; safety
agricultural/horticultural premises
    369–71
    erection/selection of equipment 371
    supplementary equipotential
        bonding 370
    switchgear and control gear 371
alarm, security and emergency
    installations 39
alternator 248–9
    construction of 83–4
Amicus (AEEU) 50
ammeter 106, 107, 108
ampere 96

apparent power 273, 280, 281–2
area classification 323–4, 325
area (measurement) 79, 81
armoured cables 358
asbestos dust safety 215
assembly drawing 62
atoms 92–3
average value 249
    bar charts 64, 233–4
    bath/shower installations 316, 361–7
    electric shock prevention 364, 365
    fixed current using equipment 366–7
    supplementary equipotential
        bonding 364–5
    switchgear and control gear 365–6
    wiring systems 365
    zones 361–4

B

battery 96
block diagram 60
body protection 14
bolts 166
bonding 140
    labelling 194–5
boom lift 183–4
break-glass contacts 346–7
BS 88 HBC fuses 303–4
BS 1361/1362 cartridge fuses 303
BS 3036 rewireable fuses 301–2
BS 7671 Wiring Regulations 6–7, 122, 312
    appendices 317–18
    Guidance Notes 313
    history and role 312–13
    Part 1 scope and principles 313–14
    Part 2 definitions 315
    Part 3 assessment of characteristics
        315
    Part 4 protection for safety 315
    Part 5 selection and erection

        equipment 315–16
    Part 6 inspection and testing 316
    Part 7 special installations/locations
        316–17
    plan and style 313
BS EN 60079 326
BS EN 60617 electrical symbols 66
    building management and control
        installations 40
    building services 39
    Building Standards 317
    building and structural engineering 38
    burns, treatment for 31

C

cable basket 356
cable jointing 41
cable ladder 356
cable runs 163
cable selection 135–9, 358
    ambient temperature 138, 178
    corrosive substances 139, 178
    damage by animals 139, 178
    and direct sunlight 139, 178
    electrolytic action 139
    environmental conditions 138–9, 318
    and moisture 139, 178
cable support and bends 174–6
    spacing regulations 176
    support regulations 175
cable tray 356
cables 159–63
    armoured/braided 358
    Cat5 357
    chasing 172–3
    colours 129–30, 318
    communication 357–8
    components 120–1, 135–9
    computer installations 40
    conductor material 135–6